Systems & Control: Foundations & Applications

A.N. Krasovskii
N.N. Krasovskii

Control Under Lack of Information

Birkhäuser
Boston • Basel • Berlin

N.N. Krasovskii
Institute of Mathematics
and Mechanics
Ekaterinburg GSP
384 Russia

A.N. Krasovskii
Faculty of Mathematics
and Mechanics
Ural State University
620083 Ekaterinburg
Russia

Library of Congress Cataloging-in-Publication Data

Krasovskii, A.N. (Andrei Nikolaevich)
Control under lack of information / A.N. Krasovskii, N.N. Krasovskii.
p. cm. -- (Systems & control)
Includes bibliographical references and index.
ISBN-13: 978-1-4612-7583-1 e-ISBN-13: 978-1-4612-2568-3
DOI: 10.1007/978-1-4612-2568-3
paper)
1. Control theory. I. Krasovskii, N.N. (Nikolai Nikolaevich)
II. Title III. Series.
QA402.3.K635 1994 94-26447
629.8'312--dc20 CIP

Printed on acid-free paper

Birkhäuser
Softcover reprint of the hardcover 1st edition 1995

ISBN-13: 978-1-4612-7583-1
Typeset and reformatted by Texniques, Inc. from authors disks.
Printed and bound by Quinn-Woodbine, Woodbine, NJ.

9 8 7 6 5 4 3 2 1

CONTENTS

PREFACE

The mathematical theory of control, essentially developed during the last decades, is used for solving many problems of practical importance. The efficiency of its applications has increased in connection with the refinement of computer techniques and the corresponding mathematical software. Real-time control schemes that include computer-realized blocks are, for example, attracting ever more attention. The theory of control provides abstract models of controlled systems and the processes realized in them. This theory investigates these models, proposes methods for solving the corresponding problems and indicates ways to construct control algorithms and the methods of their computer realization.

The usual scheme of control is the following: There is an object F whose state at every time instant t is described by a *phase variable* x. The object is subjected to a *control* action u. This action is generated by a control device U. The object is also affected by a *disturbance* v generated by the environment. The information on the state of the system is supplied to the generator U by the *informational* variable y. The mathematical character of the variables x, u, v and y are determined by the nature of the system. For example, in problems of control of an aeroplane, x may denote the collection of the coordinates and the speeds of all its essential parts; u may be the collection of the variables that characterize the thrust and the position of the steers; v may correspond to the influence of air-streams to the aeroplane; and y may be the indications of the measuring devices.

The term *control* and the scheme considered above have a very general and uncertain character. Almost every real process can be treated as a controlled process. For example, calculations on a computer can also be treated as a process of control for this computer with the help of the program. A game of chess on behalf of one or another player may also be considered as control of the position x on the chess board by the corresponding player U. The actions of the opponent player may then be considered as the actions of the environment V. And so on.

This book is devoted to problems of control in dynamical systems in situations of uncertain information about the disturbance or in conditions of conflict. We focus on a concrete class of problems of the contemporary theory of control. We consider the systems whose evolution is described by ordinary differential equations and study the processes within a given time interval $t_0 \leq t \leq \vartheta$. It is assumed that a quality index γ of the control process, which is chosen according to the nature of the considered problem, is given. Let the process be considered beginning from the initial time moment $t_* \in [t_0,\ \vartheta]$. Then the quality index γ is a certain functional of the realizations of the motion $x[t]$, of the control actions $u[t]$ and the disturbances $v[t]$ on the time interval $t_* \leq t \leq \vartheta$. The aim of the control is to make the quality index γ as small as possible (or as large as possible). We prefer to formalize the problem as that of *minimizing* γ.

We assume that the solution of the problem is aggravated by the following circumstance: Namely, because of the lack of information about the disturbance, one cannot predict uniquely the reaction of the system to the control action. Therefore, the problem consists in the choice of control that ensures the best possible result even in conditions of the most unfavourable possible disturbance. Problems of this type are studied in the framework of the theory of differential games [37], [41], [48], [79], [88], [111], [118], [122], [125], [140], [144], [147].

The central notion of the theory developed in this book is the notion of ensured result. We call an *ensured result* ρ for a fixed method (law) of control U, the upper bound for the values of the quality index γ that can occur when this method is used. We come to the problem on a control *law* that provides the smallest possible value ρ. Of course, this problem can be strictly formalized only on the basis of rigorous definitions of admissible laws of control U and disturbances v. We use a definition of *feedback control* law based on information about the current value of the variable $y[t]$. And this information variable (information image) $y[t]$ employed by us is either the current value of the phase variable $x[t]$ or the *history of the motion* $x[\tau]$, $\tau \leq t$. For example, in the case of the landing of an aeroplane, the quality index γ can be an estimation of deviations of the coordinates of the aeroplane and its speed from the required ideal values. It is natural to desire to find the method of control (control law) for which the ensured estimate ρ^0 of these deviations will be the smallest among the estimates ρ that are guaranteed by various other reasonable methods of control.

Our considerations are based on the conception developed in Ekaterinburg (Russia), and the ideology of this monograph reflects our convictions as the authors of the proposed material. We have done our best to formulate the problems without unnecessary complications but reasonably rigorously. We establish the existence of solutions, study properties of these solutions and indicate realizable methods for forming the optimal control. The theory is illustrated by model examples which provide a transparent qualitative picture of the optimal processes and give concrete numerical procedures. We have tended to give the solution of all these examples in a very detailed form so that the readers can follow our calculations.

The specific features of our approach are that (1) attention is focused on control problems for dynamical systems governed by ordinary differential equations, and (2) the system is controlled in conditions of essential lack of information about the dynamical disturbances. This approach presupposes special attention to the character of the system's informational image whose current values give the basis for constructing the control actions. We present a hierarchy of functionals γ being optimized and a corresponding hierarchy of information images that ensure effective optimization of the guaranteed control result.

Due consideration is given to the development and justification of the original method for constructing strategies proposed by the authors. The method is connected with the ideas of the *program stochastic synthesis* [57], [65], [76], [77], [143], [144] and at present is described fragmentarily only in articles [61]–[66], [68], [78], [80], [98], [99]. In this monograph we have attempted to give a readable and sufficiently rigorous presentation of this method. For typical cases, the effective algorithms are presented, including description of their implementation with the help of computers. The theoretical considerations are illustrated by concrete examples with computer simulation of the process. The principal theorems are formulated in accordance with the usual mathematical standards. Although the proofs of these theorems are outlined rather briefly, references to more detailed proofs in the publications of the authors are given.

Let us give a brief outline of the contents. We start with two model examples. With the help of these examples we illustrate the basic concepts of the motion of the object, the quality indices, the strategies, the control laws, the ensured results, the optimal strategies and the optimal ensured results. Also, the schemes of control that respond to these notions are described. For these examples we give the results of the computer simulation for the considered controlled processes. The optimal algorithms of control that form the basis for this simulation are constructed according to the general methods studied in the book.

Chapters I, II and III are devoted to the concept of the *pure (positional) strategies* of control and to the corresponding concepts of the positional quality index γ, the algorithm of feedback control and the ensured result. It is emphasized that the considered problem of control has a reasonable solution in the class of pure strategies only in the case when the so-called *saddle point condition in the small game* is fulfilled. Here we explain the connection of the considered control problem with the theory of games and, first of all, with the theory of *differential games.* The basic concepts of the *saddle point* and the *value of the game* are defined. A saddle point is understood here as a pair of the optimal pure strategies. Existence theorems for the value of the considered differential game and the saddle point are established. We show that the required optimal strategies can be constructed in principle on the basis of the value of the game by the help of the so-called *extremal shift* to the *accompanying points* (see [55], [62], [77]).

Further in these chapters a *classification* of the functionals (quality indices) γ is given. This classification takes into account the information image that enables us to construct proper feedback control in the corresponding classes of strategies. This classification begins with the *positional functionals.* In the processes of feedback control with these positional quality indices, sufficient information is given by the information image that is the current *state* of the phase vector of the controlled system. The classification ends with *quasi-positional functionals* for which the sufficient information image is the *history* of the motion of the controlled system *extended* also by some auxiliary informational variables.

As was mentioned above, attention is paid mostly to the original constructive method for the computation of the optimal ensured result and for constructing the optimal strategies or the approximating optimal control laws. This method is connected with the idea of the program stochastic synthesis. However, this method is presented and justified in this book in an independent form. We call it the method of the *upper convex hulls* in *auxiliary program constructions.* We consider these constructions in detail for the case of two basic positional functionals $\gamma_{(1)}$ and $\gamma_{(2)}$. The first of these functionals is the functional of an integral type that estimates the motion on the whole time interval $t_* \leq t \leq \vartheta$. The second functional $\gamma_{(2)}$ estimates the maximum deviation of the motion from a desired ideal realization on the time interval $[t_*, \ \vartheta]$. Then we outline the way to adapt these constructions to all the other considered classes of the quasi-positional functionals $\gamma_{(i)}$.

Let us remark here that the choice of functionals considered in this book is explained by the fact that the proposed scheme of construction of optimal strategies has much in common, when applied to the control processes, with these quality indices. Really, in the considered cases there are elements of one and the same type. Among *these* elements are the upper convex hulls for some functions that arise in the proposed schemes.

The part of the book devoted to *pure* strategies is concluded by a detailed analysis of several examples of model problems with the different kinds of positional and quasi-positional functionals that determine the corresponding quality indices. These models are sufficiently simple. We have chosen these examples because the corresponding auxiliary constructions and algorithms of optimal feedback control prove to be sufficiently transparent.

The second part of the book, Chapter IV, is devoted to the so-called *mixed strategies.* The realization of these strategies involves the use of certain probability mechanisms. It should be noted that the optimal control based on these strategies guarantees the result with probability as close to unit as desired. We deal with mixed strategies instead of pure ones in order to be able to consider also the cases when the saddle point condition in the small game is not valid. In this part of the book the mathematical statement of the problem in the considered class of mixed strategies is given and the existence of a solution is established. As in the first part of the book, attention is paid mostly to an effective method of constructing the optimal mixed strategies based on the method of the upper convex hulls. Here this method also provides algorithms that can be effectively realized on computer. We note that in the case of mixed strategies it is essential that a certain *y-model-leader* and *z-model-leader* are introduced in the scheme of control (see [61], [62], [79]). At the same time, the algorithms of control for these leaders are constructed on the basis of abstract accompanying models. Similar to the case of pure strategies these algorithms use the method of extremal shift to the accompanying points. In this chapter the general theory is also illustrated by a detailed analysis of model problems.

Let us note that in the book, along with the optimal control aimed to provide a result not worse than the a priori optimal ensured result, we also consider the strategies that form the *optimal disturbance*. This notion has the following sense: These strategies are optimal from the point of view of the imaginary opponent or the environment that is assumed to act with the purpose of maximizing the value of the quality index. The considered scheme is such that if one of the sides deviates from its optimal actions but the other side abides by its optimal actions, then the result of control turns out to be better for the second side than the a priori value of the game.

Essential development of the theory of differential games dates from the early 1960s. This development was promoted by the achievements of the mathematical theory of control and by the demands of practice. At present, the theory of control under lack of information is tightly connected with many branches of mechanics and mathematics. At the same time there remain many unclear questions, both in fundamental theory and in the domain of constructing effective procedures for forming strategies of control and their computer realization.

The examples considered have the form of rather simple models that belong to theoretical mechanics. If the reader is interested in evolutionary systems in other fields of application, then the corresponding differential equations should be constructed. The book is intended for specialists in the fields of mechanics, mathematics, informatics and their applications. It can be used as a textbook for students and postgraduates specializing in control theory and its applications. It can also be useful for engineers, physicists, biologists, and economists who apply mathematical methods of control theory to their fields. Reading this book requires the standard engineering education that corresponds to the first two years training of university students in mathematics, mechanics or physics. Everything that exceeds these limits is accompanied by bibliographical references.

The theory considered in this book has been developed in many countries, and its progress has been determined by the activity of many scientists. Here we have the possibility of naming only some investigators whose publications are closely connected with this book: T. Basar,

R. Bellman, A. Bensousan, V. Boltyanskii, A. Bryson, A. Chentsov, F. Chernous'ko, M. Crandall, R. Elliot, W. Flemming, A. Friedman, R. Gamkrelidze, Y. Ho, R. Isaacs, N. Kalton, A. Kurzhanskii, A. Kleimenov, A. Kryazhimskii, J. Lin, P. Lions, A. Melikyan, E. Miscenko, C. Marchal, V.M. Nikolskii, Ju. Osipov, V. Patsko, N. Petrov, L. Petrosyan, L. Pontryagin, B. Pshenichnii, E. Roxin, A. Subbotin, N. Subbotina, V. Tret'akov, V. Ushakov, P. Varaiya, J. Warga and others.

The authors are particularly grateful to their colleagues S.A. Brykalov, M.D. Lokshin, N.Yu. Lukoyanov, L.V. Petrak, and T.N. Reshetova. It was their help that made the publication of this book possible. We are also thankful to Christopher I. Byrnes and A.B. Kurzhanski for their support to publish our book. Finally, we wish to thank the entire staff of Birkhäuser Boston for their help in the production process of this book.

Chapter I
Pure strategies for positional functionals

This chapter begins with two model examples of the problems considered and solved in this book. The main content of this chapter is the introduction of the basic notions that enable us to formulate the problems of control under lack of information. We obtain the problem of control with the purpose of optimizing the ensured result. This gives the game problem on minimax of a chosen index.

1. A model problem

Let us begin with a model problem and consider a typical scheme of a *controlled dynamic system* that evolves in time t.

Let the controlled plant be a material point M with variable mass m. The point M moves on the plane $\{r_1, r_2\}$. Let $r[t] = \{r_1[t], r_2[t]\}$ be a radius-vector of the point at the time moment t. The process of control may evolve within a given time interval $t_0 \leq t \leq \vartheta$. The motion begins at some initial time moment $t_* \in [t_0, \ \vartheta)$ from an initial state $r[t_*] = \{r_{1*}, r_{2*}\} = r_*$ with a given initial speed $\dot{r}[t_*] = \{\dot{r}_{1*}, \dot{r}_{2*}\} = \dot{r}_*$. Here and below, the dot over a letter denotes the first derivative with respect to time t, i.e., $\dot{r} = dr/dt$. The motion is finished at the fixed time moment $t = \vartheta$. Let the point M be driven by a reactive control force $F_u[t]$ and also be subjected to a central force $F_c[t]$, a friction $F_f[t]$ and a disturbance $F_v[t]$. Assume that $F_u[t] = \{F_{1u}[t], F_{2u}[t]\}$ is equal to the vector $K\dot{m}[t]u[t]$. Here $m[t]$ is the mass of the point M varying according to the chosen program law $m[t]$, $t_* \leq t \leq \vartheta$. The control action-vector $u[t] = \{u_1[t], u_2[t]\}$ is proportional to the vector of the relative speed of the reactive mass, K is constant. Let the control action $u = \{u_1, u_2\}$ be constrained by the conditions

$$\left\{ u = \begin{bmatrix} u_1 \\ u_2 \end{bmatrix} : \ \frac{u_1^2}{a^2} + \frac{u_2^2}{b^2} \leq 1 \right\} \tag{1.1}$$

where $a > 0$ and $b > 0$ are given constants.

As a rule, we use the following description of sets. In the braces we write the notation for the element of this set, and then after the colon we indicate the conditions that define the set, i.e., the conditions for the elements of this set.

The set (1.1) describes abilities of the regulator U. Let us assume now that the admissible realization of control actions $u[t]$, $t_* \leq t < \vartheta$ must be a piecewise-continuous function of time t. Here when we speak about the realization of control actions $u[t]$, $t_* \leq t < \vartheta$ we do not mean that the values of these control actions $u[t]$ are assigned at the initial time moment t_* a priori as a determined program or as a definite function for all future time moments $t \in (t_*, \vartheta)$. In the considered case the values $u[t]$ (1.1) can be assigned as the *feedback control actions* depending on the information about the current states of the moving controlled plant. The feedback control of this type is the topic of this book. However, even if we use feedback control, after the completion of the process at time moment ϑ we obtain the control actions that were realized as a known function of t. We call this function $u[t]$, $t_* \leq t < \vartheta$ the *realization* of *control actions*, or shortly, "control actions".

Assume now that the central force $F_c[t] = \{F_{1c}[t], F_{2c}[t]\}$ attracting the point M to the origin $\{r_1 = 0,\ r_2 = 0\}$ and the friction force $F_f[t] = \{F_{1f}[t], F_{2f}[t]\}$ are proportional to the vectors $r[t]$ and $\dot{r}[t]$ respectively. Let the force $F_v[t] = \{F_{1v}[t], F_{2v}[t]\}$ of irregular *disturbance* be proportional to the disturbance $v = \{v_1, v_2\}$ which is constrained by the condition

$$\left\{ v = \begin{bmatrix} v_1 \\ v_2 \end{bmatrix} : \ \frac{v_1^2}{c^2} + \frac{v_2^2}{d^2} \leq 1 \right\} \tag{1.2}$$

where $c > 0$ and $d > 0$ are given constants. Condition (1.2) describes abilities of the device V that generates the disturbance. We note that the possible realization $v[t]$, $t_* \leq t < \vartheta$ must be a piecewise-continuous function of time t.

Thus we consider the motion of the point M with the variable mass $m[t]$ in the plane $\{r_1, r_2\}$ subjected to the following forces (see Figure 1.1):

$$F_u[t] = K\dot{m}[t]u[t] = \begin{bmatrix} K\dot{m}[t]u_1[t] \\ K\dot{m}[t]u_2[t] \end{bmatrix} \tag{1.3}$$

$$F_c[t] = \beta[t]r[t] = \begin{bmatrix} \beta[t]r_1[t] \\ \beta[t]r_2[t] \end{bmatrix} \tag{1.4}$$

$$F_f[t] = \alpha[t]\dot{r}[t] = \begin{bmatrix} \alpha[t]\dot{r}_1[t] \\ \alpha[t]\dot{r}_2[t] \end{bmatrix} \tag{1.5}$$

$$F_v[t] = Nv[t] = \begin{bmatrix} Nv_1[t] \\ Nv_2[t] \end{bmatrix}. \tag{1.6}$$

Here $K > 0$ and $N > 0$ are given constants, $\alpha[t]$ and $\beta[t]$ are given functions.

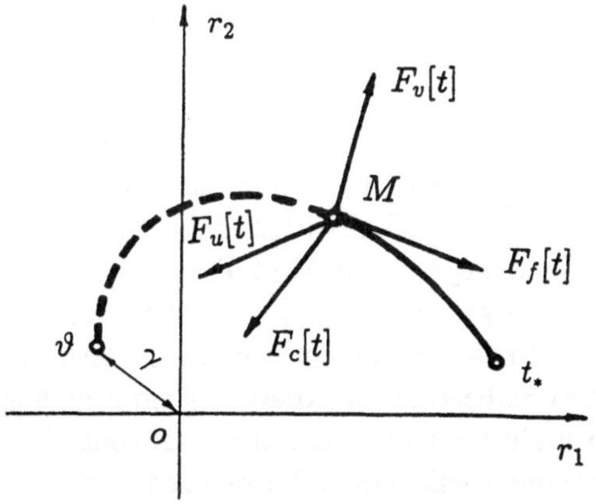

Figure 1.1

The motion $r[t]$, $t_* \leq t \leq \vartheta$ of the point M satisfies the Meshcherskii differential equation

$$\ddot{r} = \frac{\alpha[t]}{m[t]}\dot{r} + \frac{\beta[t]}{m[t]}r + K\frac{\dot{m}[t]}{m[t]}u + \frac{N}{m[t]}v, \qquad t_* \leq t < \vartheta. \tag{1.7}$$

We suppose in this example that the mass m of the point M varies according to the program law

$$m[t] = m_* \exp\{-\lambda(t - t_*)\} \tag{1.8}$$

where $m_* > 0$ and $\lambda > 0$ are given constants.

The problem consists in constructing a feedback control law U^0 with the purpose of guaranteeing the smallest possible distance between the point M and the origin $\{r_1 = 0,\ r_2 = 0\}$ at given time moment $t = \vartheta$. The *quality index* of the process of control has the following form:

$$\gamma = (r_1^2[\vartheta] + r_2^2[\vartheta])^{1/2}. \tag{1.9}$$

The formalized definitions of the admissible feedback control laws U used in this book will be given below. Here we restrict ourselves only to a remark that actually a feedback control law U is a rule that assigns the current control action $u[t]$ on the basis of the current state of the controlled system. The expression "U^0 guarantees the smallest possible distance" means the following. We assume that at any current time moment t the future values of the disturbance $v[\tau]$, $\tau > t$ are unknown to us. And more than that, we assume that at any current time instant t the current value $v[t]$ of the disturbance is also unknown to us . We know only that the restriction (1.2) is valid. Thus, the considered problem will be solved under lack of information about the disturbances v. Therefore, we would like to find a feedback control law U^0 that guarantees the smallest possible value of γ (1.9) under the assumption that the most unfavourable disturbances $v[t]$, $t_* \leq t < \vartheta$ may be realized. In this model example the most unfavourable disturbances $v_U[t]$ for a control law U are the ones that give the largest value γ (1.9), i.e., give the largest distance between M and the origin $\{r_1 = 0,\ r_2 = 0\}$ at the termination time moment ϑ in response to the chosen control law U. These disturbances can be constructed and realized in an experiment on a high-speed computer.

Of course, in reality, when a real object is controlled, the most unfavourable disturbances will hardly be realized. But this fact does not diminish the usefulness of the criterion based on a guaranteed result in the proposed approach. We shall see that optimal algorithms of feedback control proposed in this book possess the following property. If the disturbances $v[t]$, $t_* \leq t < \vartheta$ are not the most unfavourable, then the proposed method of optimal control considerably improves the result of control in comparison with an a priori ensured result.

According to the above, we say that the value ρ_u is an ensured or guaranteed result for a fixed *control law* U and for an initial state $\{t_*, r_*, \dot{r}_*\}$ of the system (1.7) if ρ_u is a supremum of the values γ (1.9) that can be realized by this method of control U. Thus, it seems natural to consider a problem about the *optimal* control law U^0 that gives the smallest value $\rho^0 = \rho_u^0$ of the *ensured result* ρ_u among all other admissible control laws U. We may call this value ρ^0 the *optimal ensured result*. It should be noted that the value ρ^0 for any initial state $\{t_*, r_*, \dot{r}_*\}$ of the controlled system (1.7) is to be calculated before the time moment t_*, which is the initial moment of a process of control. We shall consider a problem about the control law U^0 that gives an *optimal* ensured result ρ^0 for a quality index γ (1.9). So to say, the control actions $u[t]$ are aimed to minimize the value γ, and the actions of disturbance $v[t]$ can be aimed to maximize γ. The rigorous mathematical formulation of the previous phrase should include a definition of the admissible control law U. It will be given below.

Here we do not discuss the real meaning of the considered mechanical model problem. We accept the description (1.3)–(1.6) of the forces $F[t]$ as given.

Let us now define more accurately the statement of the problem.

We assume that we can control the following way. We can choose a vector-function

$$u(t, r, \dot{r}, \varepsilon) = \{u_1(t, r, \dot{r}, \varepsilon),\ u_2(t, r, \dot{r}, \varepsilon)\} \tag{1.10}$$

with the arguments t, $r = \{r_1, r_2\}$, $\dot{r} = \{\dot{r}_1, \dot{r}_2\}$ and also a parameter $\varepsilon > 0$. The function $u(\cdot)$ (1.10) must satisfy condition (1.1) for any possible $t \in [t_0,\ \vartheta)$, $r \in \mathbf{R}^2$, $\dot{r} \in \mathbf{R}^2$, and $\varepsilon > 0$.

Choose any partition $\Delta\{t_i^{(u)}\}$ of the time interval $[t_*,\ \vartheta]$ such that

$$\Delta\{t_i^{(u)}\} = \{t_1^{(u)} = t_*,\quad t_i^{(u)} < t_{i+1}^{(u)},\quad i = 1, \ldots, k_u,\quad t_{k_u+1}^{(u)} = \vartheta\} \tag{1.11}$$

where k_u is some natural number.

We note that the parameter $\varepsilon = \varepsilon_u$ is not an informational argument in (1.10). We call it a parameter of accuracy. It is chosen by us before the beginning of the process of control and then we fix it for all the future time moments t in the time interval $[t_*,\ \vartheta]$.

The chosen triple $\{u(t, r, \dot{r}, \varepsilon), \varepsilon_u, \Delta\{t_i^{(u)}\}\}$ defines a control law U. We write this in symbolic form

$$U = \{u(\cdot), \varepsilon_u, \Delta\{t_i^{(u)}\}\}. \tag{1.12}$$

As usual we call the triple $\{t, r, \dot{r}\}$ the position of the controlled system (1.7) at time t, and a function

$$u(\cdot) = \{u(t, r, \dot{r}, \varepsilon),\quad t \in [t_0,\ \vartheta),\quad r \in \mathbf{R}^2,\quad \dot{r} \in \mathbf{R}^2,\quad \varepsilon > 0\} \tag{1.13}$$

will be called a pure positional strategy.

We say that the control law U (1.12) corresponds to the strategy $u(\cdot)$ (1.13). The control law U is realized in the following way. Let us call the regulator U the device that realizes the control law U. Suppose that in the progress of the process we have achieved the moment $t_i^{(u)} \in \Delta\{t_i^{(u)}\}$, $i = 1, \ldots, k_u$. Here $\Delta\{t_i^{(u)}\}$ is a partition (1.11). We assume that the precise values of $r[t_i^{(u)}]$ and $\dot{r}[t_i^{(u)}]$ become available to the regulator U immediately, i.e., we are aware of the position $\{t_i^{(u)}, r[t_i^{(u)}], \dot{r}[t_i^{(u)}]\}$ of the system (1.7) at time moment $t_i^{(u)}$. Also we admit that the regulator U can instantly compute the vector

$$u[t_i^{(u)}] = u(t_i^{(u)}, r[t_i^{(u)}], \dot{r}[t_i^{(u)}], \varepsilon_u) \tag{1.14}$$

and create the reactive force

$$F_u[t] = F_u[t_i^{(u)}],\quad t_i^{(u)} \leq t < t_{i+1}^{(u)} \tag{1.15}$$

where $F_u[t]$ is given by (1.3). If the object M is observed and controlled with the help of a high-speed computer, which is included in regulator U, then this idealization seems to be an admissible one. We assume also that in the time interval $t_i \leq t < t_{i+1}$ some admissible disturbances $v[t]$ (1.2) act together with the control actions $u[t]$ (1.14). These disturbance actions are formed by circumstances that are uncontrollable by us.

A scheme of the process of control is shown in Figure 1.2.

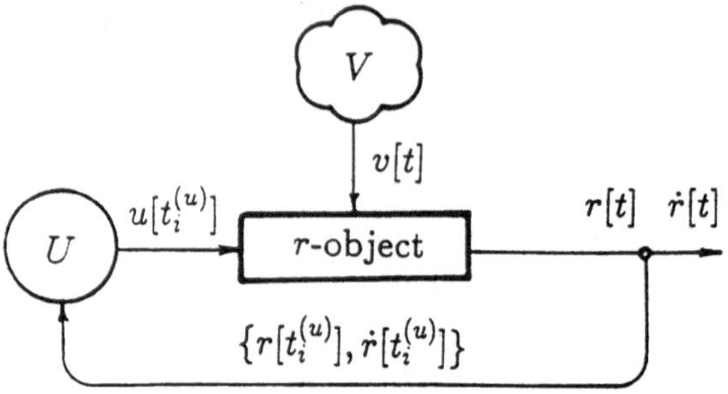

Figure 1.2

In this Figure the symbol U designates the control device or the regulator that corresponds to the control law U (1.12). The symbol V designates the environment or the device which generates the disturbances.

Let this procedure be realized step-wise on the intervals $t_i^{(u)} \leq t < t_{i+1}^{(u)}$, $i = 1, \ldots, k_u$ from the initial moment $t_1^{(u)} = t_*$, till termination at the moment $t_{k_u+1}^{(u)} = \vartheta$. Here $t_i^{(u)}$, $i = 1, \ldots, k_u + 1$, are the moments from the chosen partition $\Delta\{t_i^{(u)}\}$ (1.11), which is an element of the control law U (1.12). At the initial moment $t_1^{(u)} = t_*$ we have the initial conditions

$$r[t_1^{(u)}] = r[t_*] = r_* = \begin{bmatrix} r_{1*} \\ r_{2*} \end{bmatrix} \tag{1.16}$$

$$\dot{r}[t_1^{(u)}] = \dot{r}[t_*] = \dot{r}_* = \begin{bmatrix} \dot{r}_{1*} \\ \dot{r}_{2*} \end{bmatrix}. \tag{1.17}$$

Further, the values $r[t_i^{(u)}]$ and $\dot{r}[t_i^{(u)}]$, $i = 2, \ldots, k_u + 1$, are obtained recurrently according to the following step-by-step differential equation:

$$\ddot{r}[t] = \frac{\alpha[t]}{m[t]}\dot{r}[t] + \frac{\beta[t]}{m[t]}r[t] +$$

$$+K\frac{\dot{m}[t]}{m[t]}u(t_i^{(u)}, r[t_i^{(u)}], \dot{r}[t_i^{(u)}], \varepsilon_u) +$$

$$+\frac{N}{m[t]}v[t], \quad t_i^{(u)} \leq t < t_{i+1}^{(u)}, \quad i = 1, \ldots, k_u. \tag{1.18}$$

Therefore, in accordance with (1.8) we have

$$\ddot{r}[t] = \frac{\exp\{\lambda\,(t - t_*)\}}{m_*}\{\alpha[t]\dot{r}[t] + \beta[t]r[t] + Nv[t]\} -$$

$$-\lambda K u(t_i^{(u)}, r[t_i^{(u)}], \dot{r}[t_i^{(u)}], \varepsilon_u),$$
$$t_i^{(u)} \le t < t_{i+1}^{(u)}, \quad i = 1, \ldots, k_u. \tag{1.19}$$

Then after the termination of the whole process the realization of the disturbances

$$\{v[t],\ \ t_* \le t < \vartheta\} = \{v[t],\ \ t_i^{(u)} \le t < t_{i+1}^{(u)},\ \ i = 1, \ldots, k_u\} \tag{1.20}$$

will be obtained, too.

The given initial position $\{t_*, r_*, \dot{r}_*\}$, the chosen control law U (1.12) and the realization $v[t]$, $t_* \le t < \vartheta$, (1.20) define the motion $r[t]$, $t_* \le t \le \vartheta$, uniquely, so we obtain

$$\{r[t],\ \ t_* \le t \le \vartheta\} = \{r[t],\ \ t_i^{(u)} \le t \le t_{i+1}^{(u)},\ \ i = 1, \ldots, k_u,$$
$$r[t_1^{(u)}] = r_*, \quad \dot{r}[t_1^{(u)}] = \dot{r}_*\} \tag{1.21}$$

where $r[t]$, $t_i^{(u)} \le t \le t_{i+1}^{(u)}$ are solutions of differential equation (1.19).

Let us also introduce the following designation. Let a function be defined in some set of values of the argument. At the same time we keep in mind this function as a whole. Then we designate this function in the following way. We substitute the point instead of the letter that designates the argument. This designation was used above in (1.13). For example, a function $f[t]$ determined in a time interval $\tau_* \le t \le \tau^*$ is designated as

$$f[\tau_*[\cdot]\tau^*] = \{f[t],\ \tau_* \le t \le \tau^*\} \tag{1.22}$$

and a function determined in a half-interval $\tau_* \le t < \tau^*$ is denoted by the symbol

$$f[\tau_*[\cdot]\tau^*) = \{f[t],\ \tau_* \le t < \tau^*\}. \tag{1.23}$$

For example, a realization of disturbances $v[t]$, $t_* \le t < \vartheta$ will be designated by

$$v[t_*[\cdot]\vartheta) = \{v[t],\ t_* \le t < \vartheta\} \tag{1.24}$$

and a realization $r[t]$, $t_* \le t \le \vartheta$ of motion by

$$r[t_*[\cdot]\vartheta] = \{r[t],\ t_* \le t \le \vartheta\}. \tag{1.25}$$

Let us return now to (1.21). With regard to (1.22), (1.25) the notation (1.21) can be represented in the form

$$r_{U,v}[t_*[\cdot]\vartheta] = \{r_{U,v}[t_i^{(u)}[\cdot]t_{i+1}^{(u)}], \quad i = 1, \ldots, k_u;$$
$$r_{U,v}[t_*[t_1^{(u)}]\vartheta] = r_*, \quad \dot{r}_{U,v}[t_*[t_1^{(u)}]\vartheta] = \dot{r}_*\}. \tag{1.26}$$

Here the lower indices U and v show that motion $r_{U,v}[\cdot]$ is formed by the control law U (1.12) together with the realization of disturbances $v[t_*[\cdot]\vartheta)$ (1.20), (1.24).

The motion $r_{U,v}[t_*[\cdot]\vartheta]$ (1.26) determines the value of the vector $r_{U,v}[\vartheta] = \{r_{1_{U,v}}[\vartheta], r_{2_{U,v}}[\vartheta]\}$ such that

$$r_{U,v}[\vartheta] = r_{U,v}[t_*[t_{k+1}^{(u)}]\vartheta] \tag{1.27}$$

where $t_{k+1}^{(u)} = \vartheta$ is the termination time moment for the considered process. With regard to (1.9), a value of the quality index

$$\gamma = \left(r_{1_{U,v}}^2[\vartheta] + r_{2_{U,v}}^2[\vartheta]\right)^{1/2} \tag{1.28}$$

is determined.

Thus, the value γ (1.28) is determined uniquely by the initial position $\{t_*, r_*, \dot{r}_*\}$, the chosen feedback control law U (1.12) and the realization of disturbances $v[t_*[\cdot]\vartheta)$ (1.20). In symbolic form it can be expressed as

$$\gamma = \gamma(t_*, r_*, \dot{r}_*;\ U;\ v[t_*[\cdot]\vartheta)). \tag{1.29}$$

The realization $v[t_*[\cdot]\vartheta)$ is uncontrollable by us, when we choose the control law U (1.12) we can say only that the value γ (1.9) that can be realized will not exceed a quantity

$$\rho(t_*, r_*, \dot{r}_*;\ U) = \sup_{v[t_*[\cdot]\vartheta)} \gamma(t_*, r_*, \dot{r}_*;\ U;\ v[t_*[\cdot]\vartheta)). \tag{1.30}$$

Here supremum is calculated with respect to all admissible piecewise-continuous realizations of disturbances $v[t_*[\cdot]\vartheta)$ (1.2), (1.24).

We call a quantity $\rho(\cdot, U)$ (1.30) an ensured result for the quality index γ (1.9) and for a chosen control law U (1.12). It can be seen from (1.30) that the value $\rho(\cdot, U)$ satisfies the following two properties:

1. For any admissible realization of disturbances $v[t_*[\cdot]\vartheta)$ the value $\gamma(t_*, r_*, \dot{r}_*;\ U;\ v[t_*[\cdot]\vartheta))$ (1.29) satisfies the inequality

$$\gamma(t_*, r_*, \dot{r}_*;\ U;\ v[t_*[\cdot]\vartheta)) \leq \rho(t_*, r_*, \dot{r}_*;\ U) \tag{1.31}$$

for the initial position $\{t_*, r_*, \dot{r}_*\}$.

2. For any number $\eta > 0$ there exists an admissible realization of disturbances $v_\eta[t_*[\cdot]\vartheta)$ such that an inequality

$$\gamma(t_*, r_*, \dot{r}_*;\ U;\ v_\eta[t_*[\cdot]\vartheta)) > \rho(t_*, r_*, \dot{r}_*;\ U) - \eta \tag{1.32}$$

holds.

It is naturally to raise a problem about such optimal control law

$$U^0 = \{u^0(\cdot);\ \varepsilon_u^0;\ \Delta^0\{t_i^{(u)}\}\} \tag{1.33}$$

that the value $\rho(\cdot, U^0)$ is as small as possible, i.e.,

$$\rho(t_*, r_*, \dot{r}_*;\ U^0) = \min_U \rho(t_*, r_*, \dot{r}_*;\ U) \tag{1.34}$$

for the given initial position $\{t_*, r_*, \dot{r}_*\}$.

Let us suppose that such a control law U^0 (1.33) exists. It means that if we use this law U^0 (1.33), then we shall be guaranteed that at time moment ϑ the controlled point M will be in the ρ_u^0-neighbourhood of origin $\{r_1 = 0,\ r_2 = 0\}$. And this will be valid even if the realization of disturbances $v^0[t_*[\cdot]\vartheta)$ is most unfavourable. Such a realization $v^0[t_*[\cdot]\vartheta)$, if it exists, is a maximizing function that solves a maximum problem (1.30). Under our assumption no one admissible law U (1.12) can guarantee the inequality $\gamma \leq \rho_u$ where $\rho_u < \rho_u^0$.

However, in the cases considered here the optimal control law U^0 (1.33), (1.34) usually does not exist. Let us explain this fact. An admissible control law U (1.12) is determined by three components: by the function $u(\cdot) = u(t, r, \dot{r}, \varepsilon)$, by the parameter $\varepsilon_u > 0$ and by the partition $\Delta\{t_i^{(u)}\}$. If we change the parameter ε_u or change the partition $\Delta\{t_i^{(u)}\}$ (1.11) then in accordance with (1.12) we change the law U. Because of that the value of the ensured result $\rho(\cdot;\ U)$ (1.30) can be changed too. Usually, the smaller the time steps $t_{i+1} - t_i$, $i = 1, \dots, k$, of the partition $\Delta\{t_i\}$ (1.11), the smaller the value $\rho(t_*, r_*, \dot{r}_*;\ U)$ is. And so it occurs that no partition $\Delta^0\{t_i\}$ (1.11) gives the optimal law U^0 (1.34). Thus, it is natural to define the ensured result $\rho(t_*, r_*, \dot{r}_*;\ u(\cdot))$ only for the strategy $u(\cdot)$ (1.13) and to raise a problem about the optimal strategy $u^0(\cdot)$.

For a strategy $u(\cdot)$ (1.13) and for the initial position $\{t_*, r_*, \dot{r}_*\}$ we call the quantity

$$\rho(t_*, r_*, \dot{r}_*;\ u(\cdot)) = \overline{\lim_{\varepsilon \to 0}}\ \lim_{\delta \to 0} \sup_{\Delta_\delta} \rho(t_*, r_*, \dot{r}_*;\ U_\delta) \tag{1.35}$$

an ensured result.

Here symbol Δ_δ where $\delta > 0$ denotes the partition $\Delta\{t_i^{(u)}\}$ (1.11) that satisfies the condition

$$\max_i\ (t_{i+1}^{(u)} - t_i^{(u)}) \leq \delta. \tag{1.36}$$

In (1.35) $\rho(\cdot;\ U_\delta)$ is a value (1.30) where $U = U_\delta$. Here U_δ is a law U (1.12) corresponding to the fixed strategy $u(\cdot)$ and to the partition $\Delta_\delta\{t_i^{(u)}\}$ (1.36).

The definition of the quantity $\rho(\cdot;\ u(\cdot))$ (1.35) means the following.

For an arbitrary $\eta > 0$ there exist $\varepsilon(\eta;\ t_*, r_*, \dot{r}_*) > 0$ and $\delta(\eta, \varepsilon;\ t_*, r_*, \dot{r}_*) > 0$ such that the inequality

$$\gamma(r_{U_{\delta},\, v}[\vartheta]) \leq \rho(t_*, r_*, \dot{r}_*;\ u(\cdot)) + \eta \tag{1.37}$$

holds for any motion $r_{U_\delta, v}[t_*[\cdot]\vartheta]$ (1.26) generated by the strategy $u(\cdot)$ starting from the initial position $\{t_*, r_*, \dot{r}_*\}$ provided $\varepsilon_u \leq \varepsilon(\eta;\ t_*, r_*, \dot{r}_*)$ and $\delta < \delta(\eta, \varepsilon_u;\ t_*, r_*, \dot{r}_*)$.

The strategy $u^0(\cdot) = u^0(t, r, \dot{r}, \varepsilon)$ is optimal if

$$\rho(t_*, r_*, \dot{r}_*;\ u^0(\cdot)) = \min_{u(\cdot)} \rho(t_*, r_*, \dot{r}_*;\ u(\cdot)) = \rho_u^0(t_*, r_*, \dot{r}_*) \tag{1.38}$$

for every initial position $\{t_*, r_*, \dot{r}_*\}$ in some fixed region G in the space $\{t, r, \dot{r}\}$. In our example we can assume $G = \{t, r, \dot{r} :\ t_0 \leq t \leq \vartheta,\ | r | < \infty,\ | \dot{r} | < \infty\}$.

For the initial position $\{t_*, r_*, \dot{r}_*\}$ we call the quantity $\rho_u^0(t_*, r_*, \dot{r}_*)$ the optimal ensured result.

Thus, the definitions (1.35)–(1.38) mean that in the considered model problem the algorithm of control $U^0_{\varepsilon,\delta}$, which is based on the optimal strategy $u^0(\cdot)$ (1.38), guarantees the following result. The object M controlled by $U^0_{\varepsilon,\delta}$ is in the $(\rho_u^0 + \eta)$-neighbourhood of origin $\{r_1 = 0,\ r_2 = 0\}$ at the time moment $t = \vartheta$, if a parameter $\varepsilon > 0$ and a partition $\Delta_\delta\{t_i^{(u)}\}$ satisfy the conditions $\varepsilon \leq \varepsilon(\eta)$ and $\delta \leq \delta(\varepsilon, \eta)$. On the other hand, in Section 9 we show that for any $u(\cdot)$ (1.13), any position $\{t_*, r_*, \dot{r}_*\}$ and an arbitrary law of control U that corresponds to $u(\cdot)$ there exists a realization of disturbances $v_\eta[t_*[\cdot]\vartheta)$ such that $\gamma_{U,v_\eta} > \rho(\cdot; u^0(\cdot)) - \eta$.

Of course the values $\varepsilon(\eta)$ and $\delta(\eta, \varepsilon)$ can depend on the initial position $\{t_*, r_*, \dot{r}_*\}$. We call the optimal strategy uniform in the given region G if we can choose the values $\varepsilon(\eta)$ and $\delta(\eta, \varepsilon)$ independent of initial position $\{t_*, r_*, \dot{r}_*\} \in G$.

We have discussed this model problem in order to give some image of the mathematical problems of control that will be studied and solved in this book. It will be proved that the optimal strategy $u^0(\cdot) = u^0(t, r, \dot{r}, \varepsilon)$ (1.38) exists. Also some effective methods for calculating the value of ensured result $\rho_u^0(t_*, r_*, \dot{r}_*)$ will be given further in Chapters I and II. We shall not discuss here the construction of the optimal strategy $u^0(\cdot) = u^0(t, r, \dot{r}, \varepsilon)$ and the calculation of the value $\rho_u^0(t, r, \dot{r})$. Now we give only the results of the simulation for the corresponding process of control. It was simulated on a computer IBM PC/AT for the following parameters in the previous equations and relations:

$$t_* = t_0 = 0.00, \qquad \vartheta = 4.00, \tag{1.39}$$

$$\alpha(t) = 0.10, \ \ \beta(t) = 4.00, \ \ K = 40.00, \ \ \lambda = 0.20, \ \ N = 2.40, \ \ m_* = 1.00. \tag{1.40}$$

In the conditions (1.1), (1.2) we take

$$a = 6.00, \qquad b = 2.00, \qquad c = 2.00, \qquad d = 4.00. \tag{1.41}$$

The initial data were the following

$$r_* = \{0.00, 2.00\}, \qquad \dot{r}_* = \{2.00, 1.00\}. \tag{1.42}$$

For this initial position $\{t_*, r_*, \dot{r}_*\}$ (1.42) the value $\rho_u^0(t_*, r_*, \dot{r}_*) = 0.79$. In Figure 1.3 we depict in the plane $\{r_1, r_2\}$ the corresponding realization $r_{u^0,v^0}[t_*[\cdot]\vartheta]$ generated by the strategy $u^0(\cdot)$ and by the realization of disturbances $v^0[t_*[\cdot]\vartheta)$ which can be called optimal. Let us explain the term "optimal disturbances". Here and below we call disturbances $\{v^0[t], \ t_* \le t < \vartheta\}$ optimal if they provide the largest possible value γ. These optimal disturbances v can be constructed also by some feedback law V^0 based on some optimal strategy $v^0 = v^0(t, r, \dot{r}, \varepsilon)$. According to the above we call the strategy $v^0(\cdot)$ optimal if it ensures the largest possible value γ for any control actions aimed to minimize this value. The optimal strategy $v^0(\cdot)$ exists too. The construction of this maximizing realization of disturbances $v^0[t_*[\cdot]\vartheta)$ will be given further in Chapters I and II. In our example, the parameter $\varepsilon > 0$ and the time step $\delta = t_{i+1}^{(u)} - t_i^{(u)}$ are

$$\varepsilon = 0.02, \ \ \delta = 0.01. \tag{1.43}$$

In the considered case, the experiment gives the quantity γ close to $\rho_u^0(t_*, r_*, \dot{r}_*)$ as it should be expected

$$\gamma = (r_{1_{u^0,v^0}}[\vartheta] + r_{2_{u^0,v^0}}[\vartheta])^{1/2} = 0.78 \approx \rho_u^0(t_*, r_*, \dot{r}_*) = 0.79. \tag{1.44}$$

We note also that the value $\rho_u^0(t_*, r_*, \dot{r}_*)$ was calculated in advance, i.e., at the initial time moment t_* or before it for the parameters (1.40), (1.41) and for the initial values (1.39), (1.42). But the value γ (1.44) was found only at the termination moment ϑ for the corresponding realization of the motion $r_{u^0,v^0}[t_*[\cdot]\vartheta]$.

In Figure 1.4 we show the motion $r_{u^0,v}[t_*[\cdot]\vartheta]$ generated by the optimal strategy $u^0(\cdot)$ and by some non-optimal disturbances $v[t_*, [\cdot]\vartheta)$. We obtain the value $\gamma = 0.36$, that is essentially smaller than $\rho_u^0(t_*, r_*, \dot{r}_*)$.

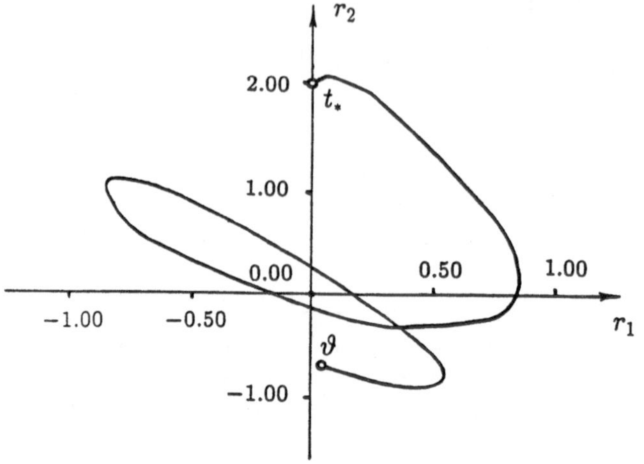

Figure 1.3

In Figure 1.5 we show the motion $r_{u,v^0}[t_* [\cdot]\vartheta]$ generated by the optimal strategy $v^0(\cdot)$ and by control actions $u[t]$ produced by some not optimal strategy $u(\cdot)$. We obtain the value $\gamma = 17.28$, which is essentially larger than $\rho_u^0(t_*, r_*, \dot{r}_*)$.

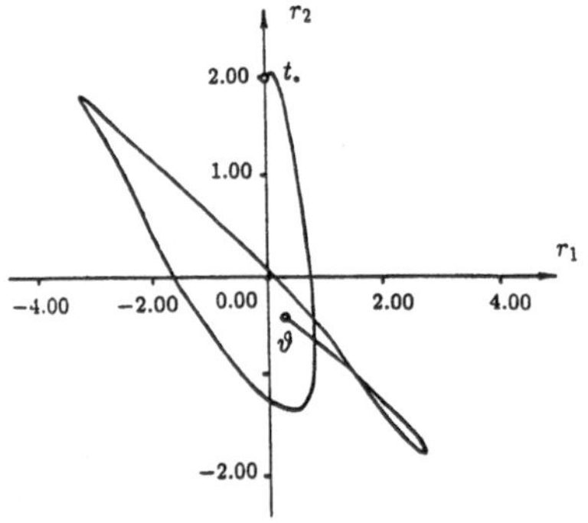

Figure 1.4

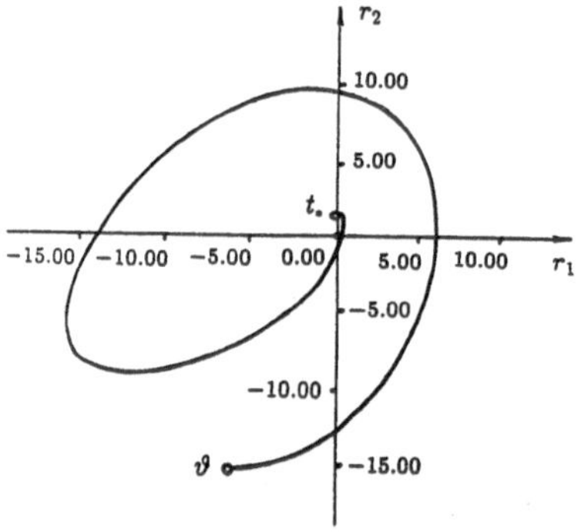

Figure 1.5

2. A model problem for mixed strategies

Let us consider one more problem that also can be solved by the method developed in this book. It is an encounter problem for two controlled objects (see Figure 2.1).

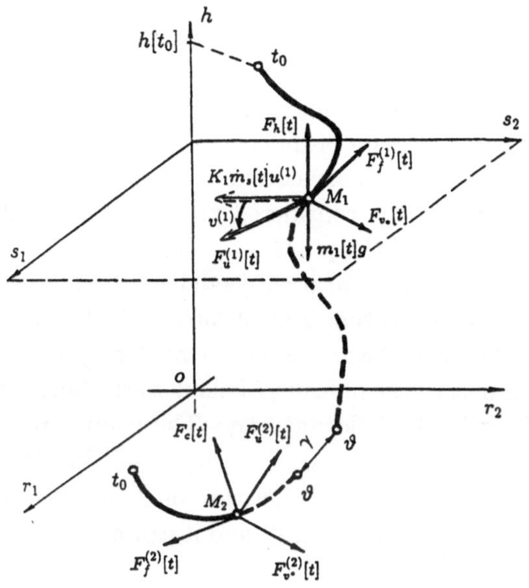

Figure 2.1

These objects are interpreted as material points M_1 and M_2 with variable masses m_1 and m_2 respectively. The first object M_1 moves in the 3-dimensional space. It is going down to the surface of earth from a given starting height $h_0 = h[t_0]$. The second object M_2 moves in the horizontal plane on the earth's surface. The aim of the objects M_1 and M_2 is to be as close to each other as possible at the moment of landing. Let the dynamics of the objects M_1 and M_2 be the following.

We assume that the object M_1 is moving down to the earth in conditions of reactive damping. Its vertical coordinate $h[t]$ changes according to a given program law. Namely, let $h[t]$ be subject to the Meshcherckii differential equation

$$m_1[t]\ddot{h} = -\alpha_1 \dot{h} + K_h[t]\dot{m}_{1h}[t] - \mid g \mid m_1[t], \quad t_0 \le t \le \vartheta \tag{2.1}$$

with a fixed starting instant of time t_0, a given starting height $h[t_0] = h_0$ and a given starting speed $\dot{h}[t_0] = \dot{h}_0$.

Here $\dot{m}_{1h}[t]$ is a component of the full speed $\dot{m}_1[t]$ of the decrease of the variable mass $m_1[t]$. This component $\dot{m}_{1h}[t]$ determines the vertical reactive damping force $F_h[t]$ applied to the object M_1. In (2.1) $K_h[t] < 0$ is a given function of t, $\alpha_1 > 0$ is a given constant and $\mid g \mid$ is the modulus of the acceleration of gravity g. The variable mass m_1 of the object M_1

varies according to the chosen law $m_1[t]$, $t_0 \leq t \leq \vartheta$. The component $\dot{m}_{1h}[t]$ of the speed $\dot{m}_1[t]$ varies according to the given program, too.

If the values $\{t_0, h_0, \dot{h}_0\}$ are given, we can calculate the time moment $t = \vartheta$ when the object M_1 reaches the earth. This moment ϑ is defined by the corresponding solution of differential equation (2.1) under condition $h[\vartheta] = 0$. Therefore, we can assume that the moment ϑ (which is the terminal moment for the considered process of control) has been found before the beginning of the process of control.

Let the object M_1 be also subjected to the action of the following horizontal forces: a reactive control force $F_u^{(1)}[t]$, a friction force $F_f^{(1)}[t]$ and an irregular disturbance (a side wind force) $F_{v_*}^{(1)}[t]$. Vectors of all these forces at every time moment t belong to the horizontal plane $\{s_1, s_2\}$ and are defined in the following way: The control reactive force $F_u^{(1)}[t]$ is generated by the control vector $u^{(1)}[t] = \{u_1^{(1)}[t], u_2^{(1)}[t]\}$ together with some "swing" disturbance $v^{(1)}$. Let the conditions be such that the norm of the vector $u^{(1)}[t]$ is equal to unit and it can be one of the four following vectors in the plane $\{s_1, s_2\}$:

$$\Big\{u^{(1)} = \begin{bmatrix} u_1^{(1)} \\ u_2^{(1)} \end{bmatrix} : \; u_{[1]}^{(1)} = \begin{bmatrix} 1 \\ 0 \end{bmatrix}, \; u_{[2]}^{(1)} = \begin{bmatrix} 0 \\ 1 \end{bmatrix}, \; u_{[3]}^{(1)} = \begin{bmatrix} -1 \\ 0 \end{bmatrix}, \; u_{[4]}^{(1)} = \begin{bmatrix} 0 \\ -1 \end{bmatrix}\Big\} \tag{2.2}$$

(see Figure 2.2).

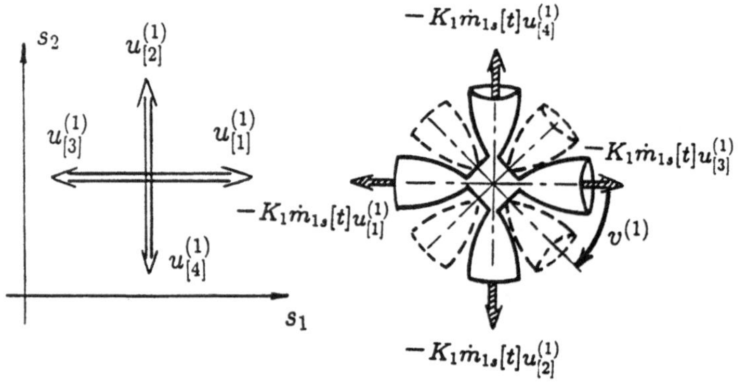

Figure 2.2

Let the modulus of the vector $F_u^{(1)}[t]$ be equal to the value $|\, K_1\dot{m}_{1s}[t] \,|$ and the vector $F_u^{(1)}[t]$ be turned with respect to the vector $u^{(1)}[t]$ on the swing angle of disturbance $v^{(1)}[t]$. Besides that, the disturbance $v^{(1)}$ satisfies the condition

$$\{v^{(1)} : \quad -\alpha_1^0 \leq v^{(1)} \leq \alpha_2^0\} \tag{2.3}$$

where $\alpha_1^0 > 0$ and $\alpha_2^0 = \alpha_1^0$ are given restrictions. Thus, the control force $F_u^{(1)}[t]$ is defined by the equality

$$F_u^{(1)} = K\dot{m}_{1s}[t]u_{v(1)}^{(1)} \tag{2.4}$$

where $K < 0$ is a given constant, and the symbol $u_{v(1)}^{(1)}$ denotes the vector $u^{(1)}$ that is turned on the angle $v^{(1)}$, i.e.,

$$u_{v(1)}^{(1)} = \begin{bmatrix} u_1^{(1)} \cos v^{(1)} - u_2^{(1)} \sin v^{(1)} \\ u_1^{(1)} \sin v^{(1)} + u_2^{(1)} \cos v^{(1)} \end{bmatrix}. \tag{2.5}$$

We assume that the angle $v^{(1)}$ is measured counter-clockwise if we look from above. Suppose that the friction force $F_f^{(1)}[t]$ is proportional to the vector of horizontal speed $\dot{s}[t] = \{\dot{s}_1[t], \dot{s}_2[t]\}$ of the object M_1, i.e.,

$$F_f^{(1)}[t] = -\alpha_1 \dot{s}[t] \tag{2.6}$$

where $\alpha_1 > 0$ is the same constant as in (2.1).

The irregular disturbance force or the force of the side wind $F_{v_*}^{(1)}[t]$ is defined by the equality

$$F_{v_*}^{(1)}[t] = N_1 v_*[t] \tag{2.7}$$

where $N_1 > 0$ is a given constant, and vector $v_*[t] = \{v_{1*}[t], v_{2*}[t]\}$ at every time moment t satisfies the restriction

$$\left\{ v_* = \begin{bmatrix} v_{1*} \\ v_{2*} \end{bmatrix} : \ v_{1*}^2 + v_{2*}^2 \le 1 \right\}. \tag{2.8}$$

In addition, at every time moment t the vector $v_*[t]$ (2.8) can be directed arbitrary in the plane $\{s_1, s_2\}$. Thus, the motion of the point M_1 in the horizontal plane $\{s_1, s_2\}$ is described by the Meshcherckii differential equation

$$m_1[t]\ddot{s} = -\alpha_1 \dot{s} + K\dot{m}_{1s}[t]u_{v(1)}^{(1)} + N_1 v_*, \quad t_0 \le t < \vartheta. \tag{2.9}$$

Here $\dot{m}_{1s}[t]$ is a component of the full speed $\dot{m}_1[t]$ of the decrease of the variable mass $m_1[t]$. This value $\dot{m}_{1s}[t]$ defines the horizontal reactive control force $F_u^{(1)}[t]$ according to (2.4). We have $\dot{m}_1[t] = \dot{m}_{1s}[t] + \dot{m}_{1h}[t]$, where $\dot{m}_{1h}[t]$ is the quantity in (2.1). Therefore, the motion of the object M_1 in the variables $h,\ s_1,\ s_2$ is described by the system of differential equations (2.1) and (2.9).

Let the object M_2 move on the earth in the horizontal plane $\{r_1, r_2\}$. This object M_2 is subjected to the action of the following horizontal forces: a control reactive force $F_u^{(2)}[t]$, a central force $F_c^{(2)}[t]$, attracting the point M_2 to (or pushing it from)the origin $\{r_1 = 0,\ r_2 = 0\}$, a friction $F_f^{(2)}[t]$ and a disturbance $F_{v*}^{(2)}[t]$. Let the forces that act on the object M_2 be analogous to the corresponding forces which act on the object M that was considered above in Section 1. Then we have

$$F_u^{(2)}[t] = K_2\dot{m}_2[t]u^{(2)}[t], \quad K_2 < 0 \ - \ \text{const}$$

$$\left\{u^{(2)} = \begin{bmatrix} u_1^{(2)} \\ u_2^{(2)} \end{bmatrix} : \ u_1^{(2)^2} + u_2^{(2)^2} \le 1\right\} \tag{2.10}$$

$$F_f^{(2)}[t] = -\alpha_2\dot{r}[t], \quad \alpha_2 > 0 \ - \ \text{const} \tag{2.11}$$

$$F_c^{(2)}[t] = \beta r[t], \quad \beta \ - \ \text{const} \tag{2.12}$$

$$F_{v*}^{(2)}[t] = N_2 v^*[t], \quad N_2 > 0 \ - \ \text{const}$$

$$\left\{v^* = \begin{bmatrix} v_1^* \\ v_2^* \end{bmatrix} : \ v_1^{*^2} + v_2^{*^2} \le 1\right\}. \tag{2.13}$$

The vectors $u^{(2)}[t]$ and $v^*[t]$ at every time moment t can be directed arbitrary in the plane $\{r_1, r_2\}$. The mass m_2 of the point M_2 changes according to the chosen program law $m_2[t]$, $t_0 \le t \le \vartheta$.

The motion of the object M_2 is described by the Meshcherckii differential equation

$$m_2[t]\ddot{r} = -\alpha_2\dot{r} + \beta r + K_2\dot{m}_2[t]u^{(2)} + N_2 v^*, \quad t_0 \le t < \vartheta. \tag{2.14}$$

We consider a situation when the objects M_1 and M_2 are driven by their control reactive forces $F_u^{(1)}[t]$ and $F_u^{(2)}[t]$ with the aim of being as close to each other as possible at the time moment ϑ. Then the quality index for the considered process of control can be chosen as the distance between the points M_1 and M_2 at the terminal time moment $t = \vartheta$, i.e.,

$$\gamma = \left((r_1[\vartheta] - s_1[\vartheta])^2 + (r_2[\vartheta] - s_2[\vartheta])^2\right)^{1/2}. \tag{2.15}$$

In addition, the process of control will be considered in the time interval $t_* \le t \le \vartheta$, where $t_* \in [t_0,\ \vartheta)$ is some initial time moment, i.e., t_* is the initial moment for the process of control. (The process is presented in Figure 2.1).

Let us discuss the use of the terms “control” and “disturbance” in this problem. We select two collections of the actions from the set of all forces applied to the united control plant that consists of two points M_1 and M_2.

The first collection comprises all the actions subjected to the "commanders" of the points M_1 and M_2. These actions should be aimed to make the value γ as small as possible. We call them the control actions and denote them by u. In the considered example we have the united control actions $u = \{u^{(1)}, u^{(2)}\}$. The second collection consists of all the actions that are not subjected to the "commanders" of the plant $\{M_1, M_2\}$. These actions might hinder the purposes of the "commanders". These actions may be generated by the environment or by some opponents, so to say. Thus, these actions might be aimed at maximizing the value γ. We call them the disturbances and denote them by v. In the considered example we have common disturbances $v = \{v^{(1)}, v_*, v^*\}$.

Below in similar cases we will always keep that in mind and will write for short: the control actions u aimed to minimize γ and disturbances v aimed to maximize γ. We have to remark that in the above two collections we do not include the regular forces generated by nature according to regular laws. In this situation it is natural to speak, as in Section 1, about guaranteeing control in conditions of uncontrollable disturbances.

The differential equation (2.1) does not contain the control actions u and the disturbances v. This equation defines only the time interval $[t_0, \ \vartheta]$ on which the process of control can be considered. In this problem of control in which the actions u and the disturbances v are aimed to minimize and maximize the quality index γ (2.15), it is sufficient to consider only the system of differential equations (2.9) and (2.14).

We confine ourselves to the case when the masses m_1 and m_2 of the objects M_1 and M_2 change according to the programs

$$m_i[t] = m_{i0}\, e^{-\lambda_i(t-t_0)}, \quad i = 1, 2 \tag{2.16}$$

where $m_{i0} > 0$ and $\lambda_i > 0, \ \ i = 1, 2$ are given constants. In addition, we assume that $m_{1h}[t]$ and $m_{1s}[t]$ in (2.1) and (2.9) are such that $m_{1h}[t] = q\, m_{1s}[t]$ for every time moment $t \in [t_0, \ \vartheta]$ where $q > 0$ is a constant. Therefore, the differential equation (2.9) takes a form

$$\ddot{s} = -\frac{\alpha_1 \dot{s}}{m_1[t]} + K_1 \lambda_1 u^{(1)}_{v^{(1)}} + \frac{N_1 v_*}{m_1[t]} \tag{2.17}$$

where $K_1 = K\,(1+q)^{-1}$, and K is the constant from (2.9).

Now we can transform the differential equations (2.9), (2.14) to the normal systems [24], [77], [79], [120], [133]. Namely, let us assume $w_1^{[r]} = r_1$, $w_2^{[r]} = \dot{r}_1$, $w_3^{[r]} = r_2$, $w_4^{[r]} = \dot{r}_2$, $w_1^{[s]} = s_1$, $w_2^{[s]} = \dot{s}_1$, $w_3^{[s]} = s_2$, $w_4^{[s]} = \dot{s}_2$.

Then

$$\begin{aligned} \dot{w}_1^{[r]} &= w_2^{[r]} \\ \dot{w}_2^{[r]} &= \frac{\beta}{m_2[t]} w_1^{[r]} - \frac{\alpha_2}{m_2[t]} + K_2\lambda_2 u_1^{(2)} + \frac{N_2 v_1^*}{m_2[t]} \\ \dot{w}_3^{[r]} &= w_4^{[r]} \\ \dot{w}_4^{[r]} &= \frac{\beta}{m_2[t]} w_3^{[r]} - \frac{\alpha_2}{m_2[t]} + K_2\lambda_2 u_2^{(2)} + \frac{N_2 v_2^*}{m_2[t]} \end{aligned} \tag{2.18}$$

$$\begin{aligned} \dot{w}_1^{[s]} &= w_2^{[s]} \\ \dot{w}_2^{[s]} &= -\frac{\alpha_1}{m_1[t]} w_4^{[s]} + K_1\lambda_1 u_{v(1)1}^{(1)} + \frac{N_1 v_{*1}}{m_1[t]} \\ \dot{w}_3^{[s]} &= w_4^{[s]} \\ \dot{w}_4^{[s]} &= -\frac{\alpha_1}{m_1[t]} w_4^{[s]} + K_1\lambda_1 u_{v(1)2}^{(1)} + \frac{N_1 v_{*2}}{m_1[t]}. \end{aligned} \tag{2.19}$$

Now subject the 4-dimensional phase vectors $w^{[r]}$ and $w^{[s]}$ to appropriate non-degenerate linear transformations $\widetilde{w}^{[l]}[t] = \left(W^{[l]}(t,\vartheta)\right)^{-1} w^{[l]}[t]$, $l = r, s$ [77], [79] where $W^{[l]}(t,\tau)$ are the fundamental matrices of solutions of the homogeneous part of the systems (2.18) and (2.19). Then we can reduce the whole 8-dimensional differential system to a suitable 4-dimensional differential system. As a result, the equations (2.9) and (2.14) are reduced to the following equations:

$$\begin{aligned} \dot{e}^{(1)} &= B^{(1)}(t) u_{v(1)}^{(1)} + C^{(1)}(t) v_*, \\ \dot{e}^{(2)} &= B^{(2)}(t) u^{(2)} + C^{(2)}(t) v^*, \end{aligned} \tag{2.20}$$

$$e^{(1)}[t] = \begin{bmatrix} \widetilde{w}_1^{[s]}[t] \\ \widetilde{w}_3^{[s]}[t] \end{bmatrix}, \quad e^{(2)}[t] = \begin{bmatrix} \widetilde{w}_1^{[r]}[t] \\ \widetilde{w}_3^{[r]}[t] \end{bmatrix} \tag{2.21}$$

where $e = \{e^{(1)}, e^{(2)}\}$ is a 4-dimensional vector that satisfies the conditions

$$e_1^{(1)}[\vartheta] = s_1[\vartheta], \quad e_2^{(1)}[\vartheta] = s_2[\vartheta], \quad e_1^{(2)}[\vartheta] = r_1[\vartheta],$$

$$e_2^{(2)}[\vartheta] = r_2[\vartheta]. \tag{2.22}$$

In (2.20) $B^{(i)}(t)$ and $C^{(i)}(t)$, $i = 1, 2$ are known matrix-functions that correspond to the parameters of the systems (2.9), (2.14), (2.16).

Now we pass to the vector $x = \begin{bmatrix} x_1 \\ x_2 \end{bmatrix}$, where

$$x_1 = e_1^{(2)} - e_1^{(1)}, \quad x_2 = e_2^{(2)} - e_2^{(1)}. \tag{2.23}$$

Then the system (2.20) takes a form

$$\dot{x} = B^{[1]}(t)u^{(1)}_{v(1)} + B^{[2]}(t)u^{(2)} + C^{[1]}(t)v_* + C^{[2]}(t)v^* \tag{2.24}$$

$$B^{[i]}(t) = \begin{pmatrix} b^{[i]}_{11}(t), \ b^{[i]}_{12}(t) \\ b^{[i]}_{21}(t), \ b^{[i]}_{22}(t) \end{pmatrix}; \quad C^{[i]}(t) = \begin{pmatrix} c^{[i]}_{11}(t), \ c^{[i]}_{12}(t) \\ c^{[i]}_{21}(t), \ c^{[i]}_{22}(t) \end{pmatrix} \tag{2.25}$$

where $B^{[i]}(t)$ and $C^{[i]}(t)$, $i = 1,2$ are known matrix-functions and $u^{(1)}_{v(1)}$, $u^{(2)}$, v_* and v^* are the vectors (2.5), (2.10), (2.8) and (2.13), respectively.

Thus we obtain the linear transformation

$$x = T_t(r, \dot{r}, s, \dot{s}) \tag{2.26}$$

which reduces the system (2.14), (2.17) to the equation (2.24).

We do not give here a more explicit description of the transformation of the variables and equations. In the present section our intention is only to illustrate the solution of the considered model problem of control in the form of its computer simulation.

With regard to (2.22), (2.23) the quality index γ (2.15) takes a form

$$\gamma = \left(x_1^2[\vartheta] + x_2^2[\vartheta]\right)^{1/2}. \tag{2.27}$$

Further, we shall denote the sets of admissible values of the control actions u and the disturbances v by the symbols P and Q, respectively. In other words, the sets P and Q describe the abilities of the regulator U and the device V that generates the disturbance. In the considered case we have

$$P = \left\{ u^{(1)} = \begin{bmatrix} u_1^{(1)} \\ u_2^{(1)} \end{bmatrix}, \quad u^{(2)} = \begin{bmatrix} u_1^{(2)} \\ u_2^{(2)} \end{bmatrix} : \ u^{(1)} \ (2.2), \ u^{(2)} \ (2.10) \right\} \tag{2.28}$$

$$Q = \left\{ v^{(1)}, \ v_* = \begin{bmatrix} v_{1*} \\ v_{2*} \end{bmatrix}, \ v^* = \begin{bmatrix} v_1^* \\ v_2^* \end{bmatrix} : \ v^{(1)} \ (2.3), \ v_* \ (2.8), \ v^* \ (2.13) \right\}. \tag{2.29}$$

The notation (2.28) means that in the considered model problem the set P consists of all the pairs of 2-dimensional vectors $u^{(1)}$ and $u^{(2)}$ that satisfy the conditions (2.2) and (2.10) respectively. The set Q (2.29) consists of all possible triples $v^{(1)}$, v_* and v^* that satisfy the conditions (2.3), (2.8) and (2.13), respectively.

Thus, the initial system of differential equations (2.9), (2.14) can be reduced by the above transformations to the following vector differential equation:

$$\dot{x} = f(t, u, v), \quad t_0 \le t < \vartheta, \quad u \in P, \quad v \in Q. \tag{2.30}$$

Here non-linear in u and v vector-function $f(t,u,v)$ is defined by the parameters of the system (2.24), u is a control action, v is a disturbance and P and Q are the sets (2.28), (2.29) respectively.

For the system (2.30) we consider the problem of the construction of the control actions u and disturbances v that are aimed to minimize and maximize, respectively, the quality index

$$\gamma = \mid x[\vartheta] \mid . \tag{2.31}$$

Here and further, the symbol $\mid x \mid$ denotes a Euclidean norm of vector $x \in \mathbf{R}^n$, i.e.,

$$\mid x \mid = (x_1^2 + \ldots + x_n^2)^{1/2} . \tag{2.32}$$

We emphasize that the system (2.30) and the quality index γ (2.31) give, generally speaking, only some suitable model for the primary controlled system (2.9), (2.14) with the index γ (2.15). However, as will be shown below, the algorithms of control that provide an effective solution of the considered problem for the x-system (2.30) with γ (2.31) can be transformed naturally for the r, s-system (2.9), (2.14) with γ (2.15), i.e., these algorithms provide an effective solution of the primary problem of control. In fact, in the scheme of the feedback control that we use it is sufficient to recalculate the informational image $\{s[t_i], \dot{s}[t_i], r[t_i], \dot{r}[t_i]\}$ into informational image $\{x[t_i]\}$ by the transformation (2.26) at every time moment t_i. In such a way we can apply the algorithm of control based on the information $\{x[t_i]\}$ to the given controlled r, s-system.

Thus, the considered problem for the system (2.30) with the quality index γ (2.31) take the form similar to the problem from Section 1 but with one essential difference. In Section 1 the controlled system (1.7) was linear in u and v. But here the system (2.30), (2.24), (2.5) is non-linear in u and v. (Speaking more accurately, the system (2.30), (2.24), (2.5) is non-linear in $u^{(1)}$ and $v^{(1)}$ and linear in $u^{(2)}$, v_*, v^*.)

Non-linearity of this type requires an approach to the solution of this problem that is quite different from the approach that was considered in Section 1.

The problem of control for the system (2.30) with the quality index γ (2.31) can be formulated and solved in one of the following classes of strategies: *pure strategies*, *counter* strategies and *mixed* strategies.

We recall [77], [79] that a pure positional strategy is understood as a law that at every time moment t assigns a control action $u[t]$ in determinate form on the basis of information about the position $\{\tau, x[\tau]\}$ that has been realized. Here either $\tau = t$, or τ is some previous time moment close to the moment t. This method of control was considered in Section 1.

The counter strategy of disturbance or the counter strategy of control are deterministic rules, too [22], [77], [79], [145], [146]. However, this kind of strategy of control is based on the following current informational data or the informational image at the time moment t. Besides information on the position $\{\tau, x[\tau]\}$ it also uses additional information about the realization $v[t]$ of the disturbance at the current time moment t.

The mixed strategy [65], [72], [79] includes the certain probability mechanism that generates the control actions. Let us emphasize, however, that at every time moment t, as in the case of pure positional strategies, the positional mixed strategy uses information only about the position $\{\tau, x[\tau]\}$.

It can be checked that in the considered case the so-called saddle point condition in a small game [48], [77], [79] is not valid for the system (2.24). It means that there exists a vector $l = \{l_1, l_2\}$ such that the equality

$$\min_{u\in P}\max_{v\in Q} \langle\, l,\ f(t,u,v)\,\rangle = \max_{v\in Q}\min_{u\in P} \langle\, l,\ f(t,u,v)\,\rangle \tag{2.33}$$

is not valid.

Here and further, the symbol $\langle a,\ b\rangle$ denotes the scalar product of two m-dimensional vectors a and b , i.e.,

$$\langle a,\ b\rangle = a^T b = a_1 b_1 + \ldots + a_m b_m \tag{2.34}$$

and the upper index T denotes transposition. We remark that all vectors in Euclidean space that are denoted by small Latin letters are understood as column vectors, i.e.,

$$a = \begin{bmatrix} a_1 \\ \vdots \\ a_m \end{bmatrix}, \quad a^T = \begin{bmatrix} a_1 \\ \vdots \\ a_m \end{bmatrix}^T = [a_1\ \ldots\ a_m]\,. \tag{2.35}$$

It is known [77], [79] that if the equality (2.33) is not true, then generally speaking there can be found such positions $\{t_*, x_*\} = \{t_*, x_{1*}, x_{2*}\}$ of the x-object (2.24) that the value of the optimal ensured result $\rho_u^0(t_*, x_*)$ in the class of pure positional strategies $u(\cdot) = u(t, x, \varepsilon)$ (see Section 1) proves to be larger than the value of the optimal ensured result $\rho_{u_v}^0(t_*, x_*)$ in the class of counter strategies $u_v(\cdot) = u(t, x, \varepsilon, v)$. However, if we use the algorithm of control based on the counter strategy, then we are restricted by a very obliging supposition of complete information about the current realization of $v[t]$. Therefore, it is natural to consider the question of a control algorithm u^0 that does not use additional information about $v[t]$ but at the same time somehow provides values of the quality index γ (2.27) that are noticeably less than the quantity $\rho_u^0(t_*, x_*)$.

Such an approach for the solution of the considered concrete model problem for the system (2.24) with γ (2.27) is the subject of this section. This approach is based on the concept of the mixed strategy. As was said above, an algorithm of control U based on this concept involves some probability mechanism. In this section we only illustrate the work of this mechanism by an example of the controlled system (2.24) with the quality index γ (2.27). The proposed scheme is such that besides the actual controlled system (2.1), (2.9), (2.14) and actual model x-object (2.24) we introduce a computer simulated y-model in the case of optimal control or a z-model if we simulate the worst disturbance. The y-model and the z-model are called "leaders" in terms of control optimization [79]. Both models are incorporated into the control loop. The evolution of all the mentioned motions is organized so that it ensures that the motion of the leader is optimal in a certain sense with respect to γ (2.27) and it ensures also stable tracking of the leader's motion by the actual motion of the x-object (2.24).

Thus, with the given controlled systems (2.1), (2.9), (2.14) and the model x-system (2.24), we associate a y-model-leader. The current state of this model is described by the vector $y[t]$. The dimension of the vector y coincides with the dimension of the phase vector x of the controlled object (2.24). The differential equation of the motion of the y-model is constructed in the following way:

Let us choose the functions

$$u_y^{(2)}(t,x,y,\varepsilon), \quad v_{*y}(t,x,y,\varepsilon), \quad v_y^*(t,x,y,\varepsilon), \tag{2.36}$$

$$p_y^*(t,x,y,\varepsilon) = \Big\{ \, p_{yl}^*(t,x,y,\varepsilon), \quad l = 1,\dots,4;$$

$$p_{yl}^*(t,x,y,\varepsilon) \geq 0, \quad \sum_{l=1}^{4} p_{yl}^*(t,x,y,\varepsilon) = 1 \, \Big\},$$

$$q_y^*(t,x,y,\varepsilon) = \Big\{ \, q_{ym}^*(t,x,y,\varepsilon), \quad m = 1,2;$$

$$q_{ym}^*(t,x,y,\varepsilon) \geq 0, \quad \sum_{m=1}^{2} q_{ym}^*(t,x,y,\varepsilon) = 1 \, \Big\} \tag{2.37}$$

where $t \in [t_0, \ \vartheta)$ and ϑ is the fixed termination time moment. For the fixed initial position $\{t_*, y_*\}$, $t_* \in [t_0, \ \vartheta]$, an assigned number $\varepsilon > 0$ and an assigned partition $\Delta\{t_i\}$ of the time interval $[t_*, \ \vartheta]$ by points t_i, $i = 1,\dots,k+1$ $(t_1 = t_*, \ t_{k+1} = \vartheta)$, the motion of the y-model is defined as the step-by-step solution of the differential equation

$$\dot{y}[t] = \sum_{l=1}^{4}\sum_{m=1}^{2} B^{[1]}(t) u^{(1)}_{[l]_{v_{[m]}}} \, p_{yl}^*(t_i, \ x[t_i], \ y[t_i], \ \varepsilon) \, q_{ym}^*(t_i, \ x[t_i], \ y[t_i], \ \varepsilon) +$$

$$+B^{[2]}(t)u_y^{(2)}(t_i,\ x[t_i],\ y[t_i],\ \varepsilon)+C^{[1]}(t)v_{*y}(t_i,\ x[t_i],\ y[t_i],\ \varepsilon)+$$
$$+C^{[2]}(t)v_y^*(t_i,\ x[t_i],\ y[t_i],\ \varepsilon),\quad t_i\le t<t_{i+1},$$
$$i=1,\dots,k,\quad y[t_1]=y_*. \tag{2.38}$$

Here $\{u_{[l]}^{(1)},\ l=1,\dots,4\}$ is a collection (2.2) of the vectors $u^{(1)}$; $v_{[1]}=-\alpha_1^0$, $v_{[2]}=\alpha_2^0$ where $-\alpha_1^0$ and α_2^0 are the swing angles in (2.3). In (2.38) $x[t_i]$ are the values of the phase vector of the controlled x-object. These values $x[t_i]$ are realized at the time moments $\tau=t_i$. We choose the functions

$$u^{(2)}(t,x,y,\varepsilon),\quad t\in[t_0,\ \vartheta),\quad x\in\mathbf{R}^2,\quad y\in\mathbf{R}^2,\quad \varepsilon>0, \tag{2.39}$$

$$p(t,x,y,\varepsilon)=\{p_l(t,x,y,\varepsilon),\quad l=1,\dots,4;$$
$$p_l(t,x,y,\varepsilon)\ge 0,\quad \sum_{l=1}^{4}p_l(t,x,y,\varepsilon)=1\}. \tag{2.40}$$

We imagine *temporary* that the motion $x[t]$, $t_*\le t\le\vartheta$ of the controlled x-object is formed in the following way:

For the fixed initial position $\{t_*,x_*\}$, for the chosen parameter $\varepsilon>0$ and for the chosen partition $\Delta\{t_i\}$, the motion $x[t]$ is determined at $t_*\le t\le\vartheta$ as the step-by-step solution of differential equation

$$\dot{x}[t]=B^{[1]}(t)u^{(1)}[t_i]_{v^{(1)}[t]}+B^{[2]}(t)u^{(2)}(t_i,x[t_i],y[t_i],\varepsilon)+$$
$$+C^{[1]}(t)v_*[t]+C^{[2]}(t)v^*[t],\quad t_i\le t<t_{i+1},\quad i=1,\dots,k,$$
$$x[t_1]=x_*. \tag{2.41}$$

Here $u^{(1)}[t_i]$ is the result of an appropriate random test for the vectors $u_{[l]}^{(1)}$, $l=1,\dots,4$ from the collection (2.2) of the control vectors $u^{(1)}$. In this test the probability of the event $u^{(1)}[t_i]=u_{[l]}^{(1)}$ is equal to $p_l(t_i,x[t_i],y[t_i],\varepsilon)$ (2.40), i.e.,

$$P(u^{(1)}[t_i]=u_{[l]}^{(1)})=p_l(t_i,x[t_i],y[t_i],\varepsilon). \tag{2.42}$$

Here and below the symbol $P(B)$ denotes the probability of the event B.

In (2.41), (2.42) $y[t_i]$ is the phase vector $y[t]$ for the y-model (2.38) at the time moment $t=t_i$. The admissible realization of disturbances

$$v[t_*[\cdot]\vartheta)=\{v^{(1)}[t]\ (2.3),\ v_*[t]\ (2.8),\ v^*[t]\ (2.13),\quad t_*\le t<\vartheta\} \tag{2.43}$$

is a realization that is formed by the environment or the device V of the supposed opponent that generates the disturbances. We assume that the actions $u[t]$ and $v[t]$ are stochastically independent in small intervals $t_i \le t < t_{i+1}$.

We imagine that the control of the x-object (2.24) is realized by the scheme presented in Figure 2.3.

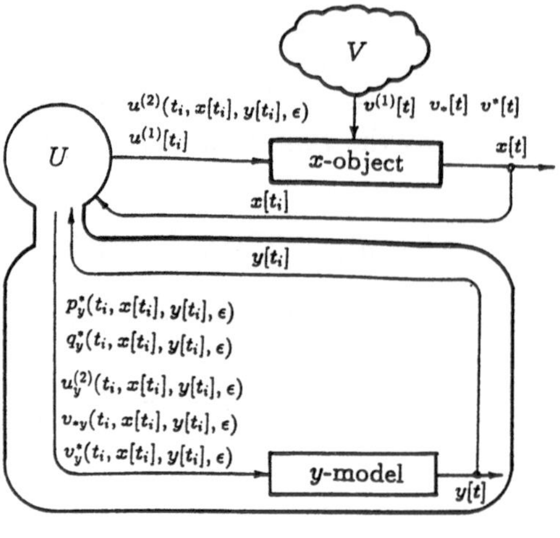

Figure 2.3

In this figure the symbol U denotes the regulator that corresponds to the control law U, and as was noted above, this regulator U includes the y-model-leader.

In the considered model problem the totality $S^u(\cdot)$ of the components

$$S^u(\cdot) = \{p(\cdot),\ \widetilde{U},\ u^{(2)}(\cdot);\ p_y^*(\cdot),\ U^*,\ q_y^*(\cdot),\ V^*,\ u_y^{(2)}(\cdot),\ v_{*y}(\cdot),\ v_y^*(\cdot)\} \tag{2.44}$$

is called the mixed strategy. Here

$$p(\cdot) = \{p_l(t,x,y,\varepsilon),\ t \in [t_0,\ \vartheta),\ x \in \mathbf{R}^2,\ y \in \mathbf{R}^2,\ \varepsilon > 0,\ l = 1,\dots,4\}, \tag{2.45}$$

$$\widetilde{U} = \{u_{[l]}^{(1)}\ (2.2),\ \ l = 1,\dots,4\}, \tag{2.46}$$

$$u^{(2)}(\cdot) = \{u^{(2)}(t,x,y,\varepsilon)\ (2.10),\quad t \in [t_0,\ \vartheta),\quad x \in \mathbf{R}^2,\ y \in \mathbf{R}^2,\ \varepsilon > 0\}, \tag{2.47}$$

$$p_y^*(\cdot) = \{p_{yl}^*(t,x,y,\varepsilon),\ \ l = 1,\dots,4\},\ \ q_y^*(\cdot) = \{q_{ym}^*(t,x,y,\varepsilon),\ \ m = 1,2\}, \tag{2.48}$$

$$U^* = \widetilde{U},\qquad V^* = \{v_{[m]},\ \ m = 1,2,\ \ v_{[1]} = -\alpha_1^0,\ \ v_{[2]} = \alpha_2^0\ (2.3)\}, \tag{2.49}$$

$$u_y^{(2)}(\cdot) = \{u_y^{(2)}(t,x,y,\varepsilon)\ (2.10)\},$$
$$v_{*y}(\cdot) = \{v_{*y}(t,x,y,\varepsilon)\ (2.8)\},\ \ v_y^*(\cdot) = \{v_y^*(t,x,y,\varepsilon)\ (2.13)\} \tag{2.50}$$

where $p(t,x,y,\varepsilon)$, $p^*_y(t,x,y,\varepsilon)$ and $q^*_y(t,x,y,\varepsilon)$ are the collections of the numbers (2.40) and (2.37), respectively.

Now let us explain the sense of the components that form the strategy $S^u(\cdot)$ (2.44). The function $p(\cdot)$ (2.45), (2.40) for every fixed $\{t, x[t], y[t], \varepsilon\}$ defines the collection of the numbers $p_l \geq 0$, $\sum_{l=1}^4 p_l = 1$, which are the probabilities of the corresponding events. The functions $p^*_y(\cdot)$ and $q^*_y(\cdot)$ (2.48), (2.37) are similar in form to the function $p(\cdot)$ (2.45), (2.40) but they are constructed for the y-model (2.38). And the functions $p^*_y(\cdot)$ and $q^*_y(\cdot)$, unlike the functions $p(\cdot)$ and $q(\cdot)$, are not used in any random tests. As we can see from equation (2.38) these functions $p^*_y(\cdot)$, $q^*_y(\cdot)$ are employed only to average the right-hand side of this equation (2.38) in deterministic way.

The control law U that corresponds to the chosen strategy $S^u(\cdot)$ (2.44) and that works in the time interval $[t_*, \ \vartheta)$ is defined as the totality of the three components (similarly as in Section 1, see (1.12)):

$$U = \{S^u(\cdot); \ \varepsilon; \ \Delta\{t_i\}\}. \tag{2.51}$$

We assume that the initial states x_* and y_* are sufficiently close to each other. In particular, in (2.38) we can take $y_* = x_*$. The prescribed control law U (2.51) together with some admissible realization of disturbances $v[t_*[\cdot]\vartheta)$ (2.43) define the motion of x-object

$$\begin{aligned} x_{U,v}[t_*[\cdot]\vartheta] = \{x_{U,v}[t_i[\cdot]t_{i+1}] = \{x_{U,v}[t], \quad t_i \leq t \leq t_{i+1}\}, \\ i = 1,\dots,k, \quad x_{U,v}[t_1] = x_*\} \end{aligned} \tag{2.52}$$

according to the scheme (2.41), (2.42). The motion of the considered y-model

$$\begin{aligned} y_{U,v}[t_*[\cdot]\vartheta] = \{y_{U,v}[t_i[\cdot]t_{i+1}] = \{y_{U,v}[t], \quad t_i \leq t \leq t_{i+1}\}, \\ i = 1,\dots,k, \quad y_{U,v}[t_1] = y_*\} \end{aligned} \tag{2.53}$$

is realized by the scheme (2.38) with the help of the computer.

We emphasize that the mechanism that forms the disturbances $v[t]$ is unknown to us. We suppose only that the swing disturbances $v^{(1)}[t]$ for every step $t_i \leq t < t_{i+1}$, $i = 1,\dots,k$ are formed under the condition of stochastic independence from the control actions $u[t] = u^{(1)}[t_i]$, $t_i \leq t < t_{i+1}$, where $u^{(1)}[t_i]$ is obtained by the corresponding stochastic test at the time moment t_i. About other components $v_*[t]$ and $v^*[t]$ of disturbances $v[t]$ (2.43) we assume the same as in Section 1.

Here the realization of extended motion $\{x_{U,v}[t_*[\cdot]\vartheta], y_{U,v}[t_*[\cdot]\vartheta]\}$ proves to be the realization of a certain stochastic process. This stochastic process will be described more strictly below in the general statement of the problem in Section 32–34. It will be done on the basis of a suitable probability space. Here we confine ourselves to a brief outline.

Depending on these or those outcomes $u[t_i]$ $(i = 1, \ldots, k)$ of the random tests for the choice of the vectors $u^{(1)}_{[l]}$ from the collection $\widetilde{U}$ (2.46) and depending on the realizations $v^{(1)}[t_i[\cdot]t_{i+1})$, $i = 1, \ldots, k$ of the random angle disturbance (2.3), we obtain this or that realization $\{x_{U,v}[t_*[\cdot]\vartheta], y_{U,v}[t_*[\cdot]\vartheta]\}$ of the random motion $\{x(\cdot), y(\cdot)\}$. The totality of all possible realizations of this motion form the stochastic process. We call it the random motion of the complex $\{x\text{-object}, y\text{-model}\}$. For this random motion we introduce an event B_ρ that consists in fulfillment of the inequality

$$\gamma = | x_{U,v}[\vartheta] | \leq \rho \tag{2.54}$$

where $\rho > 0$ is a chosen number and γ is a quality index (2.31).

To every admissible data $\{t_*, x_*, y_*;\ U, v[t_*[\cdot]\vartheta)\}$ we put into correspondence $P(B_\rho) \in [0,\ 1]$ which is the probability of the event (2.54). Then for the given extended initial position $\{t_*, x_*, y_*\}$, for the chosen feedback control law U (2.51) and for the fixed number $\nu \leq 1$ we introduce the result $\rho(t_*, x_*, y_*;\ U, \nu)$ that is ensured with the probability ν. This result is the quantity

$$\rho(t_*, x_*, y_*;\ U, \nu) = \inf\ \rho \tag{2.55}$$

where ρ denotes all the numbers for which the inequality

$$P(\gamma = | x_{U,v}[\vartheta] | \leq \rho) \geq \nu \tag{2.56}$$

holds for every admissible mechanism V that forms the disturbances $v[t_*[\cdot]\vartheta)$ (2.43).

It can be checked that among all the numbers ρ that satisfy the condition (2.56) there is the smallest number. Thus, $\rho(\cdot, U, \nu)$ is the minimal value among the numbers ρ that satisfy the condition (2.56).

For a chosen strategy $S^u(\cdot)$ (2.44) and for the initial position $\{t_*, x_*\}$ we call the quantity

$$\rho(t_*, x_*, S^u(\cdot)) = \lim_{\nu \to 1} \overline{\lim_{\varepsilon \to 0}} \lim_{\zeta \to 0} \lim_{\delta \to 0} \sup_{|x_* - y_*| \leq \zeta} \sup_{\Delta_\delta} \rho(t_*, x_*, y_*;\ U_\delta, \nu) \tag{2.57}$$

the ensured result.

It follows from the definition of the quantity $\rho(\cdot)$ (2.57) that the value $\rho(t_*, x_*, S^u(\cdot))$ is the smallest number among the numbers ρ that satisfy the following condition. For arbitrary numbers $\nu < 1$ and $\eta > 0$ there are numbers $\varepsilon(\nu, \eta) > 0$, $\zeta(\nu, \eta, \varepsilon) > 0$ and $\delta(\nu, \eta, \varepsilon, \zeta) > 0$ such that for any motion $x_{U,v}[t_*[\cdot]\vartheta]$, $x[t_*] = x_*$ generated together with the motion $y_{U,v}[t_*[\cdot]\vartheta]$, $y[t_*] = y_*$ by the control law U_δ (2.51) (which corresponds to the strategy $S^u(\cdot)$ (2.44) and $\Delta_\delta\{t_i\}: t_{i+1} - t_i \leq \delta$) and together with any mechanism V that forms the disturbances, the inequality

$$P(\gamma = | x_{U,v}[\vartheta] | \leq \rho(t_*, x_*; S^u(\cdot)) + \eta) \geq \nu \tag{2.58}$$

is valid provided

$$\varepsilon \leq \varepsilon(\nu, \eta), \quad \zeta \leq \zeta(\nu, \eta, \varepsilon), \quad | y_* - x_* | \leq \zeta, \quad \delta \leq \delta(\nu, \eta, \varepsilon, \zeta). \tag{2.59}$$

Here we speak about the realizations $x_{U,v}[t]$, $t_* \leq t \leq \vartheta$ of the stochastic process that is a random motion of x-object and also about the realizations $\{x_{U,v}[t], y_{U,v}[t], t_* \leq t \leq \vartheta\}$ of the extended random motion of the complex $\{x$-object, y-model$\}$. Therefore, strictly speaking, we must denote these realizations by symbols

$$x_{U,v}[t_*[\cdot]\vartheta, \cdot] = \{x_{U,v}[t, \omega], \quad t_* \leq t \leq \vartheta, \quad \omega \in \Omega\}, \tag{2.60}$$

$$y_{U,v}[t_*[\cdot]\vartheta, \cdot] = \{y_{U,v}[t, \omega], \quad t_* \leq t \leq \vartheta, \quad \omega \in \Omega\}. \tag{2.61}$$

Here the letter ω denotes an elementary event that defines the corresponding realization. Such a designation will be used further. But here the symbol ω is omitted in the corresponding designations.

It is important to note that the inequality

$$\gamma = | x_{U,v}[\vartheta] | \leq \rho(t_*, x_*; S^u(\cdot)) + \eta \tag{2.62}$$

is not guaranteed now for certain, but is ensured only with the probability ν. However, this probability can be made as close to 1 as desired. In fact, it suffices to fix a suitable value of a parameter $\varepsilon > 0$, choose a suitable initial state y_* and choose a suitable step $\delta > 0$ of the partition $\Delta_\delta\{t_i\}$. Therefore, the estimation $\rho(t_*, x_*; S^u(\cdot))$ (2.57) is not of the type of mathematical expectation. It means that the inequality (2.62) can be guaranteed as close to the certainty as we want.

We call the strategy $S_0^u(\cdot)$ optimal in a fixed region G in the space $\{t, x\}$ if the equality

$$\rho(t_*, x_*; S_0^u(\cdot)) = \min_{S_u(\cdot)} \rho(t_*, x_*; S^u(\cdot)) = \rho_u^0(t_*, x_*) \tag{2.63}$$

holds for every initial position $\{t_*, x_*\}$ in this region G. We call the quantity $\rho_u^0(t_*, x_*)$ an optimal ensured result for the initial position $\{t_*, x_*\}$.

Thus, using the optimal strategy $S_0^u(\cdot)$ (2.63) if it exists, we can guarantee the inequality

$$\gamma = | x_{U^0,v}[\vartheta] | \leq \rho_u^0(t_*, x_*) + \eta \tag{2.64}$$

with probability as close to 1 as we want. Here $\eta > 0$ is chosen beforehand and may be taken arbitrary small.

Now we have to explain the expression "we imagine *temporary* that the motion $x[t]$ is described by the differential equation (2.41)" (see p. 23). This expression seems somewhat vague. The sense of it is the following.

Coming back to the real r, s-object (2.9), (2.14) with quality index γ (2.15) we must explain the way in which the above construction is connected with the real control for this object. Let us discuss the following circumstance. Here the considered x-object (2.24) (described by the "imaginary" differential equation (2.41)) is auxiliary. In the real process of control for the points M_1 and M_2 the "real" value $x^*[t_i]$ of the phase vector for the auxiliary x-system that defines the control action $u[t_i]$ is computed with the help of the transformation $x^*[t_i] = T_{t_i}(r[t_i], \dot{r}[t_i], s[t_i], \dot{s}[t_i])$ of the coordinates $r[t_i]$, $s[t_i]$ and speeds $\dot{r}[t_i]$, $\dot{s}[t_i]$ of the real r, s-system (2.9), (2.14) according to the formulas (2.20), (2.23) of the transformation (2.26) mentioned above (see p. 19). At the same time, in the real process the control actions $u[t] = u[t_i]$ and the disturbances $v[t] = \{v^{(1)}[t], v_*[t], v^*[t]\}$ are supplied to the real r, s-object, which consists of two points M_1 and M_2. Thus, the actual process of control based on the optimal mixed strategy S_0^u forms the control actions $u[t] = u[t_i] = \{u^{(1)}[t_i], u^{(2)}(t_i, x^*[t_i], y[t_i], \varepsilon)\}$, $t_i \le t < t_{i+1}$ using the vectors $x^*[t_i]$ instead of the vectors $x[t_i]$ that are the solutions of (2.41). Therefore, the process of control is realized by the scheme shown in Figure 2.4.

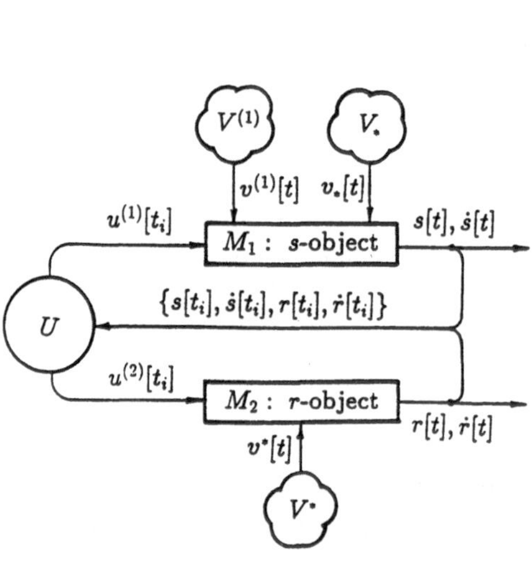

Figure 2.4

Here the regulator U contains the algorithms presented in symbolic form in Figure 2.5.

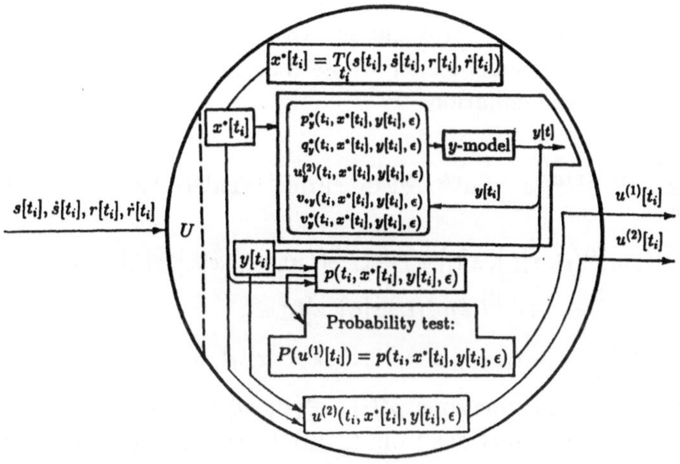

Figure 2.5

The correlation between the vectors $x[t_i]$ described by the differential equation (2.41) with some proper initial data $x[t_*] = x_*$ and the vectors $x^*[t_i]$ that we obtain transforming the vectors $s[t_i]$, $r[t_i]$, $\dot{s}[t_i]$, $\dot{r}[t_i]$ can be characterized as follows. Both vectors "almost" coincide if we assume that the system (2.41) and the r, s-system are controlled by the same actions $u[t] = u[t_i]$, $v[t]$, $t_i \leq t < t_{i+1}$. The word "almost" in the previous phrase means that for any $\eta > 0$ we can name a number δ such that

$$| x[t_i] - x^*[t_i] | \leq \eta \tag{2.65}$$

is valid for all i provided $| t_i - t_{i+1} | < \delta$ for all i. The above provides the basis for the considered scheme of control.

We would also like to note that the vectors $x[t_i]$, which are solutions of the differential equation (2.41), can be useful in computer simulation of the process of control when we test the control scheme and algorithms. In these tests we can input the same disturbance actions $v[t]$ into both the computer model of the actual r, s-system and the computer model of the x-object. The fulfilment of the inequalities (2.65) with sufficiently small η in the computer experiments is in favour of the correctness of the algorithm.

In Section 34 it will be established that the optimal mixed strategy $S_0^u(\cdot)$ exists even for a quality index γ of a more general form. Also some effective methods for calculating the value $\rho_u^0(t_*, x_*)$ (2.63) and for constructing the optimal strategy $S_0^u(\cdot)$ will be given.

Let us now look at the problem from another point of view, i.e., let us consider for the system (2.41) the problem about the disturbances $v \in Q$ (2.29) that are aimed to maximize the quality index γ (2.31). Then the problem in the class of mixed strategies is formalized in the following way:

We introduce the z-model-leader whose motion is determined by the step-by-step differential equation

$$\dot{z}[t] = \sum_{l=1}^{4}\sum_{m=1}^{2} B^{[1]}(t)u^{(1)}_{[l]_{v[m]}}\, p^*_{zl}(t_i,\ x[t_i],\ z[t_i],\ \varepsilon)\, q^*_{zm}(t_i,\ x[t_i],\ z[t_i],\ \varepsilon) +$$
$$+ B^{[2]}(t)u^{(2)}_z(t_i, x[t_i], z[t_i], \varepsilon) + C^{[1]}(t)v_{*z}(t_i, x[t_i], z[t_i], \varepsilon) +$$
$$+ C^{[2]}(t)v^*_z(t_i, x[t_i], z[t_i], \varepsilon),$$
$$t_i \le t < t_{i+1}, \quad i = 1, \dots, k, \quad z[t_1] = z_* \tag{2.66}$$

where $z \in \mathbf{R}^2$ and the functions $p^*_z(\cdot)$, $q^*_z(\cdot)$, $u^{(2)}_z(\cdot)$, $v_{*z}(\cdot)$, $v^*_z(\cdot)$ are defined similarly to the functions from (2.36)–(2.37). The only difference is that the letter y is replaced by z.

The mixed strategy $S^v(\cdot)$ is defined as a totality of the components

$$S^v(\cdot) = \{q(\cdot),\ \widetilde{V},\ v_*(\cdot),\ v^*(\cdot),\ p^*_z(\cdot),\ U^*,\ q^*_z(\cdot), V^*,\ u^{(2)}_z(\cdot),\ v_{*z}(\cdot),\ v^*_z(\cdot)\} \tag{2.67}$$

where

$$q(\cdot) = \Big\{ q(t, x, z, \varepsilon) = \{q_m(t, x, z, \varepsilon) \ge 0,\quad m = 1, 2,\quad \sum_{m=1}^{2} q_m(t, x, z, \varepsilon) = 1\},$$
$$t \in [t_0,\ \vartheta),\quad x \in \mathbf{R}^2,\quad z \in \mathbf{R}^2,\quad \varepsilon > 0 \Big\}, \tag{2.68}$$

$$\widetilde{V} = \{v_{[m]},\quad m = 1, 2,\quad v_{[1]} = -\alpha^0_1,\quad v_{[2]} = \alpha^0_2\ (2.3)\}, \tag{2.69}$$

$$p^*_z(\cdot) = \{p^*_{zl}(t, x, z, \varepsilon),\ l = 1, \dots, 4\},\quad q^*_z(\cdot) = \{q^*_{zm}(t, x, z, \varepsilon),\ m = 1, 2\},$$
$$v_*(\cdot) = \{v_*(t, x, z, \varepsilon)\ (2.8)\},\quad v^*(\cdot) = \{v^*(t, x, z, \varepsilon)\ (2.13)\} \tag{2.70}$$

and the sets U^* and V^* are defined as in (2.49).

Here we again assume *temporary* that the motion $x[t]$ of the controlled x-object (2.24) is formed in the following way: This motion $x[t]$ generated from the initial position $\{t_*, x_*\}$ by the strategy $S^v(\cdot)$ (2.67) together with any admissible control actions

$$u[t_*[\cdot]\vartheta) = \{u^{(1)}[t]\ (2.2),\quad u^{(2)}[t]\ (2.10),\quad t_* \le t < \vartheta\} \tag{2.71}$$

is defined as the step-by-step solution of the differential equation

$$\dot{x}[t] = B^{[1]}(t)u^{(1)}[t]_{v^{(1)}[t_i]} + B^{[2]}(t)u^{(2)}[t] +$$

$$+ C^{[1]}(t)v_*(t_i, x[t_i], z[t_i], \varepsilon) + C^{[2]}(t)v^*(t_i, x[t_i], z[t_i], \varepsilon), \quad t_i \leq t < t_{i+1}. \tag{2.72}$$

Here $v^{(1)}[t_i]$ is a result of an experiment of random choice of the number $v_{[m]}$, $m = 1, 2$ from the collection $\widetilde{V}$ (2.69). Therefore, we have

$$P(v^{(1)}[t_i] = v_{[m]}) = q_m(t_i, x[t_i], z[t_i], \varepsilon) \tag{2.73}$$

where $q_m(t_i, x[t_i], z[t_i], \varepsilon)$ is a number from the collection (2.68) for $t = t_i$, $x = x[t_i]$, $z = z[t_i]$. Here $z[t_i]$ is a value of the phase vector of z-model (2.66) in the time moment $t = t_i$.

For the fixed law

$$V = \{S^v(\cdot);\ \varepsilon;\ \Delta\{t_i\}\} \tag{2.74}$$

we call the quantity

$$\rho(t_*, x_*, z_*;\ V, \nu) = \sup\ \rho \tag{2.75}$$

under the condition

$$P(\gamma = \mid x_{V,u}[\vartheta] \mid \geq \rho) \geq \nu \tag{2.76}$$

an ensured result.

An ensured result for the strategy $S^v(\cdot)$ (2.67) and for the initial position $\{t_*, x_*\}$ is defined by the equality

$$\rho(t_*, x_*, S^v(\cdot)) = \lim_{\nu \to 1} \lim_{\overline{\varepsilon \to 0}} \lim_{\zeta \to 0} \lim_{\delta \to 0} \inf_{|x_* - z_*| \leq \zeta} \inf_{\Delta_\delta} \rho(t_*, x_*, z_*;\ V_\delta, \nu). \tag{2.77}$$

The strategy $S_0^v(\cdot)$ is optimal in a given region G in space $\{t, x\}$ if

$$\rho(t_*, x_*; S_0^v(\cdot)) = \max_{S^v(\cdot)} \rho(t_*, x_*;\ S_v(\cdot)) = \rho_v^0(t_*, x_*) \tag{2.78}$$

for every initial position $\{t_*, x_*\} \in G$.

We call the value $\rho_v^0(t_*, x_*)$ (2.78) an optimal ensured result for the initial position $\{t_*, x_*\}$.

So the optimal strategy $S_0^v(\cdot)$ (2.78) guarantees the fulfillment of the inequality

$$\gamma = \mid x_{V^0,u}[\vartheta] \mid \geq \rho_u^0(t_*, x_*) - \eta \tag{2.79}$$

with the probability as close to 1 as desired for any number $\eta > 0$ fixed beforehand. Here the inequality (2.79) with the desired probability is valid for any motion $x_{V^0,u}[t_*[\cdot]\vartheta]$ emanating from the initial position $\{t_*, x_*\}$ and generated by the feedback law V^0, corresponding to the optimal strategy $S_0^v(\cdot)$ (2.78), together with any admissible control actions $u[t_*[\cdot]\vartheta)$ (2.71), provided the conditions (2.59) (where $y_* = z_*$) hold. As above, we assume that the actions $u[t]$ and $v[t]$ are stochastically independent in small time intervals $[t_i,\ t_{i+1})$.

Now let us note the following.

As in the case of optimal control considered above, here the formation of the vectors $x[t_i]$ directly with the help of the differential equation (2.72) is also only "imaginary". In the real process of control the "real" value $x^*[t_i]$ of the phase vector for an auxiliary x-system that defines the actions of disturbance $v[t] = v[t_i] = \{v^{(1)}[t_i], v_*[t] = v_*(t_i, x[t_i], y[t_i], \varepsilon),\ v^*[t] = v^*(t_i, x[t_i], y[t_i], \varepsilon)\}$, $t_i \leq t < t_{i+1}$ is computed with the help of the transformation T of the coordinates $s[t_i]$, $r[t_i]$ and speeds $\dot{s}[t_i]$, $\dot{r}[t_i]$ of the real r, s-system (2.9), (2.14) according to the formulas of the transformation (2.26) mentioned on p. 19. At the same time, in the real process the action of disturbances $v_*[t] = v_*[t_i]$, $v^*[t] = v^*[t_i]$, $v^{(1)}[t] = v^{(1)}[t_i]$ and the control actions $u[t] = \{u^{(1)}[t], u^{(2)}[t]\}$ are supplied to the real r, s-object, which consists of two points M_1 and M_2. In the actual process of control in which the optimal mixed strategy S_0^v is involved, this strategy uses vectors $x^*[t_i] = T_{t_i}(s[t_i], r[t_i], \dot{s}[t_i], \dot{r}[t_i])$ instead of the vectors $x[t_i]$ that are the solutions of (2.72). Therefore, the process of control is realized by the schemes shown in Figures 2.6–2.8.

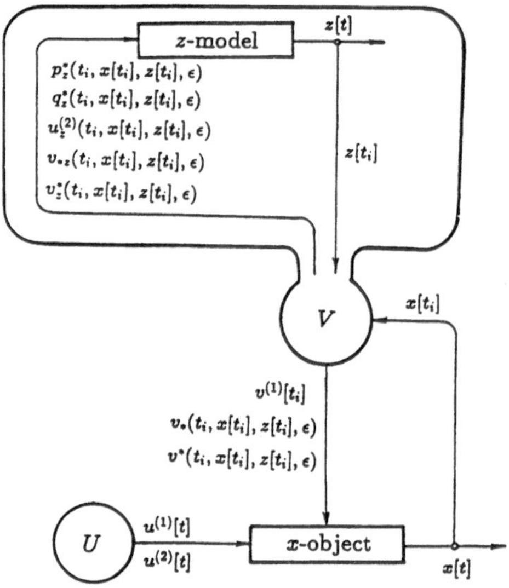

Figure 2.6

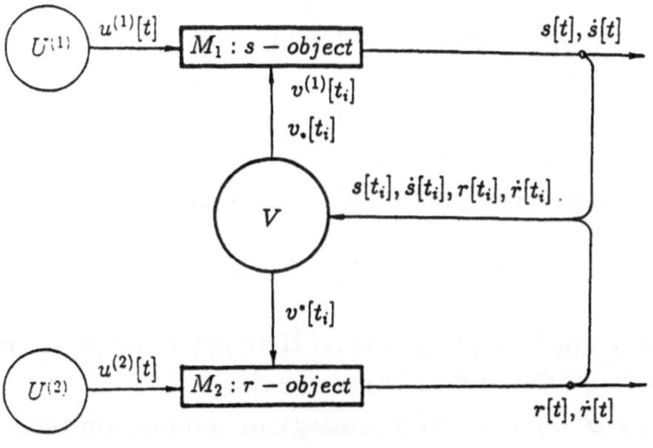

Figure 2.7

Let us remark again that the regulator V contains the algorithm depicted in symbolic form in Figure 2.8.

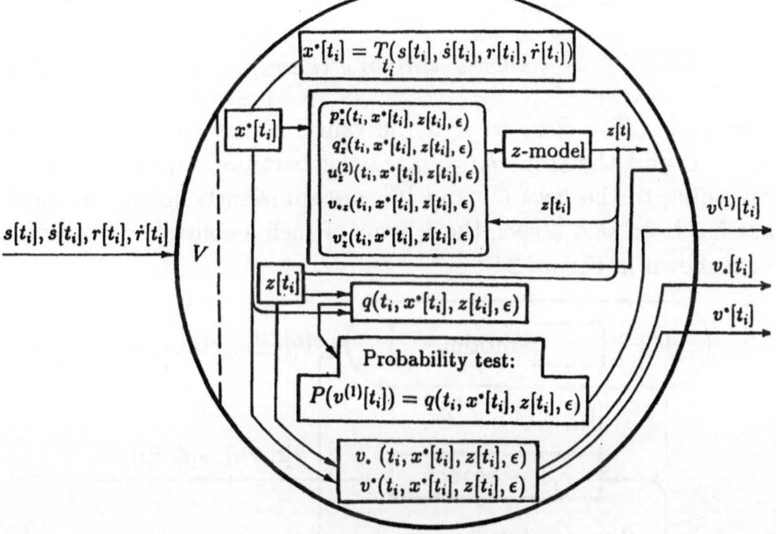

Figure 2.8

Here the connection between the vectors $x[t_i]$ and $x^*[t_i]$ is described by estimations (2.65) that are similar to those given above when the optimal mixed strategy S_0^u was considered (see the text after Figure 2.4). Thus we can justify the scheme of the process of control depicted in Figure 2.8 where the disturbance is formed in the most unfavourable way with respect to minimization of the quality index γ.

The considered problem can be formalized within the framework of the antagonistic differential game for two persons. Then using the terminology of the theory of differential games [41], [48], [79], [88] we can say that a pair of strategies $S_0^u(\cdot)$ (2.63) and $S_0^v(\cdot)$ (2.78) forms *a saddle point* $\{S_0^u(\cdot), S_0^v(\cdot)\}$ and gives the *value* $\rho^0(t_*, x_*)$ *of the game* (for the system (2.24) with the quality index (2.27)) if the equality

$$\rho_u^0(t_*, x_*) = \rho_v^0(t_*, x_*) = \rho^0(t_*, x_*) \tag{2.80}$$

holds for every position $\{t_*, x_*\} \in G$. Here $\rho_u^0(\cdot)$ and $\rho_v^0(\cdot)$ are the optimal ensured results (2.63) and (2.78) respectively.

A strategy $S^u(\cdot)$ is called a strategy of the first player and a strategy $S^v(\cdot)$ is considered as that of the second one.

So the informal description of the saddle point $\{S_0^u(\cdot), S_0^v(\cdot)\}$ and the value of the game $\rho^0(t_*, x_*)$ is the following. For any motion $x_{U^0,V^0}[t_*[\cdot]\vartheta]$ generated from any initial position $\{t_*, x_*\} \in G$ by the pair of feedback control laws U^0 and V^0, corresponding to the optimal strategies $S_0^u(\cdot)$ and $S_0^v(\cdot)$, we obtain the result

$$\gamma = | x_{U^0,V^0}[\vartheta] | \approx \rho^0(t_*, x_*) \tag{2.81}$$

with the probability close to 1, if the values of the parameters $\varepsilon_U > 0$ and $\varepsilon_V > 0$ and the steps $\delta_u = \delta_v$ of the partition $\Delta_U\{t_i\} = \Delta_V\{t_i\}$, corresponding to the laws U^0 and V^0, are sufficiently small. According to what has been said above, the scheme of such a controlled process has the form shown in Figure 2.9.

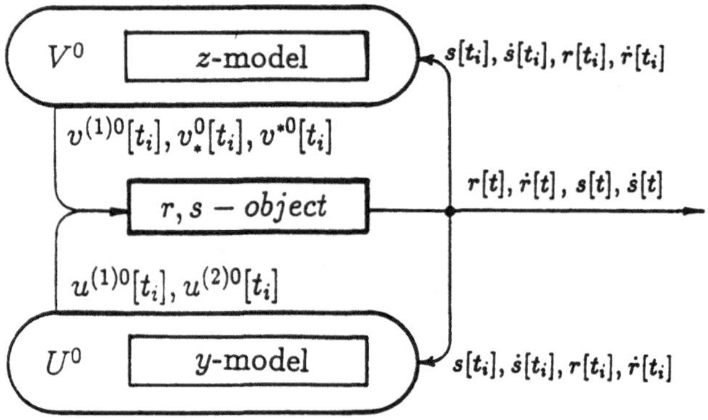

Figure 2.9

Thus, the problem consists in constructing a saddle point $\{S_0^u(\cdot), S_0^v(\cdot)\}$ and in calculating a value of the game $\rho^0(t_*, x_*)$ for any initial position $\{t_*, x_*\} \in G$.

In Section 34 it will be proved that in the considered differential game for the system (2.24) with the quality index γ (2.27), the saddle point $\{S_0^u(\cdot), S_0^v(\cdot)\}$ exists. Some procedures for constructing the optimal strategies $S_0^u(\cdot), S_0^v(\cdot)$ and for calculating the value of the game $\rho^0(t_*, x_*)$ will be given and justified. Here we say only that we construct the strategies $S_0^u(\cdot)$ and $S_0^v(\cdot)$ as the so-called extremal ones with respect to the function $\rho^0(t, x)$.

Therefore, in conclusion of this section we give only the results of the simulation for the corresponding process of control. The process was simulated on an IBM PC computer for the following values of the parameters in the equations (2.1), (2.9), (2.14), (2.17) and in the relations (2.16):

$$
\begin{gathered}
t_* = 0, \quad h_* = 10.00, \quad \dot{h}_* = 8.46, \\
\alpha_1 = 0.2, \quad \alpha_2 = 0.4, \quad \beta = 4, \\
|K_{1h}| = 160, \quad |K_1| = 40, \quad |K_2| = 25, \\
N_1 = 4, \quad N_2 = 6, \quad \alpha_1^0 = \alpha_2^0 = 1.05 \quad m_{*1} = 1, \quad m_{*2} = 1, \\
q = 0.25, \quad \lambda_1 = 0.5, \quad \lambda_2 = 0.4, \\
s_* = \{0, -2\}, \quad \dot{s}_* = \{2, 0\}, \\
r_* = \{-1, 1\}, \quad \dot{r}_* = \{0, 4\}, \\
\delta = 0.01, \quad \varepsilon = 0.01.
\end{gathered} \tag{2.82}
$$

We compute the terminal time moment $t = \vartheta = 2$ under the condition $h[\vartheta] = 0$ according to the differential equation (2.1) for the taken initial height $h_* = h_0 = 10.00$ and initial vertical speed $\dot{h}_* = \dot{h}_0 = 8.46$ of the point M_1 and for the taken parameters (2.82). For the taken initial position $\{t_*, r_*, \dot{r}_*, s_*, \dot{s}_*\}$ (2.82) the value of the game is $\rho^0(t_*, r_*, \dot{r}_*, s_*, \dot{s}_*) = 0.42$. In Figure 2.10 we depict the projection to the horizontal plane of the corresponding realizations of the motions of the points M_1 and M_2 generated by the optimal strategies $S_0^u(\cdot)$ and $S_0^v(\cdot)$ with a parameter ε and a time step $\delta = t_{i+1} - t_i$ (2.82). In this case the experiment gave the quantity γ close to $\rho^0(t_*, r_*, \dot{r}_*, s_*, \dot{s}_*)$, as should be expected:

$$
\begin{aligned}
\gamma &= ((r_1[\vartheta] - s_1[\vartheta])^2 + (r_2[\vartheta] - s_2[\vartheta])^2)^{1/2} = \\
&= 0.47 \approx \rho^0(t_*, r_*, \dot{r}_*, s_*, \dot{s}_*) = 0.42.
\end{aligned} \tag{2.83}
$$

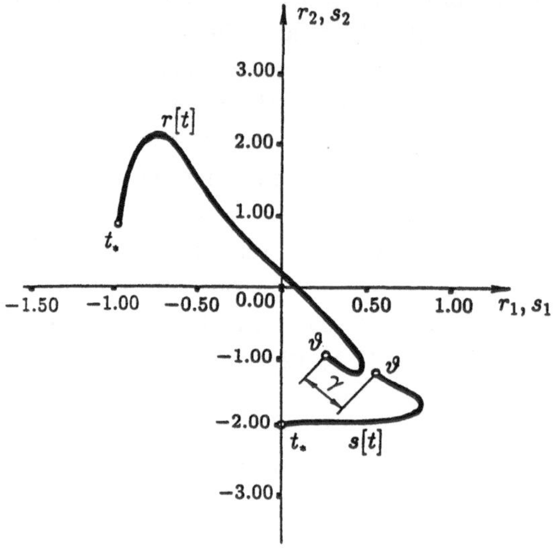

Figure 2.10

In Figure 2.11 we show the corresponding motion $x^*[t_*[\cdot]\vartheta]$.

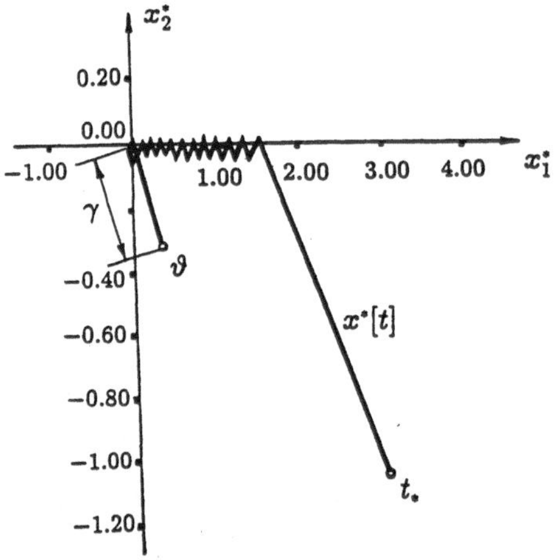

Figure 2.11

Figures 2.12 and 2.13 show the realization of the process in the case when the optimal strategy $S_0^u(\cdot)$ acts together with some mechanism V which forms the disturbance but is not based on the optimal strategy $S_0^v(\cdot)$.

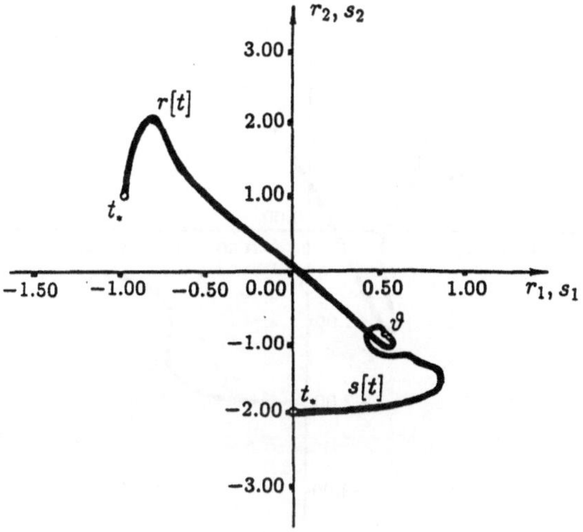

Figure 2.12

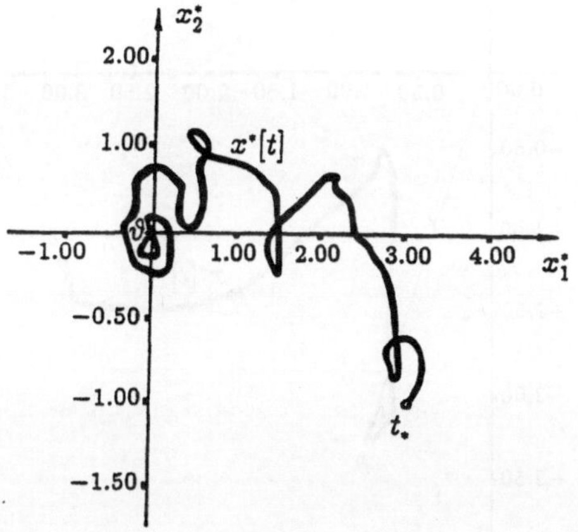

Figure 2.13

In this case the experiment gave the result $\gamma = 0.01 < \rho^0 = 0.42$. Finally in the third case some control law $U \neq U^0$, which does not use the optimal strategy $S_0^u(\cdot)$, acts together with the law V^0 based on the optimal strategy $S_0^v(\cdot)$. The corresponding realization of this process is shown in Figures 2.14 and 2.15.

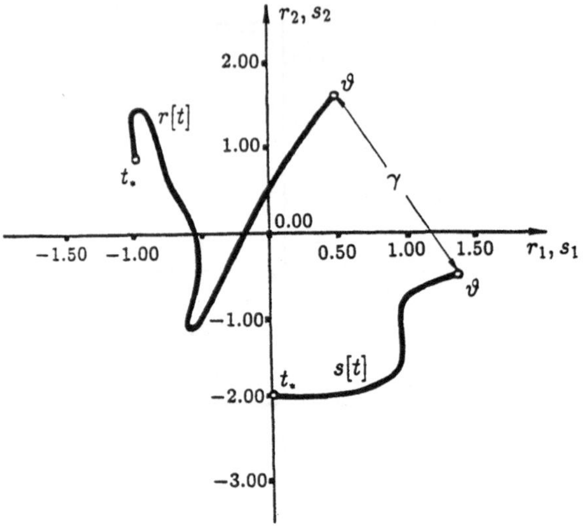

Figure 2.14

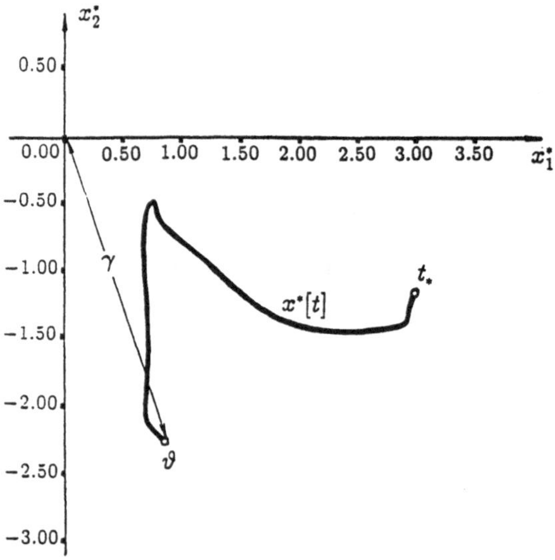

Figure 2.15

Here we have $\gamma = 2.35 > \rho^0 = 0.42$.

Thus, in the case $\{U^0, V^0\}$ we have the distance γ between the points M_1 and M_2 at the time moment ϑ close to ρ^0, i.e., the equality (2.83) holds. But if one of the players does not use his optimal strategy (see the cases $\{U^0, V\}$ or $\{U, V^0\}$), then the result of his opponent turns out to be better. In the case $\{U^0, V\}$ the distance between the points M_1 and

M_2 at the termination moment is smaller than in the first case. In the case $\{U, V^0\}$ the distance between M_1 and M_2 at the time moment ϑ is larger than in the first case.

3. Equation of motion

Assume that the motion of the controlled dynamical system is described by the ordinary vector differential equation

$$\dot{x} = f(t, x, u, v), \qquad t_0 \leq t < \vartheta \tag{3.1}$$

where x is n-dimensional phase vector, t is the time, t_0 and ϑ are fixed moments, u is an r-dimensional vector that characterizes the control action and v is an s-dimensional vector of the disturbance. For a fixed t the function $f(\cdot)$ is continuous in x, u, v; for fixed x, u, v it is piecewise-continuous in t, the points of discontinuity $f(\cdot)$ are independent of x, u, v and in these points the function $f(t, \cdot)$ is continuous on the right and has finite limits on the left in these points of discontinuity. Assume that $f(\cdot)$ satisfies a condition

$$| f(t, x, u, v) | \leq \chi (1 + | x |), \qquad \chi = \text{const} . \tag{3.2}$$

Here $| x |$ denotes the Euclidean norm of x, i.e.,

$$| x | = (x_1^2 + \ldots + x_n^2)^{\frac{1}{2}} . \tag{3.3}$$

We assume that in every bounded region G in the space $\{t, x\}$ the function $f(\cdot)$ satisfies a Lipschitz condition in x with the constant L_G, i.e.,

$$| f(t, x^{(1)}, u, v) - f(t, x^{(2)}, u, v) | \leq L_G | x^{(1)} - x^{(2)} | \tag{3.4}$$

where $\{t, x^{(i)}\} \in G$, $i = 1, 2$; $t_0 \leq t < \vartheta$.

It is assumed that admissible values of u and v are restricted by the inclusions

$$u \in P, \tag{3.5}$$

$$v \in Q. \tag{3.6}$$

Here P and Q are compact sets in Euclidean spaces $\mathbf{R}^r$ and $\mathbf{R}^s$ respectively. The sets P and Q describe abilities of the regulator U that generates the control actions and of the device V that generates the disturbances.

We will use the following terms. The value $u[t]$, the function $u[t_*[\cdot]t^*)$ and the value $v[t]$, the function $v[t_*[\cdot]t^*)$ will be called "the current control action", "the realization of control actions", "the current disturbance", and "the realization of disturbances", respectively. We will also use the shortened forms: the *control action* $u[t]$, the *control actions* $u[t_*[\cdot]t^*)$, the *disturbance* $v[t]$, the *disturbances* $v[t_*[\cdot]t^*)$.

As usual, the pair $\{t, x\}$, $t \in [t_0, \vartheta]$, $x \in \mathbf{R}^n$ will be called the *position* of the controlled system (3.1) at time t. Let some time interval $[t_*, t^*] \subset [t_0, \vartheta]$ and initial position $\{t_*, x_*\}$ be chosen. Suppose that some control actions

$$u[t_*[\cdot]t^*) = \{u[t] \in P, \ t_* \leq t < t^*\} \tag{3.7}$$

and some disturbances

$$v[t_*[\cdot]t^*) = \{v[t] \in Q, \ t_* \leq t < t^*\} \tag{3.8}$$

are realized. The actions $u[\cdot]$ (3.7) and $v[\cdot]$ (3.8) may be arbitrary Borel-measurable functions in $t \in [t_*, t^*)$.

The function

$$x[t_*[\cdot]t^*] = \{x[t], \ t_* \leq t \leq t^*\} \tag{3.9}$$

is called the *motion* of the system (3.1) generated from the initial position $\{t_*, x_*\}$ by the actions $u[\cdot]$ (3.7) and $v[\cdot]$ (3.8) if it is determined for $t_* \leq t \leq t^*$ as the solution of the differential equation

$$\dot{x}[t] = f(t, x[t], u[t], v[t]),$$

$$t_* \leq t < t^*, \quad x[t_*] = x_* \tag{3.10}$$

where $u[t]$ and $v[t]$, $t_* \leq t < t^*$ are the functions (3.7) and (3.8), respectively.

The function $x[\cdot]$ (3.9) satisfies the Lipschitz condition in t. Consequently, it is absolutely continuous.

We shall assume that the initial position $\{t_*, x_*\}$ satisfies the condition

$$\{t_*, x_*\} \in G. \tag{3.11}$$

Here G is a compact set in the space $\{t, x\}$, $t \in [t_0, \vartheta]$, $x \in \mathbf{R}^n$ which satisfies the following condition. For every $t_* \in [t_0, \vartheta]$ the intersection of G and surface $t = t_*$ is not void and for every initial position $\{t_*, x_*\} \in G$ the domain G contains all possible positions $\{t, x[t]\}$, $t \in [t_*, \vartheta]$ of the system (3.1), (3.2), (3.5), (3.6) that can be realized using all possible actions $u[\cdot]$ (3.7) and $v[\cdot]$ (3.8) and starting from a chosen position $\{t_*, x_*\}$ (3.11). Such a set G can be defined, for instance, by conditions

$$G = \{\{t, x\} : \ t \in [t_0, \vartheta], \quad | x | \leq R[t]\} \tag{3.12}$$

where

$$R[t] = (1 + R_0) \exp\{\chi (t - t_0)\} - 1, \qquad R_0 > 0 \ - \ \text{const} \tag{3.13}$$

and χ is the constant in (3.2).

4. The positional quality index

We consider minimax-maximin (with respect to u and v) problems for some quality index that determines the quality of the process of control. In a rather general case the quality index can be defined as a certain functional of the motion $x[t_*[\cdot]\vartheta]$ (3.9), (3.10) of the controlled plant (3.1) and also of the realizations $u[t_*[\cdot]\vartheta)$ (3.7) and $v[t_*[\cdot]\vartheta)$ (3.8) of control actions and disturbances respectively. Here $t_* \in [t_0, \ \vartheta)$ is the initial time moment for the process of control and ϑ is a fixed terminal time moment.

As in Sections 1 and 2 we denote the quality index by γ.

At first we restrict ourselves to the case when the quality index γ is a certain functional depending explicitly only on the motion

$$x[t_*[\cdot]\vartheta] = \{x[t], \quad t_* \leq t \leq \vartheta\} \tag{4.1}$$

of the plant (3.1). Later we also use a shortened term "functional γ", meaning the quality index

$$\gamma = \gamma(x[t_*[\cdot]\vartheta]). \tag{4.2}$$

In this book we shall consider problems of forming control actions u that are aimed at minimizing the quality index γ (4.2) and of finding disturbances v that are aimed at maximizing γ. We employ the *feedback positional control* [48], [55], [79], [88], [114]. That is, we consider a problem in which at every time moment $t \in [t_*, \ \vartheta)$ the control actions u (or the disturbances v) are formed only using the information about the current position $\{t, x[t]\}$ of the object (3.1) that is realized at this time moment t, or only using the information about some position $\{\tau, x[\tau]\}$ that is realized not long before the moment t (for example, using the information about the position $\{t_i, x[t_i]\}$, $t_i \leq t < t_{i+1}$, where t_i, $i = 1, \ldots, k+1$ are the time moments of some partition $\Delta\{t_i\} = \{t_i < t_{i+1}, \ i = 1, \ldots, k, \ t_1 = t_*, \ t_{k+1} = \vartheta\}$ of time interval $[t_*, \ \vartheta]$). This problem can be formalized within the framework of the positional differential games [37], [41], [48], [79], [88], [113], [118], [122], [125]. The exact mathematical statement of the problem will be given below in Section 6.

It is known that the problems of this kind can be solved rather naturally in some typical cases. For example, a problem of this form can be effectively solved for many cases when $\gamma = \sigma(x[\vartheta])$, i.e., when the quality index γ (4.2) depends only on the value of the phase vector $x[t]$ at the terminal time moment $t = \vartheta$, and in particular, for $\gamma = | x[\vartheta] |$ (see (1.9) and (2.27) in Sections 1 and 2). A number of concrete examples with quality indices of this type are considered in [32], [48], [79], [140], [141].

On the other hand, there are cases in which a problem formalized in this way cannot be solved sufficiently well. Let, for example, the functional γ (4.2) have a form

$$\gamma = \gamma(x[t_*[\cdot]\vartheta]) = \sum_{i=g_*}^{N} \mu_i \mid x[t^{[i]}] \mid + \max_{t_* \leq t \leq \vartheta} \mid x[t] \mid \tag{4.3}$$

$$t^{[g_*]} = \min_{t^{[i]} \geq t_*} t^{[i]}. \tag{4.4}$$

where $\mu_i > 0$ and $t^{[i]} \in [t_*,\ \vartheta]$, $i = 1, \ldots, N$ are given constants and time moments, respectively.

During the time interval $[t_*,\ t^*] \subset [t_*,\ \vartheta)$ let the motion $x[t_*[\cdot]t^*]$ be realized. Let $t^{[g^*]}$ be the time moment $t^{[i]}$ that satisfies the condition

$$t^{[g^*]} = \min_{t^{[i]} \geq t^*} t^{[i]}. \tag{4.5}$$

Denote

$$a_1 = \sum_{i=g_*}^{g^*-1} \mu_i \mid x[t^{[i]}] \mid, \qquad a_2 = \sup_{t_* \leq t < t^*} \mid x[t] \mid,$$

$$\alpha_1 = \sum_{i=g^*}^{N} \mu_i \mid x[t^{[i]}] \mid, \qquad \alpha_2 = \max_{t^* \leq t \leq \vartheta} \mid x[t] \mid . \tag{4.6}$$

With regard to (4.3)–(4.6) we can represent the functional γ (4.3) in a form

$$\gamma(x[t_*[\cdot]\vartheta]) = a_1 + \alpha_1 + \max\{a_2, \alpha_2\}. \tag{4.7}$$

Here

$$x[t_*[\cdot]\vartheta] = \{x[t_*[\cdot]t^*),\ x[t^*[\cdot]\vartheta],\ x[t_*[t^*]t^*] = x[t^*[t^*]\vartheta]\}. \tag{4.8}$$

It follows from (4.7) that information only about position $\{t^*, x[t^*]\}$ only, which is realized at the time moment t^*, is insufficient for further forming the control actions u, which should minimize $\gamma(x[t_*[\cdot]\vartheta])$ (4.3), (4.7). In fact, if we ignore the history $x[t_*[\cdot]t^*]$ of the motion $x[t_*[\cdot]\vartheta]$ (4.8), we do not know what to do in this situation. Really, if the value a_2 (4.6) is essentially larger then the value α_2 (4.6) that can occur in the time interval $[t^*,\ \vartheta]$, then in order to minimize the value γ (4.7) for the whole motion $x[t_*[\cdot]\vartheta]$ (4.8) we ought to tend to minimize the quantity α_1 (4.6). Conversely, if in the time interval $[t_*,\ t^*]$ the corresponding maximum, i.e., the value a_2 (4.6) was very small, then it can be more appropriate to try to minimize the quantity α_2 (4.6) in order to achieve the minimum of γ (4.7). So we see that information about position $\{t^*, x^* = x[t_*[t^*]t^*]\}$ is not sufficient for further effective control aimed to minimize the quantity γ (4.3), (4.7) for the whole motion $x[t_*[\cdot]\vartheta]$ (4.8).

This example of the quality index γ (4.3) prompts the following: if the functional $\gamma(x[t_*[\cdot]\vartheta])$ is such that for any history $x[t_*[\cdot]t^*)$ of the motion $x[t_*[\cdot]\vartheta]$ (4.8) this functional is monotone in $\alpha = \gamma(x[t^*[\cdot]\vartheta])$ then the hypothesis is probable that we can effectively form the control actions u by information only about the current positions $\{t, x = x[t]\}$ for minimizing the quantity $\gamma(x[t_*[\cdot]\vartheta])$.

Therefore we select the specific form of the functionals $\gamma(x[t_*[\cdot]\vartheta])$, which we call the positional functionals. Namely, the *functional* γ (4.2) is called *positional* if it can be represented in the form

$$\gamma(x[t_*[\cdot]\vartheta]) = \beta(x[t_*[\cdot]t^*), \alpha),$$

$$\alpha = \gamma(x[t^*[\cdot]\vartheta]), \quad t_* \in [t_0,\ \vartheta), \quad t^* \in (t_*,\ \vartheta] \tag{4.9}$$

where the functional $\beta(x[t_*[\cdot]t^*), \alpha)$ (for an arbitrary admissible fixed history $x[t_*[\cdot]t^*)$) is continuous and nondecreasing in α.

Now, for any possible realization $x[t_*[\cdot]t^*)$ of the history of the motion $x[t_*[\cdot]\vartheta]$ (4.8) we see that starting from the time moment t^* it is reasonable to assign the control actions u that are aimed at minimizing the quantity $\alpha = \gamma(x[t^*[\cdot]\vartheta])$ if we want to minimize the whole value $\gamma(x[t_*[\cdot]\vartheta])$. Conversely, if we are solving the problem about the actions v that tend to maximize the whole γ we ought to assign the actions v that are aimed at maximizing α (4.9). Therefore, for the functionals γ (4.2) that can be represented in the form (4.9) it can be proved [55] that the above hypothesis holds. In other words, we can successfully form the minimizing control action $u[t]$ and the maximizing disturbance $v[t]$ (in the feedback control schemes) using only the information about the *position* $\{t, x[t]\}$ (or $\{t_i, x[t_i]\}$, $t_i \leq t < t_{i+1}$) that is realized at the current time moment t (or t_i). For this reason we call *positional* the functionals γ that can be represented in the form (4.9). This definition of the positional functional γ (4.9) was introduced in the paper [55].

In particular, the following functionals are positional:

$$\gamma = \sigma(x[\vartheta]), \quad (\gamma = \mid x[\vartheta] \mid) \tag{4.10}$$

$$\gamma = \int_{t_*}^{\vartheta} \mu(t, x[t])dt + \sigma(x[\vartheta]) \tag{4.11}$$

$$\gamma = \sum_{i=g_*}^{N} \mid x[t^{[i]}] \mid, \qquad \gamma = \max_{t_* \le t \le \vartheta} \mid x[t] \mid \tag{4.12}$$

$$\gamma = \sum_{i=g_*}^{N} \mid x[t^{[i]}] - c^{[i]} \mid, \qquad \gamma = \max_{t_* \le t \le \vartheta} \mid x[t] - c[t] \mid \tag{4.13}$$

$$\gamma = \int_{t_*}^{\vartheta} \mu(t, D[t]x[t])dt + \sum_{i=g_*}^{N} \mu^{[i]}(D^{[i]}x[t^{[i]}]) \tag{4.14}$$

$$\gamma = \max \, [\sup_{t_* \le t \le \vartheta} \mu(t, D[t]x[t]), \max_{i=g_*,\dots,N} \mu^{[i]}(D^{[i]}x[t^{[i]}]) \,] \tag{4.15}$$

where μ and σ are some given continuous functions that are Lipschitz in x; $t^{[i]}$ – are given instants, $t^{[i]} \in [t_*, \ \vartheta]$; $c^{[i]}$ – are given vectors, $c^{[i]} \in \mathbf{R}^n$; $c[t]$ is a given continuous vector function.

In (4.14), (4.15), $\mu(t, D[t]x[t])$ and $\mu^{[i]}(D^{[i]}x[t^{[i]}])$ are given norm-function and norm, respectively. They are dependent on some of the components of the vector $x[t] \in \mathbf{R}^n$. These components are selected by the matrix-function $D[t]$ and the matrix $D^{[i]}$. A detailed description of the functionals γ (4.14) and γ (4.15) will be given below in Sections 10 and 11.

We note that the case of the functional γ (4.10) is well studied. Usually (4.10) is called the terminal quality index. According to the formal definition (4.9) it is a positional functional, because the equality

$$\gamma(x[t_*[\cdot]\vartheta]) = \gamma(x[t^*[\cdot]\vartheta]) = \alpha = \sigma(x[\vartheta]) \tag{4.16}$$

holds for every admissible history $x[t_*[\cdot]t^*)$, i.e., in this case we have

$$\beta(x[t_*[\cdot]t^*), \alpha) = \alpha. \tag{4.17}$$

The combined functional γ (4.11) is positional (4.9), because

$$\gamma(x[t_*[\cdot]\vartheta]) = \beta(x[t_*[\cdot]t^*), \alpha) = \int_{t_*}^{t^*} \mu(t, x[t])dt + \alpha \tag{4.18}$$

where

$$\alpha = \gamma(x[t^*[\cdot]\vartheta]) = \int_{t^*}^{\vartheta} \mu(t, x[t])dt + \sigma(x[\vartheta]). \tag{4.19}$$

For typical quality indices γ (4.12) we have, respectively,

$$\gamma(x[t_*[\cdot]\vartheta]) = \beta(x[t_*[\cdot]t^*), \alpha) = \sum_{i=g_*}^{g^*-1} \mid x[t^{[i]}] \mid + \alpha,$$

$$\alpha = \gamma(x[t^*[\cdot]\vartheta]) = \sum_{i=g^*}^{N} \mid x[t^{[i]}] \mid \tag{4.20}$$

where g^* is a number from (4.5), and

$$\gamma(x[t_*[\cdot]\vartheta]) = \beta(x[t_*[\cdot]t^*), \alpha) =$$

$$= \max[\sup_{t_* \leq t < t^*} \mid x[t] \mid, \alpha], \quad \alpha = \max_{t^* \leq t \leq \vartheta} \mid x[t] \mid . \tag{4.21}$$

It is not difficult to check that the functionals γ (4.13)–(4.15) are positional, too.

At the same time, according to what has been said above in this section (see the expressions (4.3)–(4.7)), the functional γ (4.3) is not *positional*, although for this functional γ (4.3) we have

$$\gamma = \gamma_1 + \gamma_2 \tag{4.22}$$

where γ_1 and γ_2 are the positional functionals (4.12).

According to the definition from Section 3, the motions $x[t_*[\cdot]\vartheta]$ (4.1) of the object (3.1) are absolutely continuous functions of time t. However, in Sections 8 and 9 we shall use some auxiliary functions that will be piecewise-continuous. In particular, these functions will be the motions $w[t_*[\cdot]\vartheta]$ of a suitable model and these motions will accompany the motion of the x-object (3.1). But we shall have to define the functionals γ (4.2), (4.9) for these functions also. Therefore, we shall suppose that the functionals γ (4.9) are defined on piecewise-continuous functions $x[t_*[\cdot]\vartheta] = \{x[t], \; t_* \leq t \leq \vartheta\}$ which have a finite number of points of discontinuity with finite limits. We note also that the conditions $\{t, x[t]\} \in G$ are valid for all $t \in [t_*, \; \vartheta]$. Here G is the set (3.12). We assume that the considered functions $x[t_*[\cdot]\vartheta]$ are continuous on the right. For these piecewise-continuous functions we suppose that the functionals γ (4.2), (4.9) satisfy a Lipschitz condition

$$\mid \gamma(x^{(1)}[t_*[\cdot]\vartheta]) - \gamma(x^{(2)}[t_*[\cdot]\vartheta]) \mid \leq$$

$$\leq L_* \, c(x^{(1)}[t_*[\cdot]\vartheta], x^{(2)}[t_*[\cdot]\vartheta]) \tag{4.23}$$

where L_*=const and $c(\cdot)$ is a metric

$$c(x^{(1)}[t_*[\cdot]\vartheta], x^{(2)}[t_*[\cdot]\vartheta]) =$$

$$= \sup_{t_* \leq t \leq \vartheta} \mid x^{(1)}[t] - x^{(2)}[t] \mid . \tag{4.24}$$

In conclusion of this section, let us note that later on in Chapter III we will consider certain quality indices

$$\gamma = \gamma(x[t_*[\cdot]\vartheta], u[t_*[\cdot]\vartheta), v[t_*[\cdot]\vartheta)), \tag{4.25}$$

which depended on the motion of the object (3.1) and on the realizations of control actions and disturbances.

These functionals (4.25) will not be, generally speaking, positional functionals γ (4.2), (4.9). However, in Sections 21 and 23 it will be established that for some typical quality indices of the form (4.25) and, in particular, for the functionals γ (4.25) that contain additive integral term

$$J = \int_{t_*}^{\vartheta} \varphi(t, u[t], v[t])dt \tag{4.26}$$

the feedback control problem in positional form is reasonable and can be effectively solved.

5. Pure positional strategies

In this section we define the pure positional strategies $u(\cdot), v(\cdot)$, the feedback control laws U, V (which correspond to the strategies $u(\cdot)$ and $v(\cdot)$ respectively) and the motions $x_U[\cdot]$, $x_V[\cdot]$ of the controlled x-object (3.1) generated by the laws U and V. The introduced notions are essential for the proposed formalization of differential games and minimax control problems.

According to tradition, we will say that the control actions u are created by some first player (maybe fictitious) and the disturbances v are generated by some second player (maybe fictitious, too). (The following formal constructions were illustrated on a model problem in Section 1.)

We assume that the *saddle point condition in a small game* [48], [77], [79]

$$\min_{u\in P}\max_{v\in Q} \langle\, l,\ f(t,x,u,v)\,\rangle = \max_{v\in Q}\min_{u\in P} \langle\, l,\ f(t,x,u,v)\,\rangle \tag{5.1}$$

is valid for any $\{t, x\} \in G$ (3.12), (3.13) and $l \in \mathbf{R}^n$.

Here as above (see (2.34)) the symbol $\langle l, f\rangle = l^T f$ denotes the scalar product of two n-dimensional vectors l and f. (We remind the reader that vectors are understood as column vectors. The upper index T denotes transposition.) Thus, we have

$$\langle l,\ f\rangle = [l_1 \dots l_n]\begin{bmatrix} f_1 \\ \vdots \\ f_n \end{bmatrix} = l_1 f_1 + \dots + l_n f_n. \tag{5.2}$$

We shall not discuss the condition (5.1). The meaning of the term for (5.1) and the role of this condition will be evident in subsequent chapters. Here we note only that the our methods of solving minimax control problems and our formalization of a differential game depend substantially on whether or not the considered controlled system satisfies the condition (5.1) of saddle point in a small game (see in particular Sections 1 and 2).

In the case (5.1) the problem of control actions u, which are aimed to minimize γ (4.2), (4.9), and the problem of constructing the disturbances v, which tend to maximize γ for the positional functional γ, can be formalized in the classes of *pure positional strategies* $u(\cdot)$ and $v(\cdot)$ of the first and second players, respectively.

In the general case the pure positional strategy $u(\cdot)$ of the first player (or $v(\cdot)$ of the second player) is understood as some rule which, for every admissible position $\{t, x\}$ of the object (3.1), assigns a control action u (respectively, disturbance v), also taking into account the chosen value of some auxiliary parameter ε. The pair $\{t, x[t]\}$ is called the *informational image* for the strategy $u(\cdot)$ (or $v(\cdot)$). Let us recall that strategies of this type were already used in Section 1 for a special case of the problem.

Therefore, according to tradition, let us call the function

$$u(\cdot) = \{u(t, x, \varepsilon_u) \in P, \quad \{t, x\} \in G, \quad \varepsilon_u > 0\} \tag{5.3}$$

a pure positional strategy of the first player. Here $\{t, x\}$ is the position of the system (3.1), (3.2), (5.1); P is the set (3.5); G is the set (3.12); and ε_u is a parameter. The parameter $\varepsilon_u > 0$ is not an informational argument of the strategy $u(\cdot)$ (5.3). It is a so-called *parameter of accuracy.* The sense of this parameter ε_u and its role in the solution of the problem will be evident in subsequent sections.

The feedback *control law* U, which corresponds to the strategy $u(\cdot)$ (5.3), is determined by three components: the strategy $u(\cdot)$ (5.3), the value of parameter $\varepsilon_u > 0$ and a *partition*

$$\Delta_u\{t_i^{(u)}\} = \{t_* = t_1^{(u)}, \ldots, \quad t_i^{(u)} < t_{i+1}^{(u)},$$
$$i = 1, \ldots, k_u, \quad t_{k_u+1}^{(u)} = \vartheta\} \tag{5.4}$$

of the time interval $[t_*, \vartheta]$. In (5.4) k_u is some natural number.

We write in symbolic form

$$U = \{u(\cdot);\ \varepsilon_u;\ \Delta_u\{t_i^{(u)}\}\}. \tag{5.5}$$

The motion

$$x_{U,v}[t_*[\cdot]\vartheta] = \{x_{U,v}[t], \quad t_* \leq t \leq \vartheta, \quad x_{U,v}[t_*] = x_*\} \tag{5.6}$$

generated from the given initial position $\{t_*, x_*\} \in G$ (3.12) by the control law U (5.5) together with an admissible realization of disturbances $v[t_*[\cdot]\vartheta)$ is defined in the following way. (Here the words "admissible disturbances" mean the following. The realization $v[t_*[\cdot]\vartheta)$ may be an arbitrary Borel-measurable [1], [54], [97] function in $t \in [t_*, \vartheta]$ and according to (3.6) at every time moment t the condition $v[t] \in Q$ holds. The mechanism V that forms the disturbances $v[t]$ is unknown to the first player.) If the reader prefers, he or she may assume that the admissible realizations $v[t_*[\cdot]\vartheta)$ are arbitrary piecewise-continuous functions that satisfy the condition (3.6).

Let us act on behalf of the first player. Then we have to choose the control law U. It means that we have to choose the strategy $u(\cdot)$ (5.3), to fix the parameter $\varepsilon_u > 0$ and to choose the partition $\Delta_u\{t_i^{(u)}\}$ (5.4).

To the given initial position $\{t_*, x_*\} \in G$ and parameter $\varepsilon_u > 0$ the strategy $u(\cdot)$ (5.3) assigns a control action

$$u[t_*[\cdot]t_2^{(u)}) = \{u[t] = u(t_*, x_*, \varepsilon_u) \in P, \quad t_* \le t < t_2^{(u)}\} \tag{5.7}$$

where $t_* = t_1^{(u)} \in \Delta_u\{t_i^{(u)}\}$ (5.4), $t_2^{(u)} \in \Delta_u\{t_i^{(u)}\}$.

The disturbances (or the control actions of the second player) $v[t_*[\cdot]t_2^{(u)})$ may be an arbitrary function $v[t_*[\cdot]t^*)$ (3.8) where $t^* = t_2^{(u)}$. We note that we solve the problem on behalf of the first player under lack of information about the disturbances v. So the disturbances $v[t]$, $t_* \le t < t_2^{(u)}$ are unknown to us. We know only that the restrictions $v[t] \in Q$ are valid.

This pair of realizations of the actions $u[t_*[\cdot]t_2^{(u)})$ (5.7) and $v[t_*[\cdot]t_2^{(u)})$ generates from the initial position $\{t_*, x_*\}$ on the first step $t_* \le t \le t_2^{(u)}$ a motion $x_{U,v}[t_*[\cdot]t_2^{(u)}]$ which is a solution of the differential equation

$$\dot{x}_{U,v}[t] = f(t, x_{U,v}[t], u(t_*, x_*, \varepsilon_u), v[t]),$$
$$t_* \le t < t_2^{(u)}, \quad x[t_*] = x_*. \tag{5.8}$$

Thus, a new position $\{t_2^{(u)}, x[t_2^{(u)}]\} = \{t_2^{(u)}, x_{U,v}[t_*[t_2^{(u)}]t_2^{(u)}]\} \in G$ (3.12) is obtained.

And so on step by step, the motion $x_{U,v}[t_*[\cdot]\vartheta]$ (5.6) is determined in $t_* \le t \le \vartheta$ as the step-by-step solution of the differential equation

$$\dot{x}_{U,v}[t] = f(t, x_{U,v}[t], u(t_i^{(u)}, x_{U,v}[t_i^{(u)}], \varepsilon_u), v[t]),$$
$$t_i^{(u)} \le t < t_{i+1}^{(u)}, \quad i = 1, \dots, k_u \tag{5.9}$$

with the initial conditions $x[t_1^{(u)}] = x_*$ and $x_{U,v}[t_i^{(u)}] = x_{U,v}[t_{i-1}^{(u)}[t_i^{(u)}]t_i^{(u)}]$, $i = 2, \dots, k_u$. On every half-interval $[t_i, t_{i+1})$ we obtain the control actions

$$u[t_i[\cdot]t_{i+1}) = \{u[t] = u(t_i, x_{U,v}[t_i], \varepsilon_u) \in P,$$

$$t_i^{(u)} \le t < t_{i+1}^{(u)}\} \tag{5.10}$$

generated by the strategy $u(\cdot)$ (5.3). The actions

$$v[t_i^{(u)}[\cdot]t_{i+1}^{(u)}) = \{v[t] \in Q, \quad t_i^{(u)} \le t < t_{i+1}^{(u)}\} \tag{5.11}$$

that appear in (5.9) can be any admissible disturbances in the time half-interval $[t_i,\ t_{i+1})$. And here we ignore the mechanism V that forms these actions $v[t_i^{(u)}[\cdot]t_{i+1}^{(u)})$ (5.11). In particular, the actions $v[t_i^{(u)}[\cdot]t_{i+1}^{(u)})$ (5.11) may be the "pieces" of an arbitrary admissible realization of disturbances

$$v[t_*[\cdot]\vartheta) = \{v[t] \in Q, \quad t_* \le t < \vartheta\}. \tag{5.12}$$

This realization $v[t_*[\cdot]\vartheta)$ (5.12) might be planned by the second player beforehand. But we do not know his plan. Such a scheme of the process of control is shown in Figure 5.1.

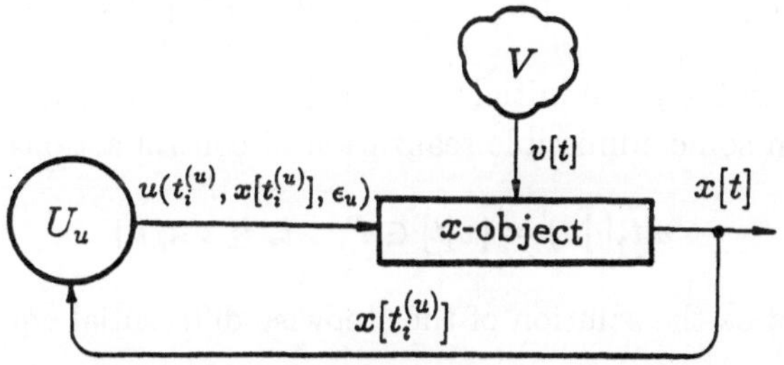

Figure 5.1

On the other hand, the mechanism V might form the disturbances $v[t_j^{(v)}[\cdot]t_{j+1}^{(v)})$ in some intervals $[t_j^{(v)},\ t_{j+1}^{(v)})$, $j = 1, \ldots, k_v$, $t_1^{(v)} = t_*$, $t_{k_v+1}^{(v)} = \vartheta$. In this case V may be a feedback law based on some pure strategy $v(\cdot)$ of the second player. This law V is determined by this strategy, some fixed parameter ε_v and some chosen partition $\Delta_v\{t_j^{(v)}\}$.

Thus, a motion $x_{U,v}[t_*[\cdot]\vartheta]$ (5.6) is a totality of the motions $x[t_i^{(u)}[\cdot]t_{i+1}^{(u)}]$ (5.9) which are continuously pasted together at the time moments $t_i^{(u)} \in \Delta_u\{t_i^{(u)}\}$ (5.4).

Now let us take the side of the second player.

The pure positional strategy $v(\cdot)$ of the second player is defined as a vector-function

$$v(\cdot) = \{v(t, x, \varepsilon_v) \in Q, \quad \{t, x\} \in G, \quad \varepsilon_v > 0\} \tag{5.13}$$

where Q is the set (3.6) and $\varepsilon_v > 0$ is a parameter of accuracy of the second player.

The law V corresponding to the strategy $v(\cdot)$ (5.13) is determined by the chosen triple: the strategy $v(\cdot)$ (5.13), the value of the parameter ε_v and the partition

$$\Delta_v\{t_i^{(v)}\} = \{t_* = t_1^{(v)}, \dots, t_i^{(v)} < t_{i+1}^{(v)},$$

$$i = 1, \dots, k_v, \quad t_{k_v+1}^{(v)} = \vartheta\} \tag{5.14}$$

where k_v is some natural number.

Thus, we have

$$V = \{v(\cdot);\ \varepsilon_v;\ \Delta_v\{t_i^{(v)}\}\}. \tag{5.15}$$

In (5.5) and (5.15) indices u and v show that the values ε_u and ε_v, the partitions $\Delta_u\{t_i^{(u)}\}$ and $\Delta_v\{t_i^{(v)}\}$, generally speaking, are different.

The motion

$$x_{V,u}[t_*[\cdot]\vartheta] = \{x_{V,u}[t], \quad t_* \le t \le \vartheta, \quad x_{V,u}[t_*] = x_*\} \tag{5.16}$$

generated from the given initial position $\{t_*, x_*\} \in G$ by the law V (5.15) together with some admissible realization of control actions

$$u[t_*[\cdot]\vartheta) = \{u[t] \in P, \quad t_* \le t < \vartheta\} \tag{5.17}$$

is determined as the solution of the stepwise differential equation

$$\dot{x}_{V,u}[t] = f(t, x_{V,u}[t], u[t], v(t_i^{(v)}, x_{V,u}[t_i^{(v)}], \varepsilon_v)),$$

$$t_i^{(v)} \le t < t_{i+1}^{(v)}, \quad i = 1, \dots, k_v \tag{5.18}$$

where $x[t_1^{(v)}] = x_*$, $x[t_i^{(v)}] = x[t_{i-1}^{(v)}[t_i^{(v)}]t_i^{(v)}]$, $i = 2, \dots, k_v$.

Concerning the realization of control actions $u[t_*[\cdot]\vartheta) = \{u[t_i^{(v)}[\cdot]t_{i+1}^{(v)}), i = 1, \dots, k_v\}$ (5.17) we can say the same as was said above about the realization $v[t_*[\cdot]\vartheta)$ (5.12). We have only to replace u by v and U by V.

The scheme of such control is shown in Figure 5.2.

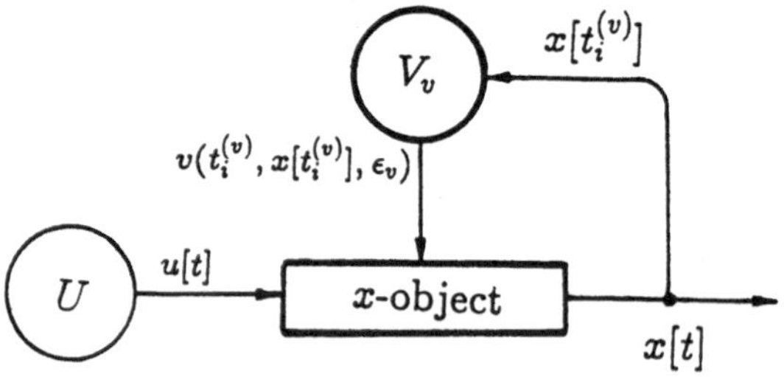

Figure 5.2

In particular, it can happen that the motion

$$x_{U,V}[t_*[\cdot]\vartheta] = \{x_{U,V}[t], \quad t_* \leq t \leq \vartheta, \quad x_{U,V}[t_*] = x_*\} \tag{5.19}$$

is generated from the given initial position $\{t_*, x_*\} \in G$ by the law U (5.5) together with the law V (5.15). We have already discussed that briefly (see p. 11). In this case the motion $x_{U,V}[t_*[\cdot]\vartheta]$ (5.19) is formed simultaneously by both schemes (5.9), (5.18). This case is shown in Figure 5.3.

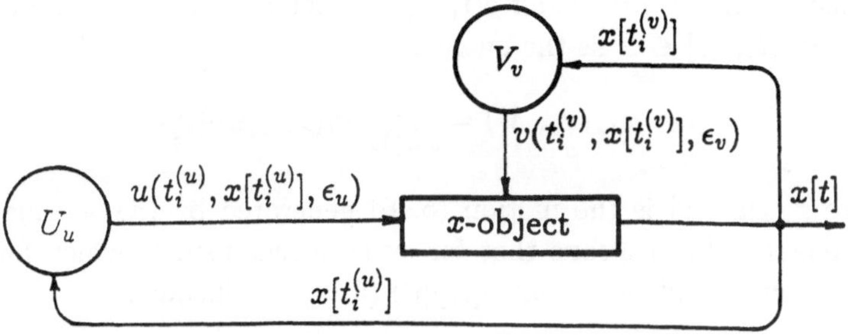

Figure 5.3

The three schemes of feedback control we have considered are shown in Figures 5.1–5.3.

6. Statement of the problem

The given initial position $\{t_*, x_*\} \in G$, the chosen control law U (5.5) and the realization of disturbances $v[t_*[\cdot]\vartheta)$ (5.12) determine a unique motion $x_{U,v}[t_*[\cdot]\vartheta]$ (5.6) of the x-object (3.1), (5.1). This motion determines in its turn the value of the functional $\gamma = \gamma(x[t_*[\cdot]\vartheta])$ (4.2), (4.9). So the data $\{t_*, x_*\}$, U, $v[t_*[\cdot]\vartheta)$ determine a unique value γ (4.2). Let us fix some control law U (5.5) and an initial position $\{t_*, x_*\} \in$ G. Then we obtain the set of all possible motions $X_{U,v}[t_*[\cdot]\vartheta]$ (5.6) that corresponds to the set of all admissible realizations $v[t_*[\cdot]\vartheta)$ (5.12) of disturbances. To this family of motions $x_{U,v}[\cdot]$ we put into correspondence the collection of the values $\gamma = \gamma(x_{U,v}[t_*[\cdot]\vartheta])$.

We call a quantity

$$\rho(t_*, x_*;\ U) = \sup_{v[t_*[\cdot]\vartheta)} \gamma(x_{U,v}[t_*[\cdot]\vartheta]) \tag{6.1}$$

an *ensured result* for the quality index γ (4.9), the chosen control law U (5.5) and the position $\{t_*, x_*\} \in$ G. In (6.1) the supremum is calculated with respect to all admissible realizations of disturbances $v[t_*[\cdot]\vartheta)$ (5.12).

The equality (6.1) means that for every mechanism V which forms the disturbances $v[t_*[\cdot]\vartheta)$ (5.12) the corresponding motion $x_{U,v}[t_*[\cdot]\vartheta]$ (5.6), (5.9) satisfies the inequality

$$\gamma(x_{U,v}[t_*[\cdot]\vartheta]) \leq \rho(t_*, x_*;\ U) \tag{6.2}$$

for any initial position $\{t_*, x_{U,v}[t_*]\} = \{t_*, x_*\} \in G$. Similarly, for the problem of actions v that are aimed to maximize the quality index γ, we define the *ensured result* $\rho(t_*, x_*;\ V)$ for a fixed law V (5.15) and a position $\{t_*, x_*\} \in G$ as the quantity

$$\rho(t_*, x_*;\ V) = \inf_{u[t_*[\cdot]\vartheta)} \gamma(x_{V,u}[t_*[\cdot]\vartheta]) \tag{6.3}$$

where $x_{V,u}[t_*[\cdot]\vartheta]$ is the motion (5.16) generated by the scheme (5.18). The equality (6.3) means that for every mechanism U which forms the realization of control actions $u[t_*[\cdot]\vartheta)$ (5.17) the inequality

$$\gamma(x_{V,u}[t_*[\cdot]\vartheta]) \geq \rho(t_*, x_*;\ V) \tag{6.4}$$

is valid.

However, we essentially use the concepts of the ensured results for the strategies $u(\cdot)$ (5.3) and $v(\cdot)$ (5.13). These quantities are defined in the following way:

Let the symbol $\Delta_\delta = \Delta\{t_i\}_\delta,\ \delta > 0$ denote a partition $\Delta\{t_i\}$ that satisfies the condition

$$\max_i\ (t_{i+1} - t_i) \leq \delta, \quad i = 1, \ldots, k\,. \tag{6.5}$$

According to (5.5) we denote the quantity $\rho(t_*, x_*;\ U)$ as

$$\rho(t_*, x_*;\ U) = \rho(t_*, x_*;\ u(\cdot), \varepsilon_u, \Delta_u\{t_i^{(u)}\})\,. \tag{6.6}$$

For a fixed strategy $u(\cdot)$ (5.3) and for the initial position $\{t_*, x_*\} \in G$ (3.12) we call the quantity

$$\begin{aligned}
&\rho(t_*, x_*;\ u(\cdot)) = \\
&\overline{\lim_{\varepsilon\to 0}}\ \lim_{\delta\to 0}\ \sup_{\Delta_\delta}\ \rho\Big(t_*, x_*;\ u(\cdot), \varepsilon_u, \Delta_u\{t_i^{(u)}\}_\delta\Big) = \\
&= \overline{\lim_{\varepsilon\to 0}}\ \lim_{\delta\to 0}\ \sup_{\Delta_\delta}\ \sup_{v[t_*[\cdot]\vartheta)} \gamma(x_{U_\delta,v}[t_*[\cdot]\vartheta])
\end{aligned} \tag{6.7}$$

an *ensured result* for this strategy $u(\cdot)$. In (6.7) U_δ is a control law U (5.5) corresponding to the parameter ε_u and the partition $\Delta_u\{t_i^{(u)}\}_\delta$ (5.4), (6.5);

and $x_{U_\delta,v}[t_*[\cdot]\vartheta]$ is a motion generated by this law U_δ; starting from the initial position $\{t_*, x_*\}$.

The definition (6.7) means the following:

Let the strategy $u(\cdot)$ (5.3) be chosen and the initial position $\{t_*, x_*\}$ be given. Then for an arbitrary sufficiently small number $\eta > 0$ there are a number $\varepsilon(\eta, t_*, x_*) > 0$ and a function $\delta(\eta, \varepsilon, t_*, x_*) > 0$ such that the inequality

$$\gamma(x_{U_\delta,v}[t_*[\cdot]\vartheta]) \leq \rho(t_*, x_*;\ u(\cdot)) + \eta \tag{6.8}$$

holds for the motion $x_{U_\delta,v}[t_*[\cdot]\vartheta]$ (5.6), (5.9) generated by the control law U_δ (5.5) corresponding to the chosen strategy $u(\cdot)$, provided

$$\varepsilon_u \leq \varepsilon(\eta, t_*, x_*), \qquad \delta \leq \delta(\eta, \varepsilon_u, t_*, x_*)\,. \tag{6.9}$$

Let us add that $\rho = \rho(t_*, x_*;\ u(\cdot))$ is the smallest of all the numbers ρ that satisfy the property (6.8), (6.9). Besides, if for the given strategy $u(\cdot)$ (5.3) and for every number $\eta > 0$ we can choose the values $\varepsilon(\eta) > 0$ and $\delta(\eta, \varepsilon) > 0$, which do not depend on the initial position $\{t_*, x_*\} \in G$, i.e., such that the inequalities

$$\varepsilon \leq \varepsilon(\eta), \qquad \delta \leq \delta(\eta, \varepsilon) \tag{6.10}$$

imply the inequality (6.8) for any initial position $\{t_*, x_*\} \in G$, then we call the ensured result $\rho(t_*, x_*;\ u(\cdot))$ *uniform* in the given region G (3.12).

We call the strategy $u^0(\cdot)$ *optimal* and call the value $\rho_u^0(t_*, x_*)$ an *optimal ensured result* of the first player, if the equality

$$\rho(t_*, x_*;\ u^0(\cdot)) = \min_{u(\cdot)} \rho(t_*, x_*;\ u(\cdot)) = \rho_u^0(t_*, x_*) \tag{6.11}$$

holds for every initial position $\{t_*, x_*\} \in G$. Here minimum is calculated with respect to all strategies $u(\cdot)$ (5.3).

The ensured result for the strategy $v(\cdot)$ (5.13) of the second player is defined by the equality

$$\rho(t, x;\ v(\cdot)) = \varliminf_{\varepsilon \to 0} \liminf_{\delta \to 0} \inf_{\Delta_\delta} \rho(t_*, x_*;\ V_\delta) =$$

$$= \varliminf_{\varepsilon \to 0} \liminf_{\delta \to 0} \inf_{\Delta_\delta} \inf_{u[t_*[\cdot]\vartheta)} \gamma(x_{V_\delta,u}[t_*[\cdot]\vartheta])\,. \tag{6.12}$$

Here $x_{V_\delta,u}[t_*[\cdot]\vartheta]$ is a motion (5.16), (5.18) generated by the law V_δ (5.15) that corresponds to the strategy $v(\cdot)$ (5.13), the parameter ε_v and the partition $\Delta_\delta = \Delta_v\{t_i^{(v)}\}_\delta$ (5.14). Here the partition Δ_δ satisfies the condition (6.5).

The following assertion stems from the definition (6.12) of the quantity $\rho(t_*, x_*;\ v(\cdot))$.

Let the strategy $v(\cdot)$ (5.13) be chosen and the initial position $\{t_*, x_*\} \in G$ be given. Then for an arbitrary number $\eta > 0$ there are $\varepsilon(\eta, t_*, x_*) > 0$ and $\delta(\eta, \varepsilon, t_*, x_*) > 0$ such that the inequality

$$\gamma(x_{V_\delta, u}[t_*[\cdot]\vartheta]) \geq \rho(t_*, x_*; v(\cdot)) - \eta \tag{6.13}$$

holds for any motion $x_{V_\delta,u}[t_*[\cdot]\vartheta]$ (5.16), (5.18) generated by the law V_δ, corresponding to the strategy $v(\cdot)$ (5.13) starting from the initial position $\{t_*, x_*\}$, provided $\varepsilon_v \leq \varepsilon(\eta, t_*, x_*)$ and $\delta \leq \delta(\eta, \varepsilon_v, t_*, x_*)$.

The value $\rho = \rho(t_*, x_*;\ v(\cdot))$ is the largest of all the numbers ρ that satisfy the condition (6.13).

The strategy $v^0(\cdot)$ and the ensured result $\rho_v^0(t_*, x_*)$ are *optimal* if the equality

$$\rho(t_*, x_*;\ v^0(\cdot)) = \max_{v(\cdot)} \rho(t_*, x_*;\ v(\cdot)) = \rho_v^0(t_*, x_*) \tag{6.14}$$

is valid for every initial position $\{t_*, x_*\} \in$ G. In (6.14) maximum is calculated over all strategies $v(\cdot)$ (5.13).

We call the optimal strategy $u^0(\cdot)$ (6.11) *uniformly optimal* in the given region G (3.12) if for the corresponding optimal ensured result $\rho_u^0(t_*, x_*)$ (6.11) the inequality (6.8) holds under the conditions (6.10), where the choice of numbers $\varepsilon(\eta) > 0$ and $\delta(\eta, \varepsilon) > 0$ can be independent of the position $\{t_*, x_*\} \in G$.

The uniformly optimal strategy $v^0(\cdot)$ (6.14) is defined similarly.

Thus, we consider the *problem* about the construction of the uniformly optimal strategies $u^0(\cdot) = u^0(t, x, \varepsilon)$ and $v^0(\cdot) = v^0(t, x, \varepsilon)$ and about the calculation of the values $\rho_u^0(t_*, x_*)$ (6.11) and $\rho_v^0(t_*, x_*)$ (6.14) (for every position in the given region G (3.12)).

The considered problem is formalized here as a differential game for the dynamical system (3.1), (3.2), (5.1) with the positional quality index $\gamma = \gamma(x[t_*[\cdot]\vartheta])$ (4.2), (4.9). Then using the terminology of the theory of the antagonistic differential games for two persons [103], [107] we say that the pair of optimal strategies $u^0(\cdot)$ (6.11) and $v^0(\cdot)$ (6.14) forms a *saddle point* $\{u^0(\cdot), v^0(\cdot)\}$ and provides the *value of the game* $\rho^0(t_*, x_*)$ if the equality

$$\rho_u^0(t_*, x_*) = \rho_v^0(t_*, x_*) = \rho^0(t_*, x_*) \tag{6.15}$$

holds for every position $\{t_*, x_*\} \in G$. In (6.15) $\rho_u^0(t_*, x_*)$ and $\rho_v^0(t_*, x_*)$ are the optimal ensured results of the first (6.11) and of the second (6.14) players, respectively.

The *problem* consists in effective construction of the value of the game $\rho^0(t_*, x_*)$ and the saddle point $\{u^0(\cdot) = u^0(t, x, \varepsilon),\ v^0(\cdot) = v^0(t, x, \varepsilon)\}$ for the considered differential game with the positional functional γ (4.2), (4.9).

The sense of the concepts of the value of the game and the saddle point is evident and is the following. If the motion $x_{U^0,V^0}[t_*[\cdot]\vartheta]$ is generated from the given initial position $\{t_*, x_*\} \in G$ according to the scheme (5.19), where $U = U^0$ and $V = V^0$ are the laws (5.5) and (5.15), which correspond to the optimal strategies $u^0(\cdot)$ (6.11) and $v^0(\cdot)$ (6.14), respectively, then the value of the functional $\gamma = \gamma(x_{U^0,V^0}[t_*[\cdot]\vartheta])$ is close to the value of the game $\rho^0(t_*, x_*)$ (6.15), provided the parameters ε_u, ε_v and the steps δ_u, δ_v of the partitions Δ_{δ_u}, Δ_{δ_v} are sufficiently small. Conversely, let the motion $x_{U^0,V}[t_*[\cdot]\vartheta]$ be generated according to the scheme (5.6), (5.9). Here $U = U^0$ is the control law (5.5) which corresponds to the optimal strategy $u^0(\cdot) \in \{u^0(\cdot), v^0(\cdot)\}$. The parameter ε_u and the step δ_u of this law U^0 are sufficiently small. Let V be some not optimal law (5.14) or any other mechanism that forms the disturbance $v[t_*[\cdot]\vartheta)$ (5.12). Then the result for the first player cannot be essentially worse than the value $\rho^0(t_*, x_*)$ of the game. That is, $\gamma(x_{U^0,V}[t_*[\cdot]\vartheta]) \leq \rho^0(t_*, x_*) + \eta$ and, as a rule, this result $\gamma(x_{U^0,V}[t_*[\cdot]\vartheta])$ is essentially better for the first player than the value $\rho^0(t_*, x_*)$ of the game, i.e., γ is essentially smaller than $\rho^0_u(t_*, x_*)$.

Similarly, for the motion $x_{U,V^0}[t_*[\cdot]\vartheta]$ if $U \neq U^0$ and V^0 is the law that corresponds to the optimal strategy $v^0(\cdot) \in \{u^0(\cdot), v^0(\cdot)\}$ and has a sufficiently small parameter ε_v and step δ_v, then the value $\gamma = \gamma(x_{U,V^0}[t_*[\cdot]\vartheta])$, as a rule, will be essentially better for the second player than the value $\rho^0(t_*, x_*)$ of the game. That means that the inequality $\gamma(x_{U,V^0}[t_*[\cdot]\vartheta]) \geq \rho^0(t_*, x_*) - \eta$ is valid. And, moreover, the value $\gamma(x_{U,V^0}[t_*[\cdot]\vartheta])$, often essentially larger than the value $\rho^0(t_*, x_*)$ of the game.

In conclusion of this section, we would like to emphasize the following. The desired optimal strategies $u^0(\cdot) = u^0(t, x, \varepsilon)$ (6.11) and $v^0(\cdot) = v^0(t, x, \varepsilon)$ (6.14) according to their definitions must be universal. Namely, we call the optimal strategy $u^0(\cdot)$ or $v^0(\cdot)$ the *positional universal strategy* in the region G if it is optimal for any initial position $\{t_*, x_*\} \in G$ (3.12). That is, for any possible current $\{t, x\} \in G$ and ε the vector $u^0(t, x, \varepsilon) \in P$ (or $v^0(t, x, \varepsilon) \in Q$) is constructed according to one and the same rule, which nevertheless was the initial position $\{t_*, x_*\}$, $t_0 \leq t_* < \vartheta$.

In [56] it is proved that there exists a solution of the problem, which is formulated here in Sections 3 to 6. The proof is based on *extremal strategies* that are constructed employing the known function $\rho^0(t, x)$ of the value of the game (6.15). These extremal strategies are connected with the concept of an *accompanying point* [77]. This method of construction of the strategies $u^0(\cdot)$ and $v^0(\cdot)$ is called the method of an *extremal shift* to the accompanying points.

It should be noted that this method of an extremal shift to the accompanying points is a modernization of the method of *extremal aiming to the stable bridge* [79].

7. An auxiliary w-object

The proposed construction of extremal strategies uses an abstract model of the given controlled object.

Let us consider at first a problem for the first player. Then we will consider a motion of some abstract $w^{[u]}$-model together with the real motion of x-object (3.1), (3.2), (5.1). The current state of this model at time $t \in [t_0,\ \vartheta]$ is determined by its n-dimensional phase vector $w^{[u]}[t]$. The phase vector $w^{[u]}[t]$ satisfies a differential equation similar to (3.1), i.e.,

$$\dot{w}^{[u]} = f(t, w^{[u]}, u_*^{[u]}, v_*^{[u]}), \quad t_0 \le t < \vartheta\,. \tag{7.1}$$

The function $f(\cdot)$ in (7.1) coincides with the function $f(\cdot)$ in (3.1), (3.2), (5.1), and the auxiliary control actions $u_*^{[u]}$ and $v_*^{[u]}$ satisfy the restrictions (3.5) and (3.6), i.e.,

$$u_*^{[u]} \in P, \qquad v_*^{[u]} \in Q\,. \tag{7.2}$$

The motion $w^{[u]}[t_*[\cdot]t^*]$, in some interval $[t_*,\ t^*]$, $t_* \in [t_0,\ \vartheta)$, $t^* \in (t_*,\ \vartheta]$ is generated from a suitable initial position $\{t_*, w_*^{[u]}\}$ by the control actions $u_*^{[u]}[t_*[\cdot]t^*)$ and $v_*^{[u]}[t_*[\cdot]t^*)$. The upper index $[u]$ in (7.1), (7.2) shows that we deal with the problem for the first player. We note that in the construction of a suitable motion $w^{[u]}[t_*[\cdot]t^*]$ of $w^{[u]}$-model (7.1), both actions $u_*^{[u]}[t_*[\cdot]t^*)$ and $v_*^{[u]}[t_*[\cdot]t^*)$ are some auxiliary *control actions* of the first player. Therefore, here $v_*^{[u]}[t_*[\cdot]t^*)$ is not uncontrollable by the first player, in contrast to the disturbance actions $v[t_*[\cdot]t^*)$ for the x-system (3.1).

Thus, the motion

$$w^{[u]}[t_*[\cdot]t^*] = \{w^{[u]}[t], \quad t_* \le t \le t^*,$$
$$w^{[u]}[t_*] = w_*^{[u]}\} \tag{7.3}$$

is determined as the solution of the differential equation

$$\dot{w}^{[u]}[t] = f(t, w^{[u]}[t], u_*^{[u]}[t], v_*^{[u]}[t]), \quad t_* \le t < t^* \tag{7.4}$$

with an initial condition $w^{[u]}[t_*] = w_*^{[u]}$ and control actions $u_*^{[u]}[t]$, $v_*^{[u]}[t]$ chosen by the first player.

We assume that every admissible motion $w^{[u]}[t_*[\cdot]t^*]$ (7.3), (7.4) satisfies the conditions

$$\{t, w^{[u]}[t]\} \in G^*, \quad t_* \leq t \leq t^* \tag{7.5}$$

where G^* is the chosen region in space $\{t, w\}$, $t \in [t_0,\ \vartheta]$, $w \in \mathbf{R}^n$. The set G^* satisfies all the properties imposed on G (3.12) and $G^{[\xi]} \in G^*$, $\xi > 0$. Here $G^{[\xi]}$ is composed from ξ-neighbourhoods of sections G_t of G (3.12) for all $t \in [t_0,\ \vartheta]$.

We note that the motions of $w^{[u]}$-model (7.1) are abstract and "imaginary". But in some cases it can be supposed that the motions of $w^{[u]}$-model are formed by a computer, which is included into the regulator U (see Figure 5.1). Then $w^{[u]}$-model can be treated as a certain computer model in contrast to the motion of the real controlled x-object (3.1). We note also that all the motions $w^{[u]}[t_*[\cdot]t^*]$ (7.3) satisfy the Lipschitz condition in t, i.e.,

$$| w^{[u]}[t''] - w^{[u]}[t'] | \leq L^*_{G^*} | t'' - t' | \tag{7.6}$$

where $t' \in [t_*,\ t^*]$, $t'' \in [t_*,\ t^*]$, $L^*_{G^*}$ is constant.

It is essential that choosing appropriate control actions $u[t]$, $v_*^{[u]}[t]$ one can ensure that the corresponding motions of the x-object and w-model are sufficiently close to each other. In this section we give a method for forming these realizations of control actions $u[\cdot]$ for the x-object (3.1) and control actions $v_*^{[u]}[\cdot]$ for the w-model (7.1).

On the other hand, if we solve the problem on behalf of the second player we introduce his w-model. This $w^{[v]}$-model is described by a differential equation similar to (7.1), i.e.,

$$\dot{w}^{[v]} = f(t, w^{[v]}, u_*^{[v]}, v_*^{[v]}), \quad t_0 \leq t < \vartheta,$$
$$u_*^{[v]} \in P, \quad v_*^{[v]} \in Q\,. \tag{7.7}$$

The motion

$$w^{[v]}[t_*[\cdot]t^*] = \{w^{[v]}[t], \quad t_* \leq t \leq t^*,$$
$$w^{[v]}[t_*] = w_*^{[v]}, \quad \{t_*, w_*^{[v]}\} \in G^*\} \tag{7.8}$$

of $w^{[v]}$-model (7.7) is determined similarly to the motion $w^{[u]}[t_*[\cdot]t^*]$ (7.3), (7.4) and the conditions (7.5), (7.6) are valid if only in (7.3)–(7.6) we substitute $[v]$ instead of $[u]$.

Being constructed for the second player, this $w^{[v]}$-model can be included into the device V (see Figure 5.2). In this case we also give a method for forming the realization of disturbance $v^{[v]}[\cdot]$ for the x-object (3.1) and the realization of control $u_*^{[v]}[\cdot]$ for the $w^{[v]}$-model (7.7), which provide the required closeness of the motions $x[\cdot]$ (3.9) and $w^{[v]}[\cdot]$ (7.8).

Let us consider again the problem on behalf of the first player. Now we shall give conditions guarantee the required closeness of the motions of x-object (3.1) and $w^{[u]}$-model (7.1).

Let us form the n-dimensional vector

$$s^{[u]} = x - w^{[u]}. \tag{7.9}$$

Let the positions $\{\tau_*, x[\tau_*]\} \in G$ and $\{\tau_*, w^{[u]}[\tau_*]\} \in G^*$, $\tau_* \in [t_0, \vartheta)$ be realized in some way. For some halfinterval $\tau_* \le t < \tau^*$, $\tau^* \in (\tau_*, \vartheta]$ we choose the control actions

$$u_e[\tau_*[\cdot]\tau^*) = \{u_e[\tau] = u_e \in P, \quad \tau_* \le t < \tau^*\} \tag{7.10}$$

for the x-object (3.1), and the control actions

$$v_{*e}^{[u]}[\tau_*[\cdot]\tau^*) = \{v_{*e}^{[u]}[\tau] = v_{*e}^{[u]} \in Q, \quad \tau_* \le t < \tau^*\} \tag{7.11}$$

for the $w^{[u]}$-model (7.1). We choose these realizations of control actions so that the vectors u_e, $v_{*e}^{[u]}$ satisfy the conditions

$$\max_{v\in Q} \langle\, s^{[u]}[\tau_*],\ f(\tau_*, x[\tau_*], u_e, v)\,\rangle =$$

$$= \min_{u\in P}\max_{v\in Q} \langle\, s^{[u]}[\tau_*],\ f(\tau_*, x[\tau_*], u, v)\,\rangle \tag{7.12}$$

$$\min_{u\in P} \langle\, s^{[u]}[\tau_*],\ f(\tau_*, w^{[u]}[\tau_*], u, v_{*e}^{[u]})\,\rangle =$$

$$= \max_{v\in Q}\min_{u\in P} \langle\, s^{[u]}[\tau_*],\ f(\tau_*, w^{[u]}[\tau_*], u, v)\,\rangle\,. \tag{7.13}$$

Generally speaking, these conditions do not determine the vectors u_e and $v_{*e}^{[u]}$ uniquely. We just choose some vector u_e (7.12) and some vector $v_{*e}^{[u]}$ (7.13) for every $\{\tau_*, x[\tau_*], w^{[u]}[\tau_*]\}$. These vectors u_e (7.12) and $v_{*e}^{[u]}$ (7.13) determine the control actions $u_e[t] = u_e[\tau_*]$, $\tau_* \le t < \tau^*$ (7.10) and $v_{*e}^{[u]}[t] = v_{*e}^{[u]}[\tau_*]$, $\tau_* \le t < \tau^*$ (7.11), which turn out to be constant on the time halfinterval $[\tau_*, \tau^*)$.

Let an admissible realization of disturbances $v[\tau_*[\cdot]\tau^*)$ (for the x-object) and an admissible realization of control actions $u_*^{[u]}[\tau_*[\cdot]\tau^*)$ (for the $w^{[u]}$-model) be formed in this or that way. We remind the reader that the admissible realizations of disturbances or control actions are arbitrary Borel-measurable functions in $t \in [\tau_*, \tau^*)$, which are restricted by inclusions $v[t] \in Q$ and $u_*^{[u]}[t] \in P$.

Now we consider the motions $x[\tau_*[\cdot]\tau^*]$ and $w^{[u]}[\tau_*[\cdot]\tau^*]$ which are defined by the differential equations

$$\dot{x}[t] = f(t, x[t], u_e, v[t]), \quad \tau_* \leq t < \tau^*, \tag{7.14}$$

and

$$\dot{w}^{[u]}[t] = f(t, w^{[u]}[t], u_*^{[u]}[t], v_{*e}^{[u]}), \quad \tau_* \leq t < \tau^*, \tag{7.15}$$

with given initial conditions $x[\tau_*]$ and $w^{[u]}[\tau_*]$, respectively.

We consider the function

$$\nu(t, x, w) = | x - w |^2 \exp\{-2\lambda (t - t_0)\} \tag{7.16}$$

which will estimate the distance between the phase vectors $x[t]$ of x-object and $w^{[u]}[t]$ of $w^{[u]}$-model at the current time moment $t \in [\tau_*, \tau^*]$. This function $\nu(\cdot)$ can be considered as some Liapunov function [2], [101].

In (7.16) $\lambda > 0$ is a constant that is not smaller than the Lipschitz constant L_{G^*} in the condition (3.4), where G is replaced by G^*.

Lemma 7.1. *For any number $\varepsilon > 0$ there is a number $\delta(\varepsilon) > 0$ such that the following holds. Let the positions $\{\tau_*, x[\tau_*]\} \in G$ and $\{\tau_*, w^{[u]}[\tau_*]\} \in G^*$ be given and the time moment τ^* satisfy the condition*

$$\tau^* - \tau_* \leq \delta(\varepsilon), \quad \tau^* \in (\tau_*, \vartheta]. \tag{7.17}$$

Let the motion $x[\tau_*[\cdot]\tau^*]$ of x-object and the motion $w^{[u]}[\tau_*[\cdot]\tau^*]$ of $w^{[u]}$-model be generated according to the schemes (7.14) *and* (7.15) *respectively. Then the inequality*

$$\nu(t, x[t], w^{[u]}[t]) \leq \nu(\tau_*, x[\tau_*], w^{[u]}[\tau_*]) + \varepsilon (t - \tau_*) \tag{7.18}$$

is valid for all $t \in [\tau_*, \tau^*]$.

Thus, Lemma 7.1 shows that the control actions $u_e[\tau_*[\cdot]\tau^*)$ (7.10) and $v_{*e}^{[u]}[\tau_*[\cdot]\tau^*)$ (7.11) which are chosen according to the conditions (7.12) and (7.13) together with any admissible disturbances $v[\tau_*[\cdot]\tau^*)$ and control actions $u_*^{[u]}[\tau_*[\cdot]\tau^*)$, generate from the initial positions $\{\tau_*, x_*\}$ and $\{\tau_*, w_*^{[u]}\}$ the motions $x[\tau_*[\cdot]\tau^*]$ and $w[\tau_*[\cdot]\tau^*]$ that are sufficiently close. The corresponding distance is estimated by the quantity $\nu(t, x[t], w[t])$ (7.16) and by the inequality (7.18).

We omit here the proof of Lemma 7.1. It is published in [77].

Let us consider now the problem on behalf of the second player.

Lemma 7.2. *For any $\varepsilon > 0$ there is $\delta(\varepsilon) > 0$ such that if for the positions $\{\tau_*, x[\tau_*]\} \in G$, $\{\tau_*, w^{[v]}[\tau_*]\} \in G^*$ the vectors v_e and $u_{*e}^{[v]}$ are defined from the conditions*

$$\min_{u\in P} \langle\ s^{[v]}[\tau_*],\ f(\tau_*, x[\tau_*], u, v_e)\ \rangle =$$

$$= \max_{v\in Q} \min_{u\in P} \langle\ s^{[v]}[\tau_*],\ f(\tau_*, x[\tau_*], u, v)\ \rangle \tag{7.19}$$

$$\max_{v\in Q} \langle\ s^{[v]}[\tau_*],\ f(\tau_*, w^{[v]}[\tau_*], u_{*e}^{[v]}, v)\ \rangle =$$

$$= \min_{u\in P} \max_{v\in Q} \langle\ s^{[v]}[\tau_*],\ f(\tau_*, w^{[v]}[\tau_*], u, v)\ \rangle \tag{7.20}$$

where

$$s^{[v]}[\tau_*] = w^{[v]}[\tau_*] - x[\tau_*] \tag{7.21}$$

then for $\tau^* \leq \tau_* + \delta(\varepsilon)$, $\tau^* \leq \vartheta$ the disturbances $v_e[t] = v_e$, $\tau_* \leq t < \tau^*$ together with any admissible control actions $u[\tau_*[\cdot]\tau^*)$ generate the motion $x[\tau_*[\cdot]\tau^*)$ and the actions $u_{*e}^{[v]}[t] = u_{*e}^{[v]}$, $\tau_* \leq t < \tau^*$ together with any $v_*^{[v]}[\tau_*[\cdot]\tau^*)$ generate the motion $w^{[v]}[\tau_*[\cdot]\tau^*]$ of $w^{[v]}$-model (7.7) for which the inequality

$$\nu(t, x[t], w^{[v]}[t]) \leq \nu(\tau_*, x[\tau_*], w^{[v]}[\tau_*]) + \varepsilon\,(t - \tau_*) \tag{7.22}$$

holds for all $t \in [\tau_*, \tau^*]$.

The proof of Lemma 7.2 we also omit here.

8. Accompanying points. An extremal shift

In this section the method of construction of the extremal strategies $u_e(\cdot) = u_e(t, x, \varepsilon)$ and $v_e(\cdot) = v_e(t, x, \varepsilon)$ is described.

Let us assume that we can construct some function $\rho(t, w)$ of the position $\{t, w\} \in G^*$ (7.5) so that the following conditions are valid:

1^0. For every fixed $t \in [t_0,\ \vartheta]$ the function $\rho(t, w)$ satisfies the Lipschitz condition in w, i.e.,

$$\mid \rho(t, w^{(1)}) - \rho(t, w^{(2)}) \mid \leq L^* \mid w^{(2)} - w^{(1)} \mid \tag{8.1}$$

where $\{t, w^{(i)}\} \in G^*$, $i = 1, 2$; L^* is a constant.

2^0. The equality

$$\gamma(w[\vartheta[\vartheta]\vartheta]) = \rho(\vartheta, w[\vartheta]) \tag{8.2}$$

holds for every $\{\vartheta, w[\vartheta]\} = \{\vartheta, w[\vartheta[\vartheta]\vartheta]\} \in G^*$.

3^0_u. Suppose a position $\{\tau_*, w_*^{[u]}\} \in G^*$, a number $\varepsilon_* > 0$, a moment $\tau^* > \tau_*$ and some admissible control actions $v_*^{[u]}[\tau_*[\cdot]\tau^*)$ are given. Then there exist an $\varepsilon^* > 0$ and the admissible control actions $u_*^{[u]}[\tau_*[\cdot]\tau^*)$, which together with $v_*^{[u]}[\tau_*[\cdot]\tau^*)$ generate from the initial position $\{\tau_*, w_*^{[u]}\}$ a motion $w^{[u]}[\tau_*[\cdot]\tau^*]$ of the $w^{[u]}$-model (7.1) such that

$$\beta(w^{[u]}[\tau_*[\cdot]\tau^*), \rho(\tau^*, w^{[u]}[\tau^*]) + 2\varepsilon^*) \leq$$

$$\leq \rho(\tau_*, w_*^{[u]}) + 2\varepsilon_* \tag{8.3}$$

where $\beta(\cdot)$ is the functional in (4.9); $w^{[u]}[\tau^*] = w^{[u]}[\tau_*[\tau^*]\tau^*]$.

The last condition 3^0_u is called the *property of u-stability* [79] for the function $\rho(t, w)$.

Let the function $\rho(t, w)$ satisfy also the following *property of v-stability* [79].

4^0_v. Suppose we are given a position $\{\tau_*, w_*^{[v]}\} \in G^*$, a number $\varepsilon_* > 0$, a moment $\tau^* > \tau_*$ and some admissible control actions $u_*^{[v]}[\tau_*[\cdot]\tau^*)$. Then there exist an $\varepsilon^* > 0$ and the admissible control actions $v_*^{[v]}[\tau_*[\cdot]\tau^*)$, which together with $u_*^{[v]}[\tau_*[\cdot]\tau^*)$ generate from the initial position $\{\tau_*, w_*^{[v]}\}$ a motion $w^{[v]}[\tau_*[\cdot]\tau^*]$ of the $w^{[v]}$-model (7.7) such that

$$\beta(w^{[v]}[\tau_*[\cdot]\tau^*), \rho(\tau^*, w^{[v]}[\tau^*]) - 2\varepsilon^*) \geq$$

$$\geq \rho(\tau_*, w_*^{[v]}) - 2\varepsilon_* \tag{8.4}$$

where $w^{[v]}[\tau^*] = w^{[v]}[\tau_*[\tau^*]\tau^*]$.

It is proved that a function $\rho(t, w)$ that satisfies the conditions 1^0, 2^0, 3^0_u and 4^0_v exists and this function is the *value of the game* $\rho^0(t, x)$ (6.15) for the considered differential game for the system (3.1), (3.2), (5.1) with the positional quality index γ (4.2), (4.9). The scheme of this proof will be given below in Section 9.

The sense of the properties of u-stability 3^0_u and v-stability 4^0_v is the following. The property of u-stability for the function $\rho(t, w)$ means that for every admissible $v_*^{[u]}[\tau_*[\cdot]\tau^*)$ there exist the control actions $u_*^{[u]}[\tau_*[\cdot]\tau^*)$ such that along the corresponding motion $w^{[u]}[\tau_*[\cdot]\tau^*]$ the function $\rho(t, w)$ actually does not increase (in the sense of the estimation (4.9)). Conversely, the property of v-stability means that for every admissible $u_*^{[v]}[\tau_*[\cdot]\tau^*)$ there exist the actions $v_*^{[v]}[\tau_*[\cdot]\tau^*)$ such that along the corresponding motion $w^{[v]}[\tau_*[\cdot]\tau^*]$ the function $\rho(t, w)$ actually does not decrease (in the sense of the estimation (4.9)).

Now we define the strategies $u_e(\cdot) = u_e(t, x, \varepsilon)$ and $v_e(\cdot) = v_e(t, x, \varepsilon)$, which are called the *extremal* strategies with respect to the function $\rho(t, w)$. These strategies are constructed in the following way:

At first we define the extremal strategy $u_e(\cdot)$ of the first player (see Section 5).

We choose some small value of the parameter $\varepsilon_u > 0$. Suppose that some position $\{t, x\} \in G$ (3.12) is fixed.

We denote by the symbol $K^{[u]}(\varepsilon_u, t, x)$ the set of positions $\{t, w^{[u]}\}$ that satisfy the conditions

$$K^{[u]}(\varepsilon_u, t, x) \subset G^* \tag{8.5}$$

$$\nu(t, x, w^{[u]}) \leq \varepsilon_u + \varepsilon_u\,(t - t_0) \tag{8.6}$$

where G^* is the set from (7.5); $\nu(\cdot)$ is the function (7.16).

According to (7.16) we have

$$K^{[u]}(\varepsilon_u, t, x) = \Big\{\{t, w^{[u]}\} : \ \{t, w^{[u]}\} \in G^*,$$

$$| \, x - w^{[u]} \, | \leq R^{[u]}(\varepsilon_u, t)\Big\} \tag{8.7}$$

where

$$R^{[u]}(\varepsilon_u, t) = (\varepsilon_u + \varepsilon_u\,(t - t_0))^{1/2} \exp\{\lambda\,(t - t_0)\}\,. \tag{8.8}$$

The set $K^{[u]}(\varepsilon_u, t, x)$ of positions $\{t, w^{[u]}\}$ is such that all the corresponding points $w^{[u]}$ are located in Euclidean space $\mathbf{R}^n$ in the sphere $K_t^{[u]}$ with the center x. The radius $R^{[u]}(\varepsilon_u, t)$ of this sphere is determined by the equality (8.8). We assume further that $\{t, x\} \in G$ (3.12), $\{t, w^{[u]}\} \in G^*$, $R^{[u]}(\varepsilon_u, t) \leq \xi$, $t_0 \leq t \leq \vartheta$, where $\xi > 0$ is the value from the definition of the set G^* (see (7.5)).

We call the point $w^{[u]0}(t, x, \varepsilon_u)$ that satisfies the conditions

$$\{t, w^{[u]0}(t, x, \varepsilon_u)\} \in K^{[u]}(\varepsilon_u, t, x) \tag{8.9}$$

and

$$\rho(t, w^{[u]0}(t, x, \varepsilon_u)) = \min_{\{t, w^{[u]}\} \in K^{[u]}(\varepsilon_u, t, x)} \rho(t, w^{[u]}) \tag{8.10}$$

an *accompanying point* [55], [77].

We sometimes treat the objects w, x, etc. as vectors and sometimes as points. We hope that this will not lead to confusion since the meaning will be clear from the context.

According to property 1^0 (8.1) of the function $\rho(t, w)$ the point $w^{[u]0}(t, x, \varepsilon_u)$ (8.10) exists. We note, however, that this point can be nonunique for fixed t, x, ε_u. We choose one of the points for every given triple t, x, ε_u.

The position $\{t, w^{[u]0}(t, x, \varepsilon_u)\}$ that corresponds to the point $w^{[u]0}(t, x, \varepsilon_u)$ is called the *accompanying position* for the position $\{t, x\}$ of x-object.

The strategy $u_e(\cdot) = u_e(t, x, \varepsilon_u)$ *extremal to the function* $\rho(t, w)$ (or simply an *extremal strategy*) is understood as some rule which to every $\varepsilon_u > 0$ and every possible position $\{t, x\} \in G$ puts into correspondence the vector $u_e = u_e(t, x, \varepsilon_u) \in P$ that satisfies the condition (7.12) (where $\tau_* = t$, $s^{[u]} = s^{[u]0}$), i.e., the condition

$$\max_{v \in Q} \langle\ s^{[u]0}(t, x, \varepsilon_u),\ f(t, x, u_e(t, x, \varepsilon_u), v)\ \rangle =$$

$$= \min_{u \in P} \max_{v \in Q} \langle\ s^{[u]0}(t, x, \varepsilon_u),\ f(t, x, u, v)\ \rangle \tag{8.11}$$

where P and Q are compacta (3.5), (3.6) and

$$s^{[u]0}(t, x, \varepsilon_u) = x - w^{[u]0}(t, x, \varepsilon_u)\,. \tag{8.12}$$

Here $w^{[u]0}(t, x, \varepsilon_u)$ is an accompanying point (8.10).

We call (8.11), (8.12) the condition of *extremal shift to the accompanying point* for u (for the first player).

The continuity of the function $f(\cdot)$ (3.1) in u, v and compactness of the sets P (3.5) and Q (3.6) imply that there exists a value $u_e = u_e(t, x, \varepsilon_u)$ that satisfies the minimax condition (8.11). But it can be nonunique. For every fixed parameter $\varepsilon_u > 0$ and position $\{t, x\}$ we choose one definite value $u_e(t, x, \varepsilon_u)$ (8.11). Thus, the extremal strategy $u_e(\cdot) = u_e(t, x, \varepsilon_u)$ is defined.

It should be noted also that $u_e(t, x, \varepsilon_u)$ can be chosen as a function Borel measurable in x. This assertion is based on the theorem [1].

Let us consider the situation on behalf of the second player, i.e., let us construct the extremal strategy $v_e(\cdot) = v_e(t, x, \varepsilon)$ for this player.

Let us choose some value of parameter $\varepsilon_v > 0$ and fix some position $\{t, x\} \in G$ of the x-object (3.1). We denote by $K^{[v]}(\varepsilon_v, t, x)$ the set of positions $\{t, w^{[v]}\}$ similar to (8.7), i.e.,

$$K^{[v]}(\varepsilon_v, t, x) = \Big\{ \{t, w^{[v]}\} :\ \{t, w^{[v]}\} \in G^*,$$

$$|\ x - w^{[v]}\ | \le (\varepsilon_v + \varepsilon_v\,(t - t_0))^{1/2} \exp\{\lambda\,(t - t_0)\} \Big\}\,. \tag{8.13}$$

We assume further that $\{t, x\} \in G$, $\{t, w^{[v]}\} \in G^*$, $R^{[v]}(\varepsilon_v, t) \le \xi$, $t_0 \le t \le \vartheta$. Here the value $R^{[v]}$ is determined similar to (8.8) by substituting u by v (see p. 62). An accompanying point $w^{[v]0}(t, x, \varepsilon_u)$ (of the second player) is defined by the conditions

$$\{t, w^{[v]0}(t, x, \varepsilon_u)\} \in K^{[v]}(\varepsilon_v, t, x) \tag{8.14}$$

and

$$\rho(t, w^{[v]0}(t,x,\varepsilon_v)) = \max_{\{t,w^{[v]}\}\in K^{[v]}(\varepsilon_v,t,x)} \rho(t, w^{[v]}) \,. \tag{8.15}$$

This point can be also nonunique for fixed $t,\ x,\ \varepsilon_v$. We choose one of the points for every given triple $t,\ x,\ \varepsilon_v$.

An extremal strategy $v_e(\cdot) = v_e(t,x,\varepsilon_v)$ is a function which to every possible $\varepsilon_v > 0$ and every position $\{t,x\}$ puts into correspondence a vector $v_e = v_e(t,x,\varepsilon_v) \in Q$ that satisfies the condition

$$\min_{u\in P} \langle\ s^{[v]0}(t,x,\varepsilon_v),\ f(t,x,u,v_e(t,x,\varepsilon_v))\ \rangle =$$

$$= \max_{v\in Q}\min_{u\in P} \langle\ s^{[v]0}(t,x,\varepsilon_v),\ f(t,x,u,v)\ \rangle \tag{8.16}$$

where

$$s^{[v]0}(t,x,\varepsilon_v) = w^{[v]0}(t,x,\varepsilon_v) - x\,. \tag{8.17}$$

Here $w^{[v]0}(t,x,\varepsilon_v)$ is an accompanying point (8.15).

As we can see, the condition (8.16) is similar to the condition (7.19) from Lemma 7.2.

Again, as in (8.11), for every given triple $t,\ x,\ \varepsilon_v$ we choose one vector $v_e(t,x,\varepsilon_v) \in Q$. As above, we note that $v_e(t,x,\varepsilon_v)$ can be chosen as a Borel-measurable function in x.

We call (8.16), (8.17) the condition of *extremal shift to the accompanying point* for v (for the second player). Thus, the extremal strategy $v_e(\cdot) = v_e(t,x,\varepsilon_v)$ is defined.

This method of forming the process of control, which uses the extremal strategies $u_e(\cdot)$ and $v_e(\cdot)$ basing on the accompanying points $w^{[u]0}$ and $w^{[v]0}$, was offered in the paper [55], and in the book [77] it was called the *extremal shift.*

It is essential that this method for constructing the extremal actions $u_e(t_i^{[u]}, x[t_i^{[u]}], \varepsilon_u)$ and $v_e(t_i^{[v]}, x[t_i^{[v]}], \varepsilon_v)$ use the values of the function $\rho(t,w)$ only in small neighbourhoods of the current positions $\{t_i, x[t_i]\}$. And it does not require calculating $\rho(t,w)$ in the whole region G^*.

The term *shift* means that actually the vector u_e (8.11) is aimed to shift the controlled x-object in the direction opposite to the vector $s^{[u]0}$ (8.12). Thus, the action u_e tends to bring the x-object closer to the accompanying point $w^{[u]0}$ (8.10). The value of the game $\rho(t, w^{[u]0})$ in this point $w^{[u]0}$ is the smallest in the set $K^{[u]}(t,x,\varepsilon_u)$ (8.7). Similarly, the extremal choice of the vector v_e (8.16) and the accompanying point $w^{[v]0}$ (8.15) pursues the contrary aim. Both cases are illustrated in Figures 8.1 and 8.2.

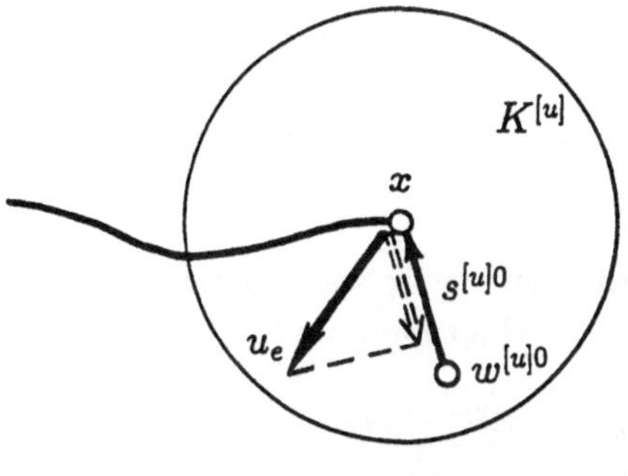

Figure 8.1

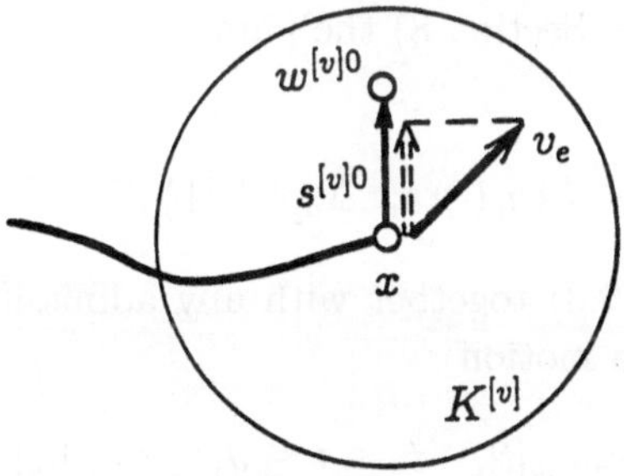

Figure 8.2

9. Existence of the solution. (Value of the game. Optimal strategies)

In this section we give an outline of the proof that the function $\rho(t, w)$, which satisfies the conditions 1^0, 2^0, 3^0_u and 4^0_v from Section 8, is the value of the game $\rho^0(t, x)$ and that this function $\rho(t, w)$ exists. Also, it will be proved that the extremal strategies $u_e(\cdot)$ and $v_e(\cdot)$, which were defined in Section 8, are the optimal ones $u^0(\cdot) = u^0(t, x, \varepsilon)$ and $v^0(\cdot) = v^0(t, x, \varepsilon)$.

We have the following:

Lemma 9.1. *Let the function $\rho(t, w)$ that satisfies the conditions 1^0, 2^0, 3^0_u (from Section 8) exist. Then for an arbitrary number $\eta > 0$ there exist $\varepsilon(\eta) > 0$ and $\delta(\eta, \varepsilon) > 0$ such that for any motion $x_{U_e,v}[t_*[\cdot]\vartheta]$ generated by the scheme* (5.9) *from an initial position $\{t_*, x_*\} \in G$ by the control law $U_e = \{u_e(\cdot);\ \varepsilon_u;\ \Delta_{u\delta}\{t_i^{(u)}\}\}$ (corresponding to the extremal* *strategy* $u_e(\cdot) = u_e(t, x, \varepsilon)$ *and* $\varepsilon_u \leq \varepsilon(\eta)$, $\max_i(t^{(u)}_{i+1} - t^{(u)}_i) \leq \delta(\eta, \varepsilon_u))$ *the inequality*

$$\gamma(x_{U_e,v}[t_*[\cdot]\vartheta]) \leq \rho(t_*, x_*) + \eta \tag{9.1}$$

holds.

Scheme of the proof. Let us fix some number $\varepsilon_u > 0$ and choose a partition $\Delta_{u_\delta}\{t^{(u)}_i\}$ (5.4) of the time interval $[t_*,\ \vartheta]$ such that

$$\max_i\ (t^{(u)}_{i+1} - t^{(u)}_i) \leq \delta(\varepsilon_u) \quad (i = 1, \ldots, k_u) \tag{9.2}$$

where $\delta(\varepsilon_u) > 0$ is the number from Lemma 7.1 chosen for the fixed $\varepsilon_u > 0$.

Let U_e be the control law U (5.5) corresponding to the extremal strategy $u_e(\cdot) = u_e(t, x, \varepsilon)$ (see Section 8) the parameter ε_u and the partition $\Delta_{u_\delta}\{t^{(u)}_i\}$ (9.2), i.e.,

$$U_e = \{u_e(\cdot); \varepsilon_u; \Delta_{u_\delta}\{t^{(u)}_i\}\}\,. \tag{9.3}$$

The control law U_e (9.3) together with any admissible disturbances $v[t_*[\cdot]\vartheta)$ (5.12) generate a motion

$$x_{U_e,v}[t_*[\cdot]\vartheta] = \{x_{U_e,v}[t], \quad t_* \leq t \leq \vartheta, \quad x_{U_e,v}[t_*] = x_*\} \tag{9.4}$$

according to the scheme (5.9) where $U = U_e$.

Let us prove that for any $\eta > 0$ we can choose $\varepsilon(\eta) > 0$ such that for the motion $x_{U_e,v}[t_*[\cdot]\vartheta]$ (9.4) the condition (9.1) holds for any initial position $\{t_*, x_*\} \in G$.

When we form the motion $x_{U_e,v}[t_*[\cdot]\vartheta]$ (9.4) in accordance with the law U_e which corresponds to the extremal strategy $u_e(\cdot)$ we obtain a collection of the accompanying points (8.5)–(8.10)

$$w^{[u]0}[t_i] = w^{[u]0}(t^{(u)}_i, x_{U_e,v}[t^{(u)}_i], \varepsilon_u), \quad i = 1, \ldots, k_u + 1\,. \tag{9.5}$$

These points satisfy the conditions

$$\rho(t^{(u)}_i, w^{[u]0}(t^{(u)}_i, x[t^{(u)}_i], \varepsilon_u)) =$$

$$= \min_{\{t_i, w^{[u]}\} \in K^{[u]}(t^{(u)}_i, x[t^{(u)}_i], \varepsilon_u)} \rho(t^{(u)}_i, w^{[u]}) \tag{9.6}$$

that follow from (8.10).

Therefore, at every time moment $t^{(u)}_i \in \Delta_{u_\delta}\{t^{(u)}_i\}$ (9.2) the accompanying point $w^{[u]0}[t_i]$ (9.5) is determined for the realized point $x_{U_e,v}[t^{(u)}_i] = x_{U_e,v}[t_*[t^{(u)}_i]t^{(u)}_i]$.

Consequently, we obtain the collection of vectors

$$s^{[u]0}[t_i] = x[t_i^{(u)}] - w^{[u]0}(t_i^{(u)}, x[t_i^{(u)}], \varepsilon_u),$$

$$i = 1, \dots, k_u + 1 . \tag{9.7}$$

According to the conditions of Lemma 7.1 the collection (9.7) determines the corresponding collection of control actions

$$v_{*e}^{[u]}[t_i^{(u)}[\cdot]t_{i+1}^{(u)}) = \{v_{*e}^{[u]}[t] = v_{*e}^{[u]}[t_i^{(u)}] \in Q, \quad t_i^{(u)} \le t < t_{i+1}^{(u)}\}$$

$$i = 1, \dots, k_u \tag{9.8}$$

for the $w^{[u]}$-model (7.1). Thus in (9.8) $v_{*e}^{[u]}[t_i^{(u)}]$ is a vector that satisfies the condition (7.13) where $\tau_* = t_i^{(u)}$.

In its turn, the collection $\{v_{*e}^{[u]}[\cdot]\}$ (9.8), together with the collection of accompanying positions $\{t_i^{(u)}, w^{[u]0}[t_i]\} \in G^*$, and with the value $\varepsilon_1^{[u]} = \varepsilon_u$ according to the property of u-stability 3_u^0, from Section 8 define the collection of control actions

$$u_*^{[u]}[t_i^{(u)}[\cdot]t_{i+1}^{(u)}) = \{u_*^{[u]}[t] \in P, \quad t_i^{(u)} \le t < t_{i+1}^{(u)}\}, \quad i = 1, \dots, k_u \tag{9.9}$$

and the collection of numbers

$$\varepsilon_{i+1}^{[u]} > 0, \quad i = 1, \dots, k_u . \tag{9.10}$$

These numbers $\varepsilon_j^{[u]}$ and $\varepsilon_{j+1}^{[u]}$ for any interval $[t_j^{(u)}, t_{j+1}^{(u)})$ correspond to the values ε_* and ε^* involved in the property 3_u^0 where $\tau_* = t_j^{(u)}$ and $\tau^* = t_{j+1}^{(u)}$.

Thus, to the motion $x_{U_e,v}[t_*[\cdot]\vartheta]$ (9.4) of the controlled object (3.1), (3.2), (5.1) we put in accordance a certain imaginary so-called *accompanying motion*

$$w_a^{[u]}[t_*[\cdot]\vartheta] = \{w_a^{[u]}[t], \quad t_* \le t \le \vartheta\} \tag{9.11}$$

of $w^{[u]}$-model (7.1) (see Figure 9.1). We mark this motion with lower index a. The motion $w_a^{[u]}[t_*[\cdot]\vartheta]$ (9.11) is formed in such a way that at the time moments $t_i^{(u)}$, $i = 2, \dots, k_u$ it can be discontinuous. On the time intervals $t_i^{(u)} \le t < t_{i+1}^{(u)}$ the "pieces" $w_a^{[u]}[t_i^{(u)}[\cdot]t_{i+1}^{(u)})$ of this motion $w_a^{[u]}[\cdot]$ (9.11) start from the accompanying points $w^{[u]0}[t_i]$ (9.5) and are generated by the control actions $u_*^{[u]}[\cdot]$ (9.9) and $v_{*e}^{[u]}[\cdot]$ (9.8). Let us denote

$$w_a^{[u]*}[t_{i+1}^{(u)}] = \lim_{t \to t_{i+1}^{(u)} - 0} w_a^{[u]}[t_i^{(u)}[t]t_{i+1}^{(u)}) . \tag{9.12}$$

This point, generally speaking, does not coincide with the next accompanying point $w^{[u]0}[t_{i+1}]$ (9.5), it may be $w_a^{[u]*}[t_{i+1}^{(u)}] \neq w_a^{[u]}[t_{i+1}^{(u)}]$.

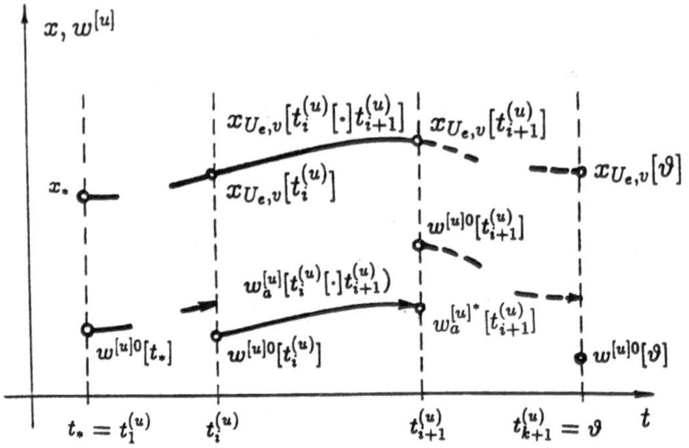

Figure 9.1

For the motions $x_{U_e,v}[t_*[\cdot]\vartheta]$ (9.4) and

$$w_a^{[u]}[t_*[\cdot]\vartheta] = \{w_a^{[u]}[t_i^{(u)}[\cdot]t_{i+1}^{(u)}), \quad i = 1, \ldots, k_u,$$

$$w_a^{[u]}[t_i^{(u)}] = w^{[u]0}[t_i^{(u)}]\} \tag{9.13}$$

on every time interval $t_i^{(u)} \leq t \leq t_{i+1}^{(u)}$ according to Lemma 7.1 and (8.5)–(8.10), the following inequalities

$$\nu(t_{i+1}^{(u)}, x_{U_e,v}[t_{i+1}^{(u)}], w_a^{[u]*}[t_{i+1}^{(u)}]) \leq$$

$$\leq \nu(t_i^{(u)}, x_{U_e,v}[t_i^{(u)}], w^{[u]0}(t_i^{(u)}, x_{U_e,v}[t_i^{(u)}], \varepsilon_u))+$$

$$+\,\varepsilon_u\,(t_{i+1}^{(u)} - t_i^{(u)}) \leq \varepsilon_u + \varepsilon_u(t_{i+1}^{(u)} - t_1^{(u)}), \qquad i = 1, \ldots, k_u \tag{9.14}$$

are valid. According to the property 3_u^0 (8.3) the inequality

$$\beta(w_a^{[u]}[t_i^{(u)}[\cdot]t_{i+1}^{(u)}),\ \rho(t_{i+1}^{(u)}, w_a^{[u]*}[t_{i+1}^{(u)}]) + 2\varepsilon_{i+1}^{[u]}) \leq$$

$$\leq \rho(t_i^{(u)}, w^{[u]0}(t_i^{(u)}, x_{U_e,v}[t_i^{(u)}], \varepsilon_u)) + 2\varepsilon_i^{[u]}, \quad i = 1, \ldots, k_u \tag{9.15}$$

holds. Here $\nu(\cdot)$ and $\beta(\cdot)$ are the functions (7.16) and (4.9), respectively; $\varepsilon_i^{[u]}$ are the numbers (9.10).

Let us show that the following induction in i holds.

If at the time moment $t_{i+1}^{(u)}$, $i = 1, \ldots, k_u$ the condition

$$\gamma(w_a^{[u]}[t_{i+1}^{(u)}[\cdot]\vartheta]) \leq \rho(t_{i+1}^{(u)}, w^{[u]0}(t_{i+1}^{(u)}, x_{U_e,v}[t_{i+1}^{(u)}], \varepsilon_u)) + 2\varepsilon_{i+1}^{[u]} \tag{9.16}$$

holds, then at the time moment $t_i^{(u)}$the inequality

$$\gamma(w_a^{[u]}[t_i^{(u)}[\cdot]\vartheta]) \le \rho(t_i^{(u)}, w^{[u]0}[t_i^{(u)}, x_{U_e,v}[t_i^{(u)}], \varepsilon_u)) + 2\varepsilon_i^{[u]} \tag{9.17}$$

is valid.

Really, since the functional γ (4.2) is *positional* (4.9) then we have the following equality

$$\gamma(w_a^{[u]}[t_i^{(u)}[\cdot]\vartheta]) = \beta(w_a^{[u]}[t_i^{(u)}[\cdot]t_{i+1}^{(u)}), \alpha) \tag{9.18}$$

where

$$\alpha = \gamma(w_a^{[u]}[t_{i+1}^{(u)}[\cdot]\vartheta]) \,. \tag{9.19}$$

Also we have the inequalities

$$\rho(t_{i+1}^{(u)}, w^{[u]0}(t_{i+1}^{(u)}, x_{U_e,v}[t_{i+1}^{(u)}], \varepsilon_u)) \le$$
$$\le \rho(t_{i+1}^{(u)}, w_a^{[u]*}[t_{i+1}^{(u)}]) \tag{9.20}$$

which follow from (9.14) and the definition (8.5)–(8.10) of the accompanying points $w^{[u]0}[t_i]$ (9.5).

Then from (9.16), (9.18) and since the functional $\beta(\cdot, \alpha)$ (9.18) is non-decreasing in α (9.19), we come to the inequality

$$\gamma(w_a^{[u]}[t_i^{(u)}[\cdot]\vartheta]) \le$$
$$\le \beta(w_a^{[u]}[t_i^{(u)}[\cdot]t_{i+1}^{(u)}), \rho(t_{i+1}^{(u)}, w_a^{[u]}[t_{i+1}^{(u)}]) + 2\varepsilon_{i+1}^{[u]}) =$$
$$= \beta(w_a^{[u]}[t_i^{(u)}[\cdot]t_{i+1}^{(u)}), \rho(t_{i+1}^{(u)}, w^{[u]0}(t_{i+1}^{(u)}, x_{U_e,v}[t_{i+1}^{(u)}], \varepsilon_u)) + 2\varepsilon_{i+1}^{[u]}) \,. \tag{9.21}$$

From the inequality (9.21) and taking into account (9.15) and (9.20)we derive the estimation (9.17). Therefore the induction (9.16), (9.17) is proved.

Besides, according to the property 2^0 of the function $\rho(\vartheta, w[\vartheta])$ from Section 8 (see (8.2)), we have the following inequality

$$\gamma(w_a^{[u]}[\vartheta[\vartheta]\vartheta]) \le \rho(\vartheta, w^{[u]0}[\vartheta]) + 2\varepsilon_{k_u+1}^{[u]} \tag{9.22}$$

for the terminal time moment $t_{k_u+1}^{(u)} = \vartheta$.

Then using the induction (9.16), (9.17), (9.22) in i from $i = k_u + 1$ to $i = 1$ we obtain that for the initial time moment $t_1^{(u)} = t_*$ the inequality

$$\gamma(w_a^{[u]}[t_*[\cdot]\vartheta) \le \rho(t_*, w^{[u]0}(t_*, x_*, \varepsilon_u) + 2\varepsilon_u \tag{9.23}$$

holds.

The inequality

$$\rho(t_*, w^{[u]0}(t_*, x_*, \varepsilon_u)) \le \rho(t_*, x_*) \tag{9.24}$$

follows from the definition (9.6) of the accompanying point $w^{[u]0}[t_1] = w^{[u]0}(t_*, x_*, \varepsilon_u)$. In fact, the inequality (9.24) holds because $\{t_*, x_*\} \in K^{[u]}(t_*, x_*, \varepsilon_u)$.

Now, according to the construction of the motion $x_{U_e,v}[t_*[\cdot]\vartheta]$ (9.4) and the accompanying motion $w_a^{[u]}[t_*[\cdot]\vartheta]$ (9.13), we obtain from (9.23), (9.24) the estimation

$$\gamma(x_{U_e,v}[t_*[\cdot]\vartheta]) \le \rho(t_*, x_*) + \eta(\varepsilon_u),$$

$$\lim_{\varepsilon_u \to 0} \eta(\varepsilon_u) = 0\,. \tag{9.25}$$

We see that this estimation follows from (9.14), the form of the function $\nu(t, x, w)$ (7.16) and the continuity property (4.23) of the functional γ with respect to the metric $c(\cdot)$ (4.24).

Lemma 9.1 is thus proved.

Lemma 9.1 shows that the extremal strategy $u_e(\cdot)$ guarantees the result $\rho(t_*, x_*)$ uniformly with respect to $\{t_*, x_*\} \in G$.

Similarly, one can prove the following:

Lemma 9.2. *Let there exist a function $\rho(t, w)$ that satisfies the conditions 1^0, 2^0 and 4^0_v from Section 8. Then for any number $\eta > 0$ one can find $\varepsilon(\eta) > 0$ and $\delta(\eta, \varepsilon) > 0$ such that for any motion $x_{V_e,u}[t_*[\cdot]\vartheta]$ generated by the scheme* (5.18) *from a given initial position $\{t_*, x_*\} \in G$ by the control law $V_e = \{v_e(\cdot);\ \varepsilon_v;\ \Delta_{v\delta}\{t_i^{(v)}\}\}$ (corresponding to the extremal strategy $v_e(\cdot) = v_e(t, x, \varepsilon)$* (8.16), (8.17), *the parameter $\varepsilon_v \le \varepsilon(\eta)$ and the partition $\Delta_{v\delta}\{t_i^{(v)}\}$* (5.14) : $\max_i(t_{i+1}^{(v)} - t_i^{(v)}) \le \delta(\eta, \varepsilon_v)$) *together with any admissible control actions $u[t_*[\cdot]\vartheta)$* (5.17) *the inequality*

$$\gamma(x_{V_e,u}[t_*[\cdot]\vartheta]) \ge \rho(t_*, x_*) - \eta \tag{9.26}$$

holds.

It follows from Lemma 9.2 that the extremal strategy $v_e(\cdot)$ guarantees the result $\rho(t_*, x_*)$ uniformly with respect to $\{t_*, x_*\} \in G$. Lemma 9.1 and Lemma 9.2. imply the following:

Theorem 9.1. *Let the function $\rho(t, w)$ satisfying the conditions 1^0, 2^0, 3^0_u and 4^0_v from Section 8 exist. Then $\rho^0(t, x) = \rho(t, x)$ is the value of the differential game for the system* (3.1), (3.2), (5.1) *with the positional* *quality index* γ (4.9). *The extremal strategies* $u_e(\cdot)$ *and* $v_e(\cdot)$ *are optimal ones* $u^0(\cdot) = u_e^0(t,x,\varepsilon)$ *and* $v^0(\cdot) = v_e^0(t,x,\varepsilon)$. *They give a uniform positional saddle point* $\{u^0(\cdot), v^0(\cdot)\}$ *in this differential game.*

It remains only to establish the existence of the function $\rho(t,w)$ that satisfies the conditions 1^0, 2^0, 3_u and 4_v^0 from Section 8. This fact was established in the paper [56] where one method of construction of this function $\rho(t,w)$ was used. This method is based on the concept of the so-called *Q-procedures* (see [22], [77]). These constructions and proofs are omitted here. They were described in detail in the book [77].

So we have the following result.

Theorem 9.2. *The differential game for the system* (3.1), (3.2), (5.1) *with the positional quality index* $\gamma(x[t_*[\cdot]\vartheta])$ (4.9) *always has a value of the game* $\rho^0(t,x)$ *and a uniform positional saddle point* $\{u^0(\cdot), v^0(\cdot)\}$ *with strategies* $u_e(\cdot) = u_e(t,x,\varepsilon)$ *and* $v_e(\cdot) = v_e(t,x,\varepsilon)$ *extremal to the function* $\rho^0(t,x)$.

We note that the strategies $u_e(\cdot)$ and $v_e(\cdot)$ are *universal* according to the definition given in Section 6. However, this method of construction of the function $\rho(t,w)$ based on Q-procedures is not sufficiently effective in practical calculations.

On the other hand, the main result of the present book is a new sufficiently constructive method for calculation of the value of the game $\rho^0(t,x)$ and for construction of the optimal strategies $u^0(\cdot) = u^0(t,x,\varepsilon)$ and $v^0(\cdot) = v^0(t,x,\varepsilon)$ and their approximations. But this method is applicable to a more restricted class of problems in the general result given by Theorem 9.2.

Let us remark the following. The above-given method for constructing optimal strategies $u^0(\cdot) = u_e(\cdot)$ and $v^0(\cdot) = v_e(\cdot)$ is strictly justified from a theoretical point of view. However, it is not always convenient in practice for the following reason. To ensure the desired estimation η (9.1), (9.26) we need sufficiently small parameters ε_u, ε_v and steps $\delta(\eta,\varepsilon_u)$, $\delta(\eta,\varepsilon_v)$. These values are sometimes too small to be realized in a practical control scheme. We can try to overcome this difficulty in the following way. When constructing the $w^{[u]}$-model we can relax the restriction on the control action $v_*^{[u]}$. Namely, we can impose the restriction

$$v_*^{[u]} \in Q_{\varepsilon_*}, \quad \varepsilon_* > 0 \tag{9.27}$$

where Q_{ε_*} is the ε_*-neighbourhood of the set Q. If we use the constructions of Sections 7 to 9 with the relaxed restriction (9.27) instead of the primary restriction (7.2), we obtain, as a rule, a desired estimate similar to (7.18) for essentially larger step $\delta(\varepsilon) > 0$ than in the case of the primary restriction (7.2). Therefore, we can ensure the desired closeness of the motion $x_{U^0,v}[t]$ and the accompanying motion $w_a^{[u]}[t]$ by taking a practically realizable step. We can say that the introduction of the restriction (9.27) stabilizes the process of control. Unfortunately, the use of the relaxed restriction (9.27) somewhat spoils the function $\rho(t,w)$, which gives the basis for constructing the control actions. Really, this function $\rho(t,w)$ is substituted by the function $\rho_{u_*}(t,w,\varepsilon_*)$ that corresponds to the restriction (9.27). As a rule, this function turns out to be larger than $\rho(t,w)$. Thus, we ensure only the estimation

$$\gamma(x_{U^0,v}[t_*[\cdot]\vartheta]) \leq \rho_{u_*}(t_*,x_*,\varepsilon_*) + \eta\,. \tag{9.28}$$

Employing the restriction (9.27) we, on the one hand, stabilize the process of control, but on the other hand, spoil the ensured result $\rho(\cdot)$. However, a proper choice of the value $\varepsilon_* > 0$ enables us to make the process sufficiently stable without spoiling the guaranteed result too much. The construction of the optimal strategy $v_e(\cdot)$ can be modified similarly. Namely, we can impose the restriction

$$u_*^{[v]} \in P_{\varepsilon_*}, \quad \varepsilon_* > 0 \tag{9.29}$$

where P_{ε_*} is the ε_*-neighbourhood of the set P. The use of this restriction instead of the primary one (7.2) again, on the one hand, stabilizes the process on behalf of the second player, but on the other hand, spoils the ensured result of this player. Namely, using this modified strategy $v^0(\cdot) = v_e(\cdot)$ he ensures only the result

$$\gamma(x_{u,V^0}[t_*[\cdot]\vartheta]) \geq \rho_{v_*}(t_*,x_*,\varepsilon_*) - \eta\,. \tag{9.30}$$

Of course, the inequality

$$\rho_{v_*}(t_*,x_*,\varepsilon_*) \leq \rho_{u_*}(t_*,x_*,\varepsilon_*) \tag{9.31}$$

holds. And, as a rule, the inequality (9.31) is strict.

We have described here the simplest method of stabilization of the process of control. Naturally, one can construct more eleborate methods of stabilization. A powerful tool for this stabilization is the Lyapunov theory of stability.

For instance, one can calculate the values of the optimal strategies $u_e(t,x,\varepsilon)$ and $v_e(t,x,\varepsilon)$ basing not on the conditions (8.11), (8.12) or (8.16), (8.17) but on the conditions (8.11) and

$$s^{[u]} = \left\{ \frac{\partial \nu(t,x,w^{[u]})}{\partial x_i}, \quad i = 1,\ldots,n \right\} \tag{9.32}$$

or (8.16) and

$$s^{[v]} = \left\{ \frac{\partial \nu(t, x, w^{[v]})}{\partial x_i}, \quad i = 1, \ldots, n \right\} \tag{9.33}$$

respectively. Here instead of the function $\nu(t, x, w)$ (7.16) one can use in (8.5), (8.6) and (8.9), (8.10) some appropriate positively definite Lyapunov function $\nu(t, x, w)$. For the function $\nu(t, x, w)$ one can take, for example, some positively definite quadratic form

$$\nu(t, x, w) = \sum_{i,j=1}^{n} \beta_{ij}(t)(x_i - w_i)(x_j - w_j)\,. \tag{9.34}$$

In particular, we obtain the conditions (7.16), (8.7), (8.8), (8.11), (8.12), (8.16), (8.17) if we take in (9.34)

$$\beta_{ij}(t) = \delta_{ij}\beta(t) = \delta_{ij}(\varepsilon + \varepsilon(t - t_0))\, e^{-2\lambda(t-t_0)} \tag{9.35}$$

where δ_{ij} is the symbol of Kroneker.

The elements of this approach to stabilization of control processes are given in [77].

Chapter II
Stochastic program synthesis of pure strategies for a positional functional

The basic result of this chapter is a new effective calculation *procedure* for the value of the game $\rho^0(t,x)$ and also the corresponding construction of the optimal strategies $u^0(\cdot) = u^0(t,x,\varepsilon)$ and $v^0(\cdot) = v^0(t,x,\varepsilon)$. These constructions were suggested and developed in [61], [64]–[66], [80], [98]. They are connected with the idea of stochastic program synthesis, which was developed in Ekatherinburg [57], [60], [61], [75], [77], [80], [144].

Below we shall consider the same differential game in the class of pure strategies as we did in Chapter I, but for certain particular cases of the system (3.1), (5.1) with some typical positional functionals γ (4.2), (4.9). For these systems and quality indices γ, a procedure for calculation of the approximating value of the game $\rho^a(t,x)$ and for construction of the approximating optimal pure strategies $u^a(\cdot)$ and $v^a(\cdot)$ will be presented and justified.

10. Differential games for typical functionals $\gamma_{(1)}$ and $\gamma_{(2)}$

Below we consider a dynamical system that is described by the differential equation

$$\dot{x} = A(t)x + f(t,u,v), \quad t_0 \le t < \vartheta \tag{10.1}$$

under the restrictions (3.5) and (3.6), i.e., $u \in P$ and $v \in Q$. Here the vectors x, u, v and the sets P, Q are defined as in Section 3. The matrix-valued function $A(t)$ and the vector function $f(t,u,v)$ are piecewise-continuous in $t \in [t_0,\ \vartheta]$ and $f(t,u,v)$, for every fixed t is continuous in u and v belonging to the sets P and Q, respectively. The points of discontinuity in t are assumed to be independent of u, v. In these points the function f is continuous in t from the right and it has the limits in t from the left in these points. For short, we call such functions piecewise-continuous in t.

The function $f(\cdot)$ in (10.1) satisfies a condition similar to (5.1), i.e.,

$$\min_{u\in P}\max_{v\in Q}\ \langle l,\ f(t,u,v)\rangle = \max_{v\in Q}\min_{u\in P}\ \langle l,\ f(t,u,v)\rangle \tag{10.2}$$

for every vector l.

In this chapter we shall consider two particular but *typical* forms of the positional functional γ (4.2), (4.9). Namely, we shall consider the functionals γ (4.14) and γ (4.15). We denote these functionals $\gamma_{(1)}$ and $\gamma_{(2)}$, respectively. So we have

$$\gamma_{(1)} = \int_{t_*}^{\vartheta} \mu(t, D_*(t)x[t])dt + \sum_{i=g_*}^{N_*} \mu^{[i]}(D_*^{[i]}x[t_*^{[i]}]), \tag{10.3}$$

$$\gamma_{(2)} = \max\left[\sup_{t_* \le t \le \vartheta} \mu(t, D_*(t)x[t]), \max_{i=g_*,\dots,N_*} \mu^{[i]}(D_*^{[i]}x[t_*^{[i]}])\right] \tag{10.4}$$

where t_* is an initial time moment for the process of control; ϑ is a fixed terminal moment for this process. Other quantities in (10.3), (10.4) are defined in the following way.

In the time interval $[t_0, \ \vartheta]$ we fix moments $t_*^{[i]}$, $i = 1,\dots,N_*$, $t_*^{[N_*]} = \vartheta$, semi-norms $\mu_*^{[i]}(x)$ and choose a semi-norm-function $\mu_*(t,x)$ piecewise-continuous in t. The considered semi-norms are constructed as follows. Let a collection of integers $\nu^{[i]} = \nu[t_*^{[i]}] \in [1, \ n]$, $i = 1,\dots,N_*$ and the collection of constant $(\nu^{[i]} \times n)$-matrices $D_*^{[i]}$, $i = 1,\dots,N_*$ be given. The semi-norm $\mu_*^{[i]}(x)$ is defined as some norm $\mu^{[i]}(D_*^{[i]}x)$. The semi-norm-function $\mu(t,x)$ is constructed as the norm-function $\mu(t, D_*(t)x)$ piecewise-continuous in t. Here $D_*(t)$ is a piecewise-constant $(\nu[t] \times n)$-matrix-function, $\nu[t] \in [1, \ n]$, $t_0 \le t \le \vartheta$. We assume that all the functions are continuous in t from the right and have the limits in t from the left, i.e., piecewise-continuous in t, as we have called them above for short.

We assume here and in similar cases below that the lines of every matrix D are linear-independent and the transformation $x^* = Dx$ determines the space $\{x^*\}$ whose dimension is equal to the number ν of lines of the matrix D.

In (10.3), (10.4) $t^{[g_*]}$ is the smallest of the moments $t_*^{[i]}$ which satisfy the condition $t_*^{[i]} \ge t_*$, i.e.,

$$g_* = \min_{t_*^{[i]} \ge t_*} i \,. \tag{10.5}$$

Of course, the functional $\gamma_{(1)}$ could be presented in the form integral of Styltiess:

$$\gamma_{(1)} = \int_{t_*}^{\vartheta} d\mu(t, D[t]x[t]).$$

However, we prefer the more explicit form of the sum (10.3).

The functionals $\gamma_{(1)}$ (10.3) and $\gamma_{(2)}$ (10.4) are *positional* according to the definition given in Section 4. That is, they can be represented in the form (4.9).

In fact, we have

$$\gamma_{(1)}(x[t_*[\cdot]\vartheta]) = \beta_{(1)}(x[t_*[\cdot]t^*), \alpha_{(1)}) =$$

$$= \int_{t_*}^{t^*} \mu(t, D_*(t)x[t])dt + \sum_{i=g_*}^{g^*} \mu^{[i]}(D_*^{[i]}x[t_*^{[i]}]) + \alpha_{(1)},$$

$$\alpha_{(1)} = \gamma_{(1)}(x[t^*[\cdot]\vartheta]) = \int_{t^*}^{\vartheta} \mu(t, D_*(t)x[t])dt + \sum_{i=g^*+1}^{N^*} \mu^{[i]}(D_*^{[i]}x[t_*^{[i]}]), \qquad (10.6)$$

$$\gamma_{(2)}(x[t_*[\cdot]\vartheta]) = \beta_{(2)}(x[t_*[\cdot]t^*), \alpha_{(2)}) =$$

$$= \max\left[\sup_{t_* \le t < t^*} \mu(t, D_*(t)x[t]), \max_{i=g_*,\dots,g^*} \mu^{[i]}(D_*^{[i]}x[t_*^{[i]}])\right], \alpha_{(2)}],$$

$$\alpha_{(2)} = \gamma_{(2)}(x[t^*[\cdot]\vartheta]) =$$

$$= \max\left[\sup_{t^* \le t \le \vartheta} \mu(t, D_*(t)x[t]), \max_{i=g^*+1,\dots,N_*} \mu^{[i]}(D_*^{[i]}x[t_*^{[i]}])\right] \qquad (10.7)$$

where $t^{[g^*]}$ is the largest of the moments $t_*^{[i]}$ satisfying the condition $t_*^{[i]} < t^*$, i.e.,

$$g^* = \max_{t_*^{[i]} < t^*} i\,. \qquad (10.8)$$

For a particular case (10.1), (10.2) of the system (3.1), (5.1) and for the positional functionals $\gamma_{(1)}$ (10.3) and $\gamma_{(2)}$ (10.4) Theorem 9.2 holds. That is, *the differential game for the system* (10.1), (10.2) *with the positional functional* $\gamma_{(i)}$ (10.3), (10.4) *has a value* $\rho^0_{(i)}(t,x)$ *and a saddle point* $\{u^0_{(i)}(\cdot) = u^0_{(i)}(t,x,\varepsilon),\ v^0_{(i)}(\cdot) = v^0_{(i)}(t,x,\varepsilon)\}$, *where* i=1,2.

11. Approximating functionals

Now for the functionals $\gamma_{(1)}$ (10.3) and $\gamma_{(2)}$ (10.4) we shall introduce some approximating functionals $\gamma_{*(1)}$ and $\gamma_{*(2)}$. These functionals are constructed in the following way:

Let

$$\Delta\{\tau_*^{[h]}\} = \{t_0 = \tau_*^{[1]}, \quad \tau_*^{[h]} < \tau_*^{[h+1]},$$

$$h = 1,\dots,M, \quad \tau_*^{[M]} = \vartheta\} \qquad (11.1)$$

be a partition of the time interval $[t_0,\ \vartheta]$. This partition (11.1) includes, in particular, all instants $t_j^0 \in [t_0,\ \vartheta]$ that divide the intervals where the matrix-function $D_*(t)$ from (10.3), (10.4) is constant. Besides, let $\Delta\{\tau_*^{[h]}\}$ also include all the moments $t_j^* \in [t_0,\ \vartheta]$ that divide the intervals of continuity of the piecewise-continuous norm-function $\mu(t, D_*(t)x)$ from (10.3), (10.4).

Let us consider at first the functional $\gamma_{(1)}$ (10.3). In this case we replace the quantity

$$\int_{\tau_*^{[h]}}^{\tau_*^{[h+1]}} \mu(\tau, D_*(\tau)x[\tau])d\tau \qquad (11.2)$$

by the quantity

$$(\tau_*^{[h+1]} - \tau_*^{[h]})\mu(\tau_*^{[h]}, D_*(\tau_*^{[h]})x[\tau_*^{[h]}]) =$$

$$= \mu(\tau_*^{[h]}, (\tau_*^{[h+1]} - \tau_*^{[h]})D_*(\tau_*^{[h]})x[\tau_*^{[h]}]))\,. \qquad (11.3)$$

We denote

$$D_{(1)}(\tau_*^{[h]}) = (\tau_*^{[h+1]} - \tau_*^{[h]})D_*(\tau_*^{[h]}). \tag{11.4}$$

In the case of the functional $\gamma_{(2)}$ (10.4) we denote

$$D_{(2)}(\tau_*^{[h]}) = D_*(\tau_*^{[h]}). \tag{11.5}$$

We select only moments $\tau_*^{[h]}$ such that $\mu(\tau_*^{[h]}, \cdot) \not\equiv 0$.

Let us take the moments $t_*^{[i]}$, $i = 1, \ldots, N_*$ in (10.3), (10.4) and the moments $\tau_*^{[h]}$, $h = 1, \ldots, M$ in (11.1) and enumerate them anew. We denote these moments by

$$t^{[1]} \leq t^{[2]} \leq \ldots \leq t^{[i]} \leq t^{[i+1]} \leq \ldots \leq t^{[N]} = \vartheta\,. \tag{11.6}$$

Assume

$$\mu^{[i]}(D^{[i]}x) = \mu(\tau_*^{[h]}, D_{(k)}(\tau_*^{[h]})x), \quad k = 1, 2 \tag{11.7}$$

if $t^{[i]} = \tau_*^{[h]} \neq t_*^{[j]}$ and

$$\mu^{[i]}(D^{[i]}x) = \mu^{[j]}(D_*^{[j]}x) \tag{11.8}$$

if $t^{[i]} = t_*^{[j]}$.

For the sake of simplicity of notation we assume also that all the moments $\tau_*^{[h]}$ and $t_*^{[j]}$ are different. Therefore, all the moments $t^{[i]}$ are also different. Otherwise the changes in the formulas (11.7), (11.8) and the formulas (11.9), (11.10), (11.11) below are obvious.

Thus, for the functional $\gamma_{(1)}$ (10.3) we obtain the functional

$$\gamma_{*(1)} = \sum_{i=g}^{N} \mu^{[i]}(D^{[i]}x[t^{[i]}]) \tag{11.9}$$

and for $\gamma_{(2)}$ (10.4) the functional

$$\gamma_{*(2)} = \max_{i=g,\ldots,N} \mu^{[i]}(D^{[i]}x[t^{[i]}]) \tag{11.10}$$

where

$$g = \min_{t_* \le t^{[i]}} i \,. \tag{11.11}$$

We call $\gamma_{*(1)}$ (11.9) and $\gamma_{*(2)}$ (11.10) *approximating functionals* to $\gamma_{(1)}$ (10.3) and $\gamma_{(2)}$ (10.4), respectively.

The functionals $\gamma_{*(1)}$ and $\gamma_{*(2)}$ are *positional*, i.e., they can be represented in the form (4.9). In fact, we have

$$\gamma_{*(1)}(x[t_*[\cdot]\vartheta]) = \beta_{*(1)}(x[t_*[\cdot]t^*), \alpha_{*(1)}) =$$

$$= \sum_{i=g}^{\widetilde{g}} \mu^{[i]}(D^{[i]}x[t^{[i]}]) + \alpha_{*(1)},$$

$$\alpha_{*(1)} = \gamma_{*(1)}(x[t^*[\cdot]\vartheta]) = \sum_{i=\widetilde{g}+1}^{N} \mu^{[i]}(D^{[i]}x[t^{[i]}]), \tag{11.12}$$

$$\gamma_{*(2)}(x[t_*[\cdot]\vartheta]) = \beta_{*(2)}(x[t_*[\cdot]t^*), \alpha_{*(2)}) =$$

$$= \max\,[\, \max_{i=g,\dots,\widetilde{g}} \mu^{[i]}(D^{[i]}x[t^{[i]}]),\ \alpha_{*(2)}\,],$$

$$\alpha_{*(2)} = \gamma_{*(2)}(x[t^*[\cdot]\vartheta]) = \max_{i=\widetilde{g}+1,\dots,N} \mu^{[i]}(D^{[i]}x[t^{[i]}]) \tag{11.13}$$

where

$$\widetilde{g} = \max_{t^{[i]} < t^*} i. \tag{11.14}$$

Consequently, for these functionals $\gamma_{*(1)}$ and $\gamma_{*(2)}$ Theorem 9.2 holds, i.e., *the differential game for the system* (10.1), (10.2) *with the functional* $\gamma_{*(i)}$ (11.9), (11.10) *has a value* $\rho^0_{*(i)}(t,x)$ *and a saddle point* $\{u^0_{*(i)}(\cdot) = u^0_{*(i)}(t,x,\varepsilon),\ v^0_{*(i)}(\cdot) = v^0_{*(i)}(t,x,\varepsilon)\}$ *where* $i = 1,2$.

Here the lower index $*$ shows that the functions $u^0_{*(i)}(\cdot)$, $v^0_{*(i)}(\cdot)$ and $\rho^0_{*(i)}$ correspond to the approximating functional $\gamma_{*(i)}$, $i = 1,2$.

The following assertion holds.

Lemma 11.1. *For each* $\zeta > 0$ *there exists a number* $\delta(\zeta) > 0$ *such that*

$$|\, \rho^0_{*(i)}(t_*, x_*) - \rho^0_{(i)}(t_*, x_*) \,| \le \zeta, \quad i = 1,2 \tag{11.15}$$

for all positions $\{t_*, x_*\} \in G$ (3.12) *if the partition step* $\delta(\Delta\{\tau_h\})$ *for* (11.1) *satisfies the condition*

$$\delta(\Delta\{\tau_h\}) = \max_h (\tau_*^{[h+1]} - \tau_*^{[h]}) \leq \delta(\zeta). \tag{11.16}$$

In (11.15) $\rho^0_{(i)}(t,x)$, $i = 1,2$ is the value of the game for the corresponding functional $\gamma_{(1)}$ (10.3) or $\gamma_{(2)}$ (10.4) and $\rho^0_{*(i)}(t,x)$, $i = 1,2$ is the value of the game for $\gamma_{*(1)}$ (11.9) or $\gamma_{*(2)}$ (11.10).

According to Lemma 11.1 the value $\rho^0_{*(i)}(t,x)$ and the optimal strategies $u^0_{*(i)}(\cdot)$ and $v^0_{*(i)}(\cdot)$, $i = 1,2$ approximate the solution of the problem from Section 3 for the system (10.1), (10.2) for the functionals $\gamma_{(1)}$ (10.3) and $\gamma_{(2)}$ (10.4).

The following sections of this chapter will be devoted to the description of an effective construction of the approximating values of the game $\rho^a_{*(1)}(t,x)$ and $\rho^a_{*(2)}(t,x)$ for the functionals $\gamma_{*(1)}$ (11.9) and $\gamma_{*(2)}$ (11.10).

In conclusion of this section we note that all the assertions given above are valid for the general case (3.1), (3.2), (5.1) of the controlled x-system. But in this chapter we consider the particular case (10.1), (10.2) of the system (3.1), (5.1) because the proposed method for calculating the value of the game $\rho^0_*(t,x)$ is most effective for the controlled system (10.1), (10.2).

12. Stochastic program maximins $\rho_{*(1)}(\cdot)$ and $\rho_{*(2)}(\cdot)$

According to the method of program stochastic synthesis we associate a stochastic w-model with the given controlled x-system (10.1), (10.2) . The current state of this model is described by a vector $w[t,\omega]$ whose evolution is subjected to the differential equation

$$\dot{w} = A(\tau)w + f(\tau,u,v), \quad t_0 \leq \tau < \vartheta \tag{12.1}$$

where $w \in \mathbf{R}^n$, $u \in P$, $v \in Q$. Here P, Q, $A(\tau)$ and $f(\tau,u,v)$ are the same as in (10.1), (10.2). And τ, $t_0 \leq \tau \leq \vartheta$ is a conceptional time variable for the model. The dot over the letter w denotes here a derivative with respect to the variable τ.

Let $\{\tau_*, w_*\} \in G^*$ be the initial position for w-model (12.1). For the interval $[\tau_*, \vartheta] \subset [t_0, \vartheta]$ we choose a partition

$$\Delta\{\tau_j\} = \{\tau_1 = \tau_*, \quad \tau_j < \tau_{j+1}, \quad j = 1,\ldots,k, \quad \tau_{k+1} = \vartheta\} \tag{12.2}$$

and some appropriate probability space $\{\Omega, \mathcal{B}, \mathbf{P}\}$ [96], [97]. We assume that the instants $t^{[i]}$ from (11.9), (11.10) are included into the partition $\Delta\{\tau_j\}$ (12.2). Let $t^{[g]}$ be the smallest of the moments $t^{[i]} \geq \tau_*$. The space $\{\Omega, \mathcal{B}, \mathbf{P}\}$ is generated by independent random variables ξ_j, $j = 1, \ldots, k$ uniformly distributed on the interval $0 \leq \xi_j \leq 1$. We can interpret that in the following way. A source Ξ of random events gives a number $\xi_j \in [0,\ 1]$ at the time moment τ_j. Further, for a fixed t all values of ξ_j $(t < \tau_j)$ are regarded as equally probable in advance. Every collection $\{\xi_1, \ldots, \xi_k\}$ of numbers $\xi_j \in [0,\ 1]$ is considered as an elementary event ω, i.e., $\omega = \{\xi_1, \ldots, \xi_k\} \in \Omega$. Hence, Ω is the unite cube in k-dimensional space $\{\xi_1, \ldots, \xi_k\}$, $\mathcal{B} = \mathcal{B}_\Omega$ is the Borel σ-algebra [96], [97] for the cube, and $\mathbf{P}(B)$ is Lebesgue measure on cube, $B \in \mathcal{B}$.

Let

$$u[\cdot] = \{u[\tau, \omega] \in P, \quad \tau_* \leq \tau < \vartheta, \quad \omega \in \Omega\} \tag{12.3}$$

and

$$v[\cdot] = \{v[\tau, \omega] \in Q, \quad \tau_* \leq \tau < \vartheta, \quad \omega \in \Omega\} \tag{12.4}$$

be the stochastic programs *nonanticipating* [97] in respect to $\xi(\tau_j) = \xi_j$, i.e.,

$$u[\tau, \omega] = u[\tau, \xi_1, \ldots, \xi_j], \quad \tau_j \leq \tau < \tau_{j+1}, \quad j = 1, \ldots, k \tag{12.5}$$

and

$$v[\tau, \omega] = v[\tau, \xi_1, \ldots, \xi_j], \quad \tau_j \leq \tau < \tau_{j+1}, \quad j = 1, \ldots, k \tag{12.6}$$

where the functions $u[\tau, \xi_1, \ldots, \xi_j]$ and $v[\tau, \xi_1, \ldots, \xi_j]$ are jointly measurable in the variables τ, $\xi_1, \ldots, \xi_j$.

Let some stochastic programs $u[\cdot]$ (12.3) and $v[\cdot]$ (12.4) be chosen. A given position $\{\tau_*, w_*\} \in G^*$, a partition $\Delta = \Delta\{\tau_j\}$ (12.2), and the chosen stochastic programs $u[\cdot]$ and $v[\cdot]$ determine a random motion of the w-model (12.1), which is the solution

$$w[\tau_*[\cdot]\tau^*, \cdot] = \{w[\tau, \omega], \quad \tau_* \leq \tau \leq \vartheta, \quad \omega \in \Omega, \quad w[\tau_*, \omega] = w_*\} \tag{12.7}$$

of the stochastic differential equation

$$\dot{w}[\tau, \omega] = A(\tau)w[\tau, \omega] + f(\tau, u[\tau, \omega], v[\tau, \omega]) \tag{12.8}$$

with the initial condition

$$w[\tau_*,\omega]=w_*,\quad \omega\in\Omega\,. \tag{12.9}$$

For the approximating functional $\gamma_{*(1)}$ (11.9) we call the quantity

$$\rho_{*(1)}(\tau_*,w_*,\Delta)=\max_{v[\cdot]}\min_{u[\cdot]}\ M\Big\{\sum_{i=g}^{N}\mu^{[i]}(D^{[i]}w[t^{[i]},\omega])\Big\} \tag{12.10}$$

the *stochastic program maximin.* Here $M\{\cdots\}$ denotes the expectation [96]. In (12.10) $\Delta=\Delta\{\tau_j\}$ is the partition (12.2), $\mu^{[i]}(\cdot)$, $D^{[i]}$, $t^{[i]}$, N are the quantities from (11.9), and also we have $w[t^{[i]},\omega]=w[\tau,\omega]$ at $\tau=t^{[i]}$, $i=g,\dots,N$ where $w[\tau_*[\cdot]\vartheta,\cdot]$ is the motion (12.7)–(12.9). In (12.10) g is the number that satisfies the condition

$$g=\min_{t^{[i]}\ge\tau_*} i\,. \tag{12.11}$$

For the second approximating functional $\gamma_{*(2)}$ (11.10) we call the *stochastic program maximin* $\rho_{*(2)}(\tau_*,w_*,\Delta)$ the quantity

$$\rho_{*(2)}(\tau_*,w_*,\Delta)=\max_{v[\cdot]}\min_{u[\cdot]}\ M\{\max_{g\le i\le N}\mu^{[i]}(D^{[i]}w[t^{[i]},\omega]\}\,. \tag{12.12}$$

Taking the appropriate dual vector random variables $l_{(1)}(\cdot)$ and $l_{(2)}(\cdot)$ we can calculate $\rho_{*(1)}(\cdot)$ (12.10) and $\rho_{*(2)}(\cdot)$ (12.12) as so-called *program extrema* $e_{*(1)}(\tau_*,w_*,\Delta)$ and $e_{*(2)}(\tau_*,w_*,\ \Delta)$, respectively. These program extrema $e_{*(1)}(\cdot)$ and $e_{*(2)}(\cdot)$ are defined in the following way.

At first we consider the case of approximating functional $\gamma_{*(1)}$ (11.9) and construct the quantity $e_{*(1)}(\tau_*,w_*,\Delta)$.

Let $l_{(1)}^{(i)}(\omega)$ be a $\nu^{[i]}$-dimensional vector random variable defined on the probability space $\{\Omega,\mathcal{B},\mathbf{P}\}$. Here $\nu^{[i]}\in[1,\ n]$ is determined by the dimension of matrix $D^{[i]}$ from (11.9), (12.10).

Let $l_{(1)}(\cdot)$ be a variable with the components $l_{(1)}^{(i)}(\omega)$, i.e., $l_{(1)}(\cdot)=\{l_{(1)}^{(i)}(\omega),\ i=g,\dots,N,\ \omega\in\Omega\}$ and let

$$\|l_{(1)}(\cdot)\|^*=\max_{i}\operatorname*{ess\,max}_{\omega}\mu^{[i]*}(l_{(1)}^{(i)}(\omega)) \tag{12.13}$$

denote the *norm* of $l_{(1)}(\cdot)$. Here $\mu^{[i]*}(l)$ is the norm conjugate to $\mu^{[i]}(l)$ where $\mu^{[i]}(\cdot)$ is the norm from (11.9), (12.10).

Denote by $X(t,\tau)$ the fundamental matrix [24, 120, 133] of the differential equation $dx/dt = A(t)x$. Let $M\{\cdots \mid \cdots\}$ denote the conditional expectation [96], [97].

The program extremum $e_{*(1)}(\tau_*, w_*, \Delta)$ is defined by the equality

$$e_{*(1)}(\tau_*, w_*, \Delta) = \sup_{\|l_{(1)}(\cdot)\|^* \leq 1} \Big[\sum_{i=g}^{N} M\left\{l_{(1)}^{(i)T}(\omega)\right\} D^{[i]} X(t^{[i]}, \tau_*) w_* +$$

$$+ M\Big\{\sum_{j=1}^{k} (\tau_{j+1} - \tau_j) \max_{v \in Q} \min_{u \in P} M\Big\{\sum_{i=d(j)}^{N} l_{(1)}^{(i)T}(\omega) *$$

$$* D^{[i]} X(t^{[i]}, \tau_j) f(\tau_j, u, v) \mid \xi_1, \ldots, \xi_j\Big\}\Big\}\Big] \qquad (12.14)$$

where

$$d(j) = \min_{t^{[i]} > \tau_j} i\,. \qquad (12.15)$$

The equality

$$\rho_{*(1)}(\tau_*, w_*, \Delta_\delta) = e_{*(1)}(\tau_*, w_*, \Delta_\delta) + O(\delta) \qquad (12.16)$$

holds. Here $\rho_{*(1)}(\cdot)$ is the stochastic program maximin (12.10); Δ_δ is a partition $\Delta_\delta\{\tau_j\}$ (12.2) satisfying the condition

$$\max_j (\tau_{j+1} - \tau_j) \leq \delta\,. \qquad (12.17)$$

The equality (12.16) is implied by the following relations.

According to the definition of the stochastic program maximin (12.10) we have

$$\rho_{*(1)}(\tau_*, w_*, \Delta) = \max_{v[\cdot]} \min_{u[\cdot]} \mu_{(1)}(D(\cdot)w(\cdot)) \qquad (12.18)$$

where $\mu_{(1)}(\cdot)$ is the corresponding norm of the stochastic vector-function, which is denoted here in the symbolic form as

$$D(\cdot)w(\cdot) = D(\tau)w[\tau, \omega], \quad \tau_* \leq \tau \leq \vartheta, \quad \omega \in \Omega\,.$$

This norm $\mu(\cdot)$ satisfies the equality

$$\mu(D(\cdot)w(\cdot)) = \max_{\|l(\cdot)\|^* \leq 1} \langle\, l(\cdot),\ D(\cdot)w(\cdot)\,\rangle \tag{12.19}$$

where the symbol $\langle \cdots, \cdots \rangle$ denotes the corresponding linear functional.

Thus, the equality

$$\rho_{*(1)}(\tau_*, w_*, \Delta) = \max_{v[\cdot]} \min_{u[\cdot]} \max_{\|l(\cdot)\|^* \leq 1} \langle\, l(\cdot),\ D(\cdot)w(\cdot)\,\rangle \tag{12.20}$$

is valid.

According to the known theorem [9], [103] we can invert here the order of operations of minimum in $u(\cdot)$ and maximum in $l(\cdot)$. After that we can change the order of operations of maximum in $v(\cdot)$ and maximum in $l(\cdot)$. However, the general theorem employed here gives us only supremum in $l(\cdot)$ and does not guarantee the existence of the corresponding maximum. Thus we obtain that

$$\rho_{*(1)}(\tau_*, w_*, \Delta_\delta) = \sup_{\|l_{(1)}(\cdot)\|^* \leq 1} \max_{v[\cdot]} \min_{u[\cdot]} \langle\, l(\cdot),\ D(\cdot)w(\cdot)\,\rangle.$$

In fact, in our cases this maximum in $l^{(1)}(\cdot)$ exists. It is not essential for us here. We omit the proof of this assertion. According to the Cauchy formula for the equation (12.8) we have

$$w[t,\omega] = X(t,\tau_*)w_* + \int_{\tau_*}^{t} X(t,\tau) f(\tau, u[\tau,\omega], v[\tau,\omega]) d\tau\,.$$

Taking into account the nonanticipating property of the stochastic programs and approximating the integrals by the corresponding integral sums and using (12.14) we obtain the equality (12.16).

Let us consider now the case of the approximating functional $\gamma_{*(2)}$ (10.4). In this case the previous constructions are repeated with the following changes:

We take

$$\|l_{(2)}(\cdot)\|^* = \underset{\omega \in \Omega}{\mathrm{essmax}} \Big[\sum_{i=g}^{N} \mu^{[i]*}(l^{(i)}_{(2)}(\omega)) \Big]\,. \tag{12.21}$$

The equality

$$\rho_{*(2)}(\tau_*, w_*, \Delta_\delta) = e_{*(2)}(\tau_*, w_*, \Delta_\delta) + O(\delta) \tag{12.22}$$

is valid where the value of the *stochastic program extremum* $e_{*(2)}(\cdot)$ is defined by the equality (12.14), where the lower index (1) is replaced by the index (2); i.e., we have

$$e_{*(2)}(\tau_*, w_*, \Delta) = \sup_{\|l_{(2)}(\cdot)\|^* \le 1} \Big[\sum_{i=g}^{N} M\left\{ l_{(2)}^{(i)T}(\omega) \right\} D^{[i]} X(t^{[i]}, \tau_*) w_* +$$

$$+M\left\{ \sum_{j=1}^{k} (\tau_{j+1} - \tau_j) \max_{v \in Q} \min_{u \in P} M\Big\{ \sum_{i=d(j)}^{N} l_{(2)}^{(i)T}(\omega) *\right.$$

$$\left. * D^{[i]} X(t^{[i]}, \tau_j) f(\tau_j, u, v) \mid \xi_1, \ldots, \xi_j \Big\}\right\}\Big]. \tag{12.23}$$

The equality (12.22) is verified similarly to (12.20). The only difference is the following. In the equality

$$\rho_{*(2)}(\tau_*, w_*, \Delta) = \max_{v[\cdot]} \min_{u[\cdot]} \mu_{(2)}(D(\cdot) w(\cdot))$$

that corresponds to the equality (12.12), the sum in i from (12.10) is replaced by max in i from (12.12). According to this, in the dual norm (12.13) max in i is replaced by the sum in i in (12.21).

Using the properties of the approximating functionals $\gamma_{*(1)}$, $\gamma_{*(2)}$ and the equalities (12.16), (12.22) we obtain the following result:

Theorem 12.1. *The value of the game $\rho^0_{*(s)}(\tau_*, x_*)$ satisfies the equality*

$$\rho^0_{*(s)}(\tau_*, x_*) = \lim_{\delta \to 0} e_{*(s)}(\tau_*, x_*, \Delta_\delta), \quad s = 1, 2 \tag{12.24}$$

where Δ_δ is a partition (12.2), (12.17).

Theorem 12.1 was given in [64],[66]. We omit here the complete proof of Theorem 12.1, since this proof is not the subject of the present book. The main aim of the present book is to give and justify an effective construction for finding the value $e_{*(s)}(\tau_*, w_*, \Delta)$, $s = 1, 2$ (12.14), (12.23).

However, after the description of these constructions we shall prove that the functions $e_{*(s)}(\tau, w, \Delta)$ obtained in this way satisfy conditions similar to the conditions 1^0, 2^0, 3^0_u and 4^0_v from Section 8 for the functionals $\gamma_{*(1)}$ (11.9) and $\gamma_{*(2)}$ (11.10). Thus, taking also into account Theorem 9.1, we obtain an independent proof of Theorem 12.1.

13. Recurrent construction for the program extremum $e_{*(1)}(\cdot)$

We consider the case of approximating functional $\gamma_{*(1)}$ (11.9) and give a procedure for finding the quantity $e_{*(1)}(\tau_*, w_*, \Delta)$ (12.14).

Denote

$$\widetilde{w}_* = X(\vartheta, t_*)w_*, \tag{13.1}$$

$$\widetilde{f}(\tau, u, v) = X(\vartheta, \tau)f(\tau, u, v), \tag{13.2}$$

$$m^{(i)} = X^T(t^{[i]}, \vartheta)D^{[i]T}M\{l^{(i)}(\omega)\}, \tag{13.3}$$

$$m_j^{(i)}(\xi_1, \ldots, \xi_j) = X^T(t^{[i]}, \vartheta)D^{[i]T}M\{l^{(i)}(\omega) \mid \xi_1, \ldots, \xi_j\}, \quad t^{[i]} > \tau_j \tag{13.4}$$

where the right-hand sides of the equalities are the corresponding terms from the expression for $e_{*(1)}(\tau_*, w_*, \Delta)$ (12.14). According to (12.14), (13.1)–(13.4) we obtain that the quantity $e_{*(1)}(\cdot)$ can be represented in the form

$$e_{*(1)}(\tau_*, w_*, \Delta)$$

$$= \max_{\|l_{(1)}(\cdot)\|^* \le 1} \Big[\sum_{i=g}^{N} M\left\{l_{(1)}^{(i)T}(\omega)\right\} D^{[i]}X(t^{[i]}, \tau_*)w_* + M\Big\{\sum_{j=1}^{k}(\tau_{j+1} - \tau_j) *$$

$$* \max_{v \in Q}\min_{u \in P}\Big(\sum_{i=d(j)}^{N} m_j^{(i)T}(\xi_1, \ldots, \xi_j)\Big)\widetilde{f}(\tau_j, u, v)\Big\}\Big]. \tag{13.5}$$

Now it is convenient to forget the original probabilistic sense of the parameters m and l. In the expressions below in this section and in similar cases in subsequent sections we deal with these parameters as with deterministic arguments of the considered functions in the corresponding domains. This interpretation of the arguments m and l applies also to the construction of the value $e_*(\cdot)$ with the help of recurrent sequences of functions $\varphi_j(\cdot)$.

According to (13.3) we introduce the vectors $l^{[i]}$ and

$$m^{[i]} = X^T(t^{[i]}, \vartheta)D^{[i]T}l^{[i]}, \quad i = 1, \ldots, N. \tag{13.6}$$

Consider two cases. In the first case $\tau_* < t^{[g]}$. (Here $t^{[g]}$ is the smallest of the time moments $t^{[i]} \geq \tau_*$, $i = 1, \dots, N$ from (10.3), (11.9), i.e., g is a number (12.11).) In this case we take

$$m^* = m_* = \sum_{i=h}^{N} m^{[i]}, \qquad h = g \tag{13.7}$$

where $m^{[i]}$ is a quantity (13.6).

In the second case $\tau_* = t^{[g]}$ we take

$$m_* = \sum_{i=h}^{N} m^{[i]}, \qquad h = g+1\,, \tag{13.8}$$

$$m^* = \sum_{i=g}^{N} m^{[i]}\,. \tag{13.9}$$

Now we can represent the value $e_{*(1)}(\cdot)$ in the form

$$e_{*(1)}(\tau_*, w_*, \Delta) = \max_{m^*, m_*} \Big[\, m^{*T} \widetilde{w}_* + \kappa_{*(1)}(\tau_*, m_*, \Delta) \Big] \tag{13.10}$$

where $\kappa_{*(1)}(\tau_*, m_*, \Delta)$ is a certain function of the indicated arguments. In order to obtain the expression for $\kappa_{*(1)}(\tau_*, m_*, \Delta)$ we offer a construction for finding the quantity $e_{*(1)}(\tau_*, w_*, \Delta)$ (13.5) which uses a recurrent procedure of taking upper convex hulls $\varphi_j(m) = \overline{\psi}_j(m)$ for some functions $\psi_j(m)$ whose argument $m \in G_\varphi$ is a deterministic vector, which substitute sums of the conditional mathematical expectations of the random vector m in (13.5). Here G_φ is some convex set. Let us recall the definition of the upper convex hull of a function $\psi(m)$ in the corresponding set of arguments.

The *upper convex hull* $\overline{\psi}(m)$, $m \in G_\varphi$ of a function $\psi(m)$ is defined as a function $\varphi(m)$ which satisfies the following conditions:

1. The function $\varphi(m)$ is *concave* [130] in $m \in G_\varphi$, i.e., for any $m' \in G_\varphi$, $m'' \in G_\varphi$ and for any number $\lambda \in [0,\ 1]$ the inequality

$$\varphi(\lambda m' + (1-\lambda)m'') \geq \lambda\varphi(m') + (1-\lambda)\varphi(m'') \tag{13.11}$$

is true.

2. The inequality

$$\psi(m) \leq \varphi(m) \tag{13.12}$$

holds for any $m \in G_\varphi$.

3. Upper convex hull $\varphi(m)$ is the minimal function which satisfies conditions 1 and 2, i.e., for any concave function $\zeta(m)$ which majorizes $\psi(m)$, $m \in G_\varphi$ the inequality

$$\varphi(m) \leq \zeta(m), \quad m \in G_\varphi \tag{13.13}$$

is true.

We use also the following property of the considered minimal concave function.

4. Let $G_\varphi \subset \mathbf{R}^n$ and $\mathcal{B}$ be the Borel σ-algebra on G_φ. For any $m \in G_\varphi$ there exists a probability measure $\eta(B \mid m)$ on the sets $B \in \mathcal{B}$ such that the equalities

$$\varphi(m) = \int_{r \in G_\varphi} \psi(r)\eta(dr \mid m) \tag{13.14}$$

and

$$m = \int_{r \in G_\varphi} r\eta(dr \mid m) \tag{13.15}$$

are true.

Actually, due to the Carathéodory theorem [130] one can choose a measure $\eta(B|m)$ concentrated on a finite collection of points $r^{[s]}(m)$. This simplifies the theoretical proofs and practical constructions.

First we describe the procedure for calculating the quantity $\kappa_{*(1)}(\tau_*, m_*, \Delta))$ in (13.10), needed for finding the quantity $e_{*(1)}(\tau_*, w_*, \Delta)$ (13.5), (13.10).

The calculation of the quantity $\kappa_{*(1)}(\cdot)$ is reduced to the following recurrent procedure:

Let $\tau_* \in [t_0, \vartheta)$ be the time moment corresponding to a fixed position $\{\tau_*, w_*\} \in G^*$. Here G^* is the domain, described in Section 7 (see (7.5)). Let $\Delta = \Delta\{\tau_j\}$ be the partition (12.2) for the time interval $[\tau_*, \vartheta]$, i.e., $\Delta = \{\tau_1 = \tau_*, \ \tau_j < \tau_{j+1}, \ j = 1, \ldots, k, \ \tau_{k+1} = \vartheta\}$.

We introduce a set

$$G_{k+1}(\tau_*) = \{m : \ m = D^{[N]T} l, \ \ \mu^{[N]*}(l) \leq 1\} \tag{13.16}$$

and a function

$$\psi_{k+1}(\tau_*, m) = 0, \quad m \in G_{k+1}(\tau_*)\,. \tag{13.17}$$

Then we have

$$\varphi_{k+1}(\tau_*, m) = \overline{\psi}_{k+1}(\tau_*, m) = 0, \qquad m \in G_{k+1}(\tau_*)\,. \tag{13.18}$$

Here the symbol $\overline{\psi}_{k+1}(\tau_*, m)$ denotes the upper convex hull of the function $\psi_{k+1}(\tau_*, m)$, $m \in G_{k+1}(\tau_*)$.

Thus, for the time moment $\vartheta = t_{k+1} \in \Delta$ we define $\varphi_{k+1}(\tau_*, m)$ as the upper convex hull of function $\psi_{k+1}(\tau_*, m)$ (13.17) in the set $m \in G_{k+1}(\tau_*)$ (13.16). We recall that $t^{[N]} = \tau_{k+1} = \vartheta$.

Let us move one step back. We assume

$$G_k(\tau_*) = G_{k+1}(\tau_*), \tag{13.19}$$

$$\Delta\psi_k(\tau_*, m) = J(\tau_k, \tau_{k+1}, m) =$$

$$= (\tau_{k+1} - \tau_k) \max_{v \in Q} \min_{u \in P} \left[m^T \widetilde{f}(\tau_k, u, v) \right], \tag{13.20}$$

$$\psi_k(\tau_*, m) = \varphi_{k+1}(\tau_*, m) + \Delta\psi_k(\tau_*, m) = \Delta\psi_k(\tau_*, m), \tag{13.21}$$

$$\varphi_k(\tau_*, m) = \overline{\psi}_k(\tau_*, m), \quad m \in G_k(\tau_*). \tag{13.22}$$

In (13.20) $\widetilde{f}(\cdot)$ is a term (13.2) in (13.5).

The justification of this step is the following. We need to maximize the expression in (13.5). Thus, we need at least to maximize the function that corresponds to the last term of the sum in j in (13.5). If we set $m = D^{[N]T} l = D^{[N]T} l^{[N]}_{(1)}(\omega)$ then according to the formula for expectation [96], [97] that uses the conditional expectations we see that for every conditional expectation

$$m = m^{(N)}_{k-1}(\xi_1, \ldots, \xi_{k-1}) = M\left\{ D^{[N]T} l^{(N)}_{(1)}(\omega) \mid \xi_1, \ldots, \xi_{k-1} \right\}$$

it is necessary to maximize the quantity

$$M\left\{ \Delta\psi_k \left(D^{[N]T}\, l^{(N)}_{(1)}(\omega) \right) \mid \xi_1, \ldots, \xi_{k-1} \right\}.$$

This maximal quantity equals the value at the point m of the upper convex hull $\varphi_k(\tau_*, m)$ of the function $\psi_k(\tau_*, m)$ for the set $G_k(\tau_*)$ (13.19), (13.16). Therefore we take $\varphi_k(\tau_*, m) = \overline{\psi}_k(\tau_*, m)$ for $m \in G_k(\tau_*)$, where $\overline{\psi}$ denotes the upper convex hull of the function ψ in the corresponding set.

The above-constructed set $G_k(\tau_*) = G_{k+1}(\tau_*)$ (13.16) and the functions $\Delta\psi_k(\tau_*, m)$ (13.20), $\psi_k(\tau_*, m)$ (13.21), $\varphi_k(\tau_*, m)$ (13.22) provide the induction base in j from $j = k$ to $j = 1$.

We assume that $j \geq 1$. Let us consider the induction step from $j + 1$ to j.

Suppose that a set $G_{j+1}(\tau_*)$ and a function $\varphi_{j+1}(\tau_*, m)$ have been constructed already for $m \in G_{j+1}(\tau_*)$.

Let us consider two cases. In the first case we have

$$\tau_{j+1} < t^{[i]}, \quad i = d(j). \tag{13.23}$$

Here $t^{[i]}$ is the moment from (11.9)) and $d(j)$ is determined by (12.15). In this case we take

$$G_j(\tau_*) = G_{j+1}(\tau_*), \tag{13.24}$$

$$\Delta\psi_j(\tau_*, m) = J(\tau_j, \tau_{j+1}, m) =$$

$$= (\tau_{j+1} - \tau_j) \max_{v \in Q} \min_{u \in P} \; m^T \tilde{f}(\tau_j, u, v), \tag{13.25}$$

$$\psi_j(\tau_*, m) = \varphi_{j+1}(\tau_*, m) + \Delta\psi_j(\tau_*, m), \quad \varphi_j(\tau_*, m) = \overline{\psi}_j(\tau_*, m) \tag{13.26}$$

for $m \in G_j(\tau_*)$. In (13.23) $d(j)$ is determined in (12.15).

In the second case we have

$$\tau_{j+1} = t^{[i]}, \quad i = d(j)\,. \tag{13.27}$$

Then the set $G_j(\tau_*)$ consists of the vectors m^*, where

$$m^* = m_* + \widehat{m}, \tag{13.28}$$

$$m_* \in G_{j+1}(\tau_*), \tag{13.29}$$

$$\widehat{m} = X^T(t^{[i]}, \vartheta) D^{[i]T} l, \quad \mu^{[i]*}(l) \leq 1, \quad i = d(j). \tag{13.30}$$

And we put

$$\psi_j(\tau_*, m^*) = \max_{m_*} [\Delta\psi_j(\tau_*, m^*) + \varphi_{j+1}(\tau_*, m_*)], \tag{13.31}$$

$$\varphi_j(\tau_*, m) = \overline{\psi}_j(\tau_*, m), \quad m \in G_j(\tau_*)\,. \tag{13.32}$$

Similar to the case $j = k$, the justification of this induction step is based on maximization of the corresponding conditional expectation of the function $\psi_j(\cdot)$ (see, for instance, [80]).

Using the induction in j from $j = k$ to $j = 1$ we get the quantity $\varphi_1(\tau_*, m)$. Here two cases are possible. In the *first case*:

$$\tau_* = \tau_1 < t^{[g]} \tag{13.33}$$

and in the *second case*:

$$\tau_* = \tau_1 = t^{[g]} \tag{13.34}$$

where g is the number (12.11).

In the first case (13.33) we obtain the set

$$G_1(\tau_*) = G^{(g)} = \Big\{m: \ m = \sum_{i=g}^{N} X^T(t^{[i]}, \vartheta) D^{[i]T} l^{[i]}, \quad \mu^{[i]*}(l^{[i]}) \le 1\Big\} \tag{13.35}$$

and the function $\varphi_1(\tau_*, m)$, $m \in G_1(\tau_*)$. This function $\varphi_1(\tau_*, m)$ determines the quantity $\kappa_{*(1)}(\tau_*, m_*, \Delta)$ in (13.10) such that

$$\kappa_{*(1)}(\tau_*, m_*, \Delta) = \varphi_1(\tau_*, m_*), \qquad m_* \in G^{(g)}. \tag{13.36}$$

In the second case (13.34) we obtain the set

$$G_1(\tau_*) = G^{(g+1)} = \Big\{m: \ m = \sum_{i=g+1}^{N} X^T(t^{[i]}, \vartheta) D^{[i]T} l^{[i]}, \quad \mu^{[i]*}(l^{[i]}) \le 1\Big\} \tag{13.37}$$

and the function $\varphi_1(\tau_*, m)$, $m \in G_1(\tau_*)$. In this case we have

$$\kappa_{*(1)}(\tau_*, m_*, \Delta) = \varphi_1(\tau_*, m_*), \quad m_* \in G^{(g+1)}. \tag{13.38}$$

The constructed quantity $\kappa_{*(1)}(\cdot)$ (13.36), (13.38) determines the value $e_{*(1)}(\tau_*, w_*, \Delta)$ (13.5), (13.10). Thus, in the first case (13.33) we have

$$m^* = m_* \in G^{(g)} \tag{13.39}$$

in (13.10).

In the second case (13.34) we take in (13.10):

$$m^* = m_* + \widehat{m}, \tag{13.40}$$

where

$$m_* \in G^{(g+1)}, \tag{13.41}$$

$$\widehat{m} = X^T(\tau_*, \vartheta)D^{[g]T}l, \quad \mu^{[g]*}(l) \leq 1\,. \tag{13.42}$$

This concludes the construction of value $e_{*(1)}(\tau_*, w_*, \Delta)$ (13.5), (13.10).

We would like to remark also that if in the considered procedure all the functions $\psi_j(m),\ m \in G_j,\ j = 1, \ldots, k$ are *concave* then we have the *regular case* [59],[79],[140] and the random vector $l_{(1)}(\cdot)$ turns out to be *determinate.* Then we come to the case of *determinate program* synthesis [71],[79], [88], [140].

Let us also note the important fact that the offered construction, which is based on the recurrent sequence of functions $\varphi_j(\cdot)$, permits us to check the required properties of *u-stability* and *v-stability* (3_u and 4_v from Section 8) for the value $e_{*(1)}$ employing the properties of the functions $\varphi_j(\cdot)$ directly. The description of these properties will be given in Sections 15 and 16.

14. Recurrent construction for the program extremum $e_{*(2)}(\cdot)$

Now we consider the case of approximating functional $\gamma_{*(2)}$ (11.10) and describe the procedure for calculation of the value $e_{*(2)}(\tau_*, w_*, \Delta)$ (12.23). Using (12.23), (13.1)–(13.4) we obtain that the quantity $e_{*(2)}(\cdot)$ has the form

$$e_{*(2)}(\tau_*, w_*, \Delta)$$

$$= \max_{\|l_{(2)}(\cdot)\|^* \leq 1} \Big[\sum_{i=g}^{N} M\left\{l_{(2)}^{(i)T}(\omega)\right\} D^{[i]}X(t^{[i]}, \tau_*)w_* + M\Big\{\sum_{j=1}^{k}(\tau_{j+1} - \tau_j) *$$

$$* \max_{v \in Q} \min_{u \in P}\Big(\sum_{i=d(j)}^{N} m_j^{(i)T}(\xi_1, \ldots, \xi_j)\Big)\widetilde{f}(\tau_j, u, v)\Big\}\Big]. \tag{14.1}$$

We now take for the value $e_{*(2)}(\cdot)$ the expression

$$e_{*(2)}(\tau_*, w_*, \Delta) = \max_{m^*, m_*, \nu}[m^{*T}\widetilde{w}_* + \kappa_{*(2)}(\tau_*, m_*, \Delta, \nu)]. \tag{14.2}$$

As in Section 13 we introduce here deterministic vectors $l^{[i]}$ and $m^{[i]}$ (see (13.6)).

We see that the quantity $e_{*(2)}(\cdot)$ (14.1) has a form similar to (13.5). The only difference is the presence of some parameter ν in the expression (14.2). The sense of this parameter $\nu \in [0,\ 1]$ will become evident from the following construction of the quantity $e_{*(2)}(\cdot)$.

Here we have to consider two cases. In the first case $\tau_* < t^{[g]}$ we have

$$m^* = m_* = \sum_{i=h}^{N} m^{[i]}, \qquad h = g,$$

$$\sum_{i=g}^{N} \mu^{[i]*}\left(l_{(2)}^{[i]} \right) = \nu = 1\,. \tag{14.3}$$

In the second case $\tau_* = t^{[g]}$ we have

$$m^* = \sum_{i=g}^{N} m^{[i]}, \quad \sum_{i=g}^{N} \mu^{[i]*}\left(l_{(2)}^{[i]} \right) = \nu = 1,$$

$$m_* = \sum_{i=h}^{N} m^{[i]}, \quad \sum_{i=h}^{N} \mu^{[i]*}\left(l_{(2)}^{[i]} \right) = \nu \in [0,\ 1] \tag{14.4}$$

where $h = g + 1$, $\mu^{[g]*}(l^{[g]}) = 1 - \nu$.

The following procedure for the construction of the quantities $\kappa_{*(2)}(\cdot)$ (14.2) and $e_{*(2)}(\cdot)$ (14.1) is similar to the procedure from Section 13. Some changes are connected mostly with the use of different norms (12.13) and (12.21) for $l_{(1)}(\cdot)$ and $l_{(2)}(\cdot)$, respectively.

In fact, in the case of the functional $\gamma_{*(1)}$ (11.9) and the corresponding value $e_{*(1)}(\cdot)$ (12.14) we have the restrictions for $l_{(1)}^{(i)}(\omega)$ in the form $\mu^{[i]*}\left(l_{(1)}^{(i)}(\omega) \right) \leq 1$ (see (12.13), (12.14)). However, in the case of the functional $\gamma_{*(2)}$ (11.10) and the corresponding value $e_{*(2)}(\cdot)$ we have the following restriction

$$\sum_{i=g}^{N} \mu^{[i]*}\left(l_{(2)}^{(i)}(\omega) \right) \leq 1 \tag{14.5}$$

for the random variable $l_{(2)}^{(i)}(\omega)$ (see (12.21), (12.23)).

Let $\nu \in [0,\ 1]$ be a parameter evaluating an admissible sum of the norms $\mu^{[i]*}\left(l^{[i]}_{(2)}\right)$ for the time interval $[\tau_j,\ \tau_{k+1}]$. Therefore we shall consider now the functions $\Delta\psi(\cdot)$, $\psi(\cdot)$ and $\varphi(\cdot)$ (similar to the ones from Section 13) of deterministic arguments τ, m and ν.

Let us assume that for the interval $[\tau_*,\ \vartheta]$ with the fixed τ_* the following equalities

$$\varphi_{k+1}(\tau_*, m, \nu) = 0, \qquad \nu \in [0,\ 1], \tag{14.6}$$

$$\Delta\psi_k(\tau_*, m, \nu) = J(\tau_k, \tau_{k+1}, m) \tag{14.7}$$

are satisfied, where the symbol $J(\tau_j, \tau_{j+1}, m)$ again as in (13.25) denotes the quantity

$$J(\tau_j, \tau_{j+1}, m) =$$
$$= (\tau_{j+1} - \tau_j) \max_{v\in Q} \min_{u\in P} m^T \widetilde{f}(\tau_j, u, v) \tag{14.8}$$

if $j \le k$. Here $\tau_j \in \Delta = \Delta\{\tau_j\}$ (12.2), $\widetilde{f}(\cdot)$ is the term (13.2).

Let

$$\psi_k(\tau_*, m, \nu) = \Delta\psi_k(\tau_*, m, \nu) \tag{14.9}$$

and

$$\varphi_k(\tau_*, m, \nu) = \overline{\psi}_k(\tau_*, m, \nu)\,. \tag{14.10}$$

As above, $\varphi_k(\tau_*, m, \nu)$ denotes the upper convex hull of the function $\psi_k(\tau_*, m, \nu)$, $m \in G_{k,\nu}(\tau_*)$ where we have now

$$G_{k,\nu}(\tau_*) = \left\{m:\ m = D^{[N]T}l;\quad \mu^{[N]*}(l) \le \nu\right\}. \tag{14.11}$$

Let us form a recurrent sequence of functions $\varphi_j(\cdot)$. We assume that $j \ge 1$. Suppose that a set $G_{j+1,\nu}(\tau_*)$ and a function $\varphi_{j+1}(\tau_*, m, \nu)$ are constructed already for $\nu \in [0,\ 1]$ and $m \in G_{j+1,\nu}(\tau_*)$, where

$$G_{j+1,\nu}(\tau_*) = \left\{m:\ \sum_{p=d(j)}^{N} X^T(t^{[p]}, \vartheta) D^{[p]T} l^{[p]},\quad \sum_{p=d(j)}^{N} \mu^{[p]*}(l^{[p]}) \le \nu\right\}. \tag{14.12}$$

Let us consider the first case in which

$$\tau_{j+1} < t^{[i]}, \qquad i = d(j). \tag{14.13}$$

Then we assume that

$$G_{j,\nu}(\tau_*) = G_{j+1,\nu}(\tau_*), \tag{14.14}$$

$$\Delta\psi_j(\tau_*, m, \nu) = J(\tau_j, \tau_{j+1}, m), \tag{14.15}$$

$$\psi_j(\tau_*, m, \nu) = \varphi_{j+1}(\tau_*, m, \nu) + \Delta\psi_j(\tau_*, m, \nu), \tag{14.16}$$

$$\varphi_j(\tau_*, m, \nu) = \overline{\psi}_j(\tau_*, m, \nu) \tag{14.17}$$

for $m \in G_{j,\nu}(\tau_*)$. In (14.15) $J(\cdot)$ is the quantity (14.8).

In the second case we suppose

$$\tau_{j+1} = t^{[i]}, \quad i = d(j). \tag{14.18}$$

Then we assume that

$$\psi_j(\tau_*, m, \nu) = \max_{m_*, \nu_*} \left[\varphi_{j+1}(\tau_*, m_*, \nu_*) + \Delta\psi_j(\tau_*, m^*, \nu)\right], \tag{14.19}$$

for $\Delta\psi_j(\tau_*, m^*, \nu)$ (14.15) where $m = m^*$ and

$$m^* = m_* + \widehat{m}, \tag{14.20}$$

$$m_* \in G_{j+1,\nu_*}(\tau_*), \quad \nu_* \leq \nu,$$

$$\widehat{m} = X^T(t^{[i]}, \vartheta) D^{[i]T} l, \quad \mu^{[i]*}(l) \leq \nu - \nu_*, \quad i = d(j). \tag{14.21}$$

Let $G_{j,\nu}(\tau_*)$ be the set of all m^* (14.20), (14.21) and let

$$\varphi_j(\tau_j, m, \nu) = \overline{\psi}_j(\tau_*, m, \nu), \quad m \in G_{j,\nu}(\tau_*). \tag{14.22}$$

The justification of this induction is similar to that in Section 13.

Using induction in j from $j = k$ to $j = 1$ we obtain the quantity $\varphi_1(\tau_*, m, \nu)$. If $\tau_* < t^{[g]}$ then the equality

$$\kappa_{*(2)}(\tau_*, m_*, \Delta, \nu) = \varphi_1(\tau_*, m_*, \nu), \quad m_* \in G_{1,\nu}(\tau_*), \quad \nu = 1 \tag{14.23}$$

holds, and if $\tau_* = t^{[g]}$ then

$$\kappa_{*(2)}(\tau_*, m_*, \Delta, \nu) = \varphi_1(\tau_*, m_*, \nu),$$

$$m_* \in G_{1,\nu}(\tau_*), \qquad \nu \leq 1 - \mu^{[g]*}(l^{[g]}) \tag{14.24}$$

is valid for $\kappa_{*(2)}(\cdot)$ (14.2).

The equalities (14.23), (14.24) can be proved using the properties of the function $\varphi(m)$, which is the upper convex hull of some function $\psi(m)$.

From (14.1), (14.20)–(14.24) we obtain the equality

$$e_{*(2)}(\tau_*, w_*, \Delta) = \max_{m^*, m_*, \nu} [m^{*T} \widetilde{w}_* + \varphi_1(\tau_*, m_*, \nu)] \tag{14.25}$$

where in the case $\tau_* < t^{[g]}$ we have the condition

$$m^* = m_* \in G_{1,\nu}(\tau_*), \qquad \nu = 1 \tag{14.26}$$

and in the case $\tau_* = t^{[g]}$ we have

$$m^* = m_* + \widehat{m}, \qquad m_* \in G_{1,\nu}(\tau_*),$$

$$\widehat{m} = X^T(\tau_*, \vartheta) D^{[g]T} l, \qquad \mu^{[g]*}(l) \leq 1 - \nu \,. \tag{14.27}$$

This concludes the construction of value $e_{*(2)}(\cdot)$ (14.1).

Here we can make remarks concerning the functions $\varphi_j(\tau_*, m, \nu)$, $m \in G_{j,\nu}(\tau_*)$ that are similar to the remarks given in the end of Section 13 about the functions $\varphi_j(\tau_*, m)$, $m \in G_j(\tau_*)$.

15. Condition of u-stability for the program extremum $e_{*(1)}(\cdot)$

The approximating functional $\gamma_{*(1)}$ (11.9) is positional (see (4.9) and (11.12)). It follows from Theorem 9.2 that a function $\rho_{*(1)}(t, w)$ is the value $\rho^0_{*(1)}(t, x)$ of the differential game for the system (10.1), (10.2) with the quality index $\gamma_{*(1)}$ (11.9) if this function $\rho_{*(1)}(t, w)$ satisfies the conditions 1^0, 2^0, 3^0_u and 4^0_v

from Section 8. It is obvious that the function $e_{*(1)}(\tau_*, w_*, \Delta)$ constructed in Sections 12 and 13 satisfies the conditions 1^0 (8.1) and 2^0 (8.2); i.e., we have

$$| e_{*(1)}(t, w^{(1)}, \Delta) - e_{*(1)}(t, w^{(2)}, \Delta) | \leq$$

$$\leq \widetilde{L} \mid w^{(2)} - w^{(1)} \mid, \quad \widetilde{L} = \text{const} \tag{15.1}$$

for fixed $t \in [t_0, \vartheta]$, $\{t, w^{(s)}\} \in G^*$ (see (7.5)), $s = 1, 2$ and the equality

$$\gamma_{*(1)}(w[\vartheta]) = e_{*(1)}(\vartheta, w[\vartheta], \Delta) \tag{15.2}$$

is true for every $\{\vartheta, w[\vartheta]\} \in G^*$.

Now we assume that the vector function $f(t, u, v)$ in (10.1), (10.2) besides the restrictions (3.5), (3.6) satisfies also the following condition. For fixed t, v the vector $f(t, u, v)$ describes a convex set when u varies in P. We would like to remark that this condition is just technical and simplifies some definitions and proofs given below. When we use it we will indicate the way to do without it.

Let us show that the function $e_{*(1)}(\tau_*, w_*, \Delta)$ (13.5), (13.10) satisfies also the following property of *u-stability*.

$3^0_{*(1)}u$. Suppose we know a domain G^* (see (7.5)), a position $\{\tau_*, w_*\} \in G^*$ and a value $\varepsilon > 0$. Then there exists a value $\delta(\varepsilon) > 0$ such that the following conditions are valid. It is important to remark that for a given bounded domain G^* the value $\delta(\varepsilon) > 0$ can be chosen independent of the initial position $\{\tau_*, w_*\} \in G^*$. Let us choose a partition $\Delta_* = \Delta_\delta = \Delta_\delta\{\tau_i\}$ (12.2), $\tau_{j+1} - \tau_j < \delta$, where $\delta \leq \delta(\varepsilon)$, moments $\tau_* \in \Delta_\delta$, $\tau^* = \tau_2 \in \Delta_\delta$ and disturbances $v[\tau_*[\cdot]\tau^*) = \{v[\tau] = v[\tau_*] \in Q,\ \tau_* \leq \tau < \tau^*\}$. Then there exist control actions $u[\tau_*[\cdot]\tau^*) = \{u[\tau] = u[\tau_*] \in P,\ \tau_* \leq \tau < \tau^*\}$ which together with $v[\tau_*[\cdot]\tau^*)$ generate from the initial position $\{\tau_*, w_*\}$ a motion $w[\tau_*[\cdot]\tau^*] = \{w[\tau],\ \tau_* \leq \tau \leq \tau^*\}$ of w-model (12.1) that comes to the state $w[\tau^*] = w^*$ such that

$$e_{*(1)}(\tau^*, w^*, \Delta^*) \leq e_{*(1)}(\tau_*, w_*, \Delta_*) + \varepsilon(\tau^* - \tau_*) \tag{15.3}$$

in the case

$$\tau_* < \tau^* \leq t^{[g]} \tag{15.4}$$

or the inequality

$$\mu^{[g]}(D^{[g]} w[t^{[g]}]) + e_{*(1)}(\tau^*, w^*, \Delta^*) \leq$$

$$\leq e_{*(1)}(\tau_*, w_*, \Delta_*) + \varepsilon(\tau^* - \tau_*) \tag{15.5}$$

holds in the case

$$\tau_* = t^{[g]}\,. \tag{15.6}$$

Here g is a number (12.11); $\mu^{[g]}(\cdot)$ is the norm in (11.9); Δ_* and Δ^* in (15.3), (15.6) are the following partitions:

$$\Delta_* = \Delta_*\{\tau_j\} = \{\tau_* = \tau_1, \quad \tau_j < \tau_{j+1}, \quad \tau_{k+1} = \vartheta\}, \tag{15.7}$$

$$\Delta^* = \Delta^*\{\tau_j^*\} = \{\tau^* = \tau_1^* = \tau_2, \quad \tau_j^* = \tau_{j+1}, \quad \tau_k^* = \tau_{k+1} = \vartheta\}\,. \tag{15.8}$$

It should be noted that the condition $3^0_{*(1)}u$ is a convenient variant of the general condition 3^0_u from Section 8 for the particular case considered here.

If the condition of the convexity of the set $\{f(t,u,v):\ u \in P\}$ given in the beginning of this section does not hold, then the differential equation (12.1) is substituted by the differential inclusion

$$\dot{w} \in A(\tau)w + H(\tau, v), \qquad t_0 \leq \tau < \vartheta \tag{15.9}$$

where $H(\tau, v)$ is the convex hull of the set $\{f(\tau,u,v):\ u \in P\}$. In accordance with this substitution of (12.1) by (15.9) in the property of u-stability we make the following changes. Instead of the existence of control actions $u[\tau_*[\cdot]\tau^*)$ we speak about the existence of a motion $w[\tau_*[\cdot]\tau^*]$ of the w-model (15.9) generated from the initial position (τ_*, w_*) by the actions $v[\tau_*[\cdot]\tau^*)$ so that this motion comes to the state $w[\tau^*] = w^*$ such that the conditions (15.3)–(15.8) are fulfilled.

Let us prove the condition $3^0_{*(1)}u$.

Proof. It was shown in Section 13 that according to (13.36), (13.38) the value $e_{*(1)}(\tau_*, w_*, \Delta)$ (13.5), (13.10) is represented in the form

$$e_{*(1)}(\tau_*, w_*, \Delta) = \max_{m^*, m_*} \Big[\ m^{*T}\widetilde{w}_* + \varphi_1(\tau_*, m_*)\ \Big]. \tag{15.10}$$

Here the vectors m^*, m_* satisfy the condition (13.39) in the case (13.33) (that is, (15.4)) or these vectors satisfy the conditions (13.40)–(13.42) in the case (13.34) (that is, (15.6)).

At first we consider the case (15.4), i.e., $\tau_* < \tau^* \leq t^{[g]}$ and in particular the case

$$\tau^* < t^{[g]} . \tag{15.11}$$

Here g is the number (12.11). We have to prove the existence of $u[\tau_*[\cdot]\tau^*)$ such that the inequality (15.3) is true.

According to (15.10), (13.39), (13.33) we have the following relations:

$$e_{*(1)}(\tau_*, w_*, \Delta_*) = \max_{m^*_{\tau_*}} \Big[\, m^{*T}_{\tau_*} \widetilde{w}_{\tau_*} + \varphi_1(\tau_*, m^*_{\tau_*}) \, \Big] =$$

$$= m^{*0T}_{\tau_*} \widetilde{w}_{\tau_*} + \varphi_1(\tau_*, m^{*0}_{\tau_*}) \tag{15.12}$$

and

$$e_{*(1)}(\tau^*, w^*, \Delta^*) = \max_{m^*_{\tau^*}} \Big[\, m^{*T}_{\tau^*} \widetilde{w}_{\tau^*} + \varphi_1(\tau^*, m^*_{\tau^*}) \, \Big] =$$

$$= m^{*0T}_{\tau^*} \widetilde{w}_{\tau^*} + \varphi_1(\tau^*, m^{*0}_{\tau^*}) . \tag{15.13}$$

Here in the considered case $\tau^* < t^{[g]}$ and therefore $\tau_* < t^{[g]}$ also we have

$$m^{*0}_{\tau_*} \in G_1(\tau_*) = G^{(g)} \tag{15.14}$$

$$m^{*0}_{\tau^*} \in G_1(\tau^*) = G^{(g)} \tag{15.15}$$

i.e., *maximizing* vectors $m^{*0}_{\tau_*}$ and $m^{*0}_{\tau^*}$ that solve the corresponding problems (15.12) and (15.13) belong to one and the same set $G^{(g)}$ (13.35).

In (15.12) and (15.13) the lower indices τ_* and τ^* emphasize that the extrema $e_{*(1)}(\tau_*, \cdot)$, $e_*(\tau^*, \cdot)$, and the corresponding vectors m^* and $\widetilde{w}$ defined in Section 13 are calculated here for the initial moments $\tau_* \in \Delta_*$ (15.7) and $\tau^* \in \Delta^*$ (15.8), respectively.

According to (15.12) and (15.13) we consider the difference

$$\Delta e_{*(1)} = e_{*(1)}(\tau^*, w^*, \Delta^*) - e_{*(1)}(\tau_*, w_*, \Delta_*) =$$

$$= m^{*0T}_{\tau^*} \widetilde{w}_{\tau^*} + \varphi_1(\tau^*, m^{*0}_{\tau^*}) - m^{*0T}_{\tau_*} \widetilde{w}_{\tau_*} - \varphi_1(\tau_*, m^{*0}_{\tau_*}) . \tag{15.16}$$

If in (15.16) instead of the vector $m^{*0}_{\tau_*} \in G^{(g)}$ that maximizes $e_{*(1)}(\tau_*, \cdot)$ (15.12) we substitute the vector $m^{*0}_{\tau^*} \in G^{(g)}$ that maximizes $e_{*(1)}(\tau^*, \cdot)$ (15.13) then we obtain the following estimation:

$$\Delta e_{*(1)} \le m_{\tau^*}^{*0T}(\widetilde{w}_{\tau^*} - \widetilde{w}_{\tau_*}) + \varphi_1(\tau^*, m_{\tau^*}^{*0}) - \varphi_1(\tau_*, m_{\tau^*}^{*0}) \,. \tag{15.17}$$

According to the notation (13.1), (13.2) and Cauchy formula [24], [79], [120], [133] the equality

$$\widetilde{w}_{\tau^*} = \widetilde{w}_{\tau_*} + \int_{\tau_*}^{\tau^*} \widetilde{f}(\tau, u[\tau], v[\tau])d\tau \tag{15.18}$$

is valid. Therefore

$$m_{\tau^*}^{*0T}(\widetilde{w}_{\tau^*} - \widetilde{w}_{\tau_*}) = m_{\tau^*}^{*0T} \int_{\tau_*}^{\tau^*} \widetilde{f}(\tau, u[\tau], v[\tau])d\tau \,. \tag{15.19}$$

For any $\varepsilon > 0$ there exists a number $\delta_\Delta(\varepsilon) > 0$ such that

$$m_{\tau^*}^{*0T}(\widetilde{w}_{\tau^*} - \widetilde{w}_{\tau_*}) \le m_{\tau^*}^{*0T}\widetilde{f}(\tau_*, u[\tau_*], v[\tau_*])(\tau^* - \tau_*) + \varepsilon(\tau^* - \tau_*) \tag{15.20}$$

provided $(\tau^* - \tau_*) \le \delta_\Delta(\varepsilon)$.

Then according to (15.17) and (15.20) the estimation

$$\Delta e_{*(1)} \le m_{\tau^*}^{*0T}\widetilde{f}(\tau_*, u[\tau_*], v[\tau_*])(\tau^* - \tau_*) +$$

$$+ \varphi_1(\tau^*, m_{\tau^*}^{*0}) - \varphi_1(\tau_*, m_{\tau^*}^{*0}) + \varepsilon(\tau^* - \tau_*) \tag{15.21}$$

is true.

The calculations (13.25), (13.26) of the quantity $\varphi_1(\cdot)$ in Section 13 imply the inequality

$$\varphi_1(\tau_*, m_{\tau^*}^{*0}) \ge \varphi_1(\tau^*, m_{\tau^*}^{*0}) + J(\tau_*, \tau^*, m_{\tau^*}^{*0}) =$$

$$= \varphi_1(\tau^*, m_{\tau^*}^{*0}) + (\tau^* - \tau_*) \max_{v\in Q} \min_{u\in P} \ m_{\tau^*}^{*0T} \widetilde{f}(\tau_*, u, v)\,. \tag{15.22}$$

Besides that we have the obvious inequality

$$\max_{v\in Q} \min_{u\in P} \ m_{\tau^*}^{*0T} \widetilde{f}(\tau_*, u, v) \geq \min_{u\in P} m_{\tau^*}^{*0T} \widetilde{f}(\tau_*, u, v[\tau_*]) \tag{15.23}$$

for every $v[\tau_*] \in Q$.

Then according to (15.21)–(15.23) we obtain the following estimation:

$$\Delta e_{*(1)} \leq (\tau^* - \tau_*)\Big[\ m_{\tau^*}^{*0T} \widetilde{f}(\tau_*, u[\tau_*], v[\tau_*]) -$$

$$- \min_{u\in P} m_{\tau^*}^{*0T} \widetilde{f}(\tau_*, u, v[\tau_*])\ \Big] + \varepsilon(\tau^* - \tau_*)\,. \tag{15.24}$$

Generally speaking, the maximizing vector $m_{\tau^*}^{*0}$ is nonunique. The set $\{m_{\tau^*}^{*0}\}$ of the maximizing vectors $m_{\tau^*}^{*0}$ for $e_{*(1)}(\tau^*, \cdot)$ (15.13) is a convex compactum. It follows from the property 1 (13.11). The set $\{f(\cdot, u)\}$ of vectors $\widetilde{f}(\cdot, u)$ where $u \in P$ is a convex compactum, too. It follows from the properties of the vector-function $f(t, u, v)$ (10.1) stipulated in the beginning of this section. Then similarly to [77, 79] we can employ a *fixed point theorem* (see, for example [9]). According to this theorem we establish that there exists a *pair* $\{m_{\tau^*}^{*0}, u^0[\tau_*]\}$, $m_{\tau^*}^{*0} \in G^{(g)}$, $u^0[\tau_*] \in P$ which satisfies the following condition

$$\min_{u\in P} m_{\tau^*}^{*0T} \widetilde{f}(\tau_*, u, v[\tau_*]) = m_{\tau^*}^{*0T} \widetilde{f}(\tau_*, u^0[\tau_*], v[\tau_*]) \tag{15.25}$$

and at the same time the pair of the realizations of the actions $v[\tau] = v[\tau_*] \in Q$, $\tau_* \leq \tau < \tau^*$ and $u[\tau\} = u^0[\tau_*] \in P$, $\tau_* \leq \tau < \tau^*$ generates from the initial position $\{\tau_*, w_*\}$ the motion $w[\tau_*[\cdot]\tau^*]$ and the corresponding state $\widetilde{w}_{\tau^*} = X(\vartheta, \tau^*)w[\tau^*]$ such that the vector $m_{\tau^*}^{*0}$ in the pair $\{m_{\tau^*}^{*0}, u^0[\tau_*]\}$ maximizes $e_*(\tau^*, w^*, \Delta^*)$, where $w^* = w[\tau^*]$. According to (15.24), (15.25) the condition (15.3) is valid in the case (15.4) where $\tau^* < t^{[g]}$. This proves the property of u-stability $3^0_{*(1)}u$ in the particular case (15.4), when the condition (15.11) holds.

In the other case when in (15.4) $\tau^* = t^{[g]}$ we prove the condition (15.3) similarly. The only difference is that we take into account a transformation of the constructions for the sets $G_j(\tau^*)$ and $G_j(\tau_*)$. Namely, in the case $\tau^* = t^{[g]}$ according to (15.10), (13.34), (13.40)–(13.42) (where we substitute τ_* by τ^*)

we have the equalities

$$e_{*(1)}(\tau^*, w^*, \Delta^*) = \max_{m^*_{\tau^*},\, m_{*\tau^*}} \Big[m^{*T}_{\tau^*} \widetilde{w}_{\tau^*} + \varphi_1(\tau^*, m_{*\tau^*}) \Big] =$$

$$= m^{*0T}_{\tau^*} \widetilde{w}_{\tau^*} + \varphi_1(\tau^*, m^0_{*\tau^*}) \tag{15.26}$$

where due to (13.40)–(13.42) the vectors $m^*_{\tau^*}$ and $m_{*\tau^*}$ satisfy the following conditions

$$m^*_{\tau^*} = m_{*\tau^*} + \widehat{m}, \qquad m_{*\tau^*} \in G_1(\tau^*) = G^{(g+1)} \tag{15.27}$$

where

$$\widehat{m} = X^T(\tau^*, \vartheta) D^{[g]T} l, \qquad \mu^{[g]*}(l) \le 1\,. \tag{15.28}$$

We denote by

$$G^{(g)} = \{m^*_{\tau^*}\} \tag{15.29}$$

the set of the vectors $m^*_{\tau^*}$ (15.27), (15, 28), i.e., $G^{(g)}$ is the set (13.37).

In the considered case $\tau^* = t^{[g]}$ we have $\tau_* < t^{[g]}$ (since $\tau_* = \tau_1 < \tau_2 = \tau^*$) and so, according to (13.33), (13.39), the quantity $e_{*(1)}(\tau_*, w_*, \Delta)$ is defined by the quantity (15.12) under the conditions (15.14).

According to (15.26), (15.12) we consider the difference

$$\Delta e_{*(1)} = e_{*(1)}(\tau^*, w^*, \Delta^*) - e_{*(1)}(\tau_*, w_*, \Delta_*) =$$

$$= \Big[m^{*0T}_{\tau^*} \widetilde{w}_{\tau^*} + \varphi_1(\tau^*, m^0_{*\tau^*}) \Big] - \Big[m^{*0T}_{\tau_*} \widetilde{w}_{\tau_*} + \varphi_1(\tau_*, m^{*0}_{\tau_*}) \Big] \tag{15.30}$$

where

$$m^{*0}_{\tau^*} \in G^{(g)}, \qquad m^0_{*\tau^*} \in G_1(\tau^*) = G^{(g+1)},$$
$$m^{*0T}_{\tau_*} \in G_1(\tau_*) = G^{(g)}\,. \tag{15.31}$$

Here $G^{(g+1)}$ is the set (13.37).

Now in (15.30), similarly to the previous case of the equality (15.16), instead of the vector $m^{*0}_{\tau_*} \in G^{(g)}$ we substitute the vector $m^{*0}_{\tau^*} \in G^{(g)}$. We obtain the estimation

$$\Delta e_{*(1)} \le m^{*0T}_{\tau^*} (\widetilde{w}_{\tau^*} - \widetilde{w}_{\tau_*}) + \varphi_1(\tau^*, m^0_{*\tau^*}) - \varphi_1(\tau_*, m^{*0}_{\tau^*}) \tag{15.32}$$

and then according to (15.20) the inequality

$$\Delta e_{*(1)} \le m^{*0T}_{\tau^*} \widetilde{f}(\tau_*, u[\tau_*], v[\tau_*])(\tau^* - \tau_*) +$$

$$+\varphi_1(\tau^*, m^0_{*\tau^*}) - \varphi_1(\tau_*, m^{*0}_{\tau^*}) + \varepsilon(\tau^* - \tau_*) \tag{15.33}$$

is true for $\tau^* - \tau_* \leq \delta_\Delta(\varepsilon)$.

Taking into account the constructions (13.27)–(13.32) in Section 13 (where $\tau_j = \tau_*,\ \tau_{j+1} = \tau^*$) we have the following equality:

$$\varphi_1(\tau_*, m^*) = \bar{\psi}_1(\tau_*, m^*), \qquad m^* \in G_1(\tau_*)\,. \tag{15.34}$$

Here

$$\psi_1(\tau_*, m^*) = \max_{m_*}\Big[\, \Delta\psi_1(\tau_*, m^*) + \varphi_2(\tau_*, m_*) \,\Big] \tag{15.35}$$

where, in accordance with (13.25),

$$\Delta\psi_1(\tau_*, m^*) = J(\tau_1, \tau_2, m^*) =$$

$$= (\tau^* - \tau_*) \max_{v\in Q} \min_{u\in P}\ m^{*T} \widetilde{f}(\tau_*, u, v) \tag{15.36}$$

and

$$m^* = m_* + \widehat{m}, \qquad m_* \in G_2(\tau_*) = G^{(g+1)}, \qquad \widehat{m} = X^T(t^{[g]}, \vartheta) D^{[g]T} l,$$

$$\mu^{[g]*}(l) \leq 1, \quad G_1(\tau_*) = G^{(g)} = \{m^*\}\,. \tag{15.37}$$

Here $G^{(g)}$ and $G^{(g+1)}$ are the sets (13.35) and (13.37).

We have the following equality

$$\varphi_2(\tau_*, m_*) = \varphi_1(\tau^*, m_*), \qquad m_* \in G^{(g+1)} \tag{15.38}$$

since $G_2(\tau_*) = G^{(g+1)}$ and $G_1(\tau^*) = G^{(g+1)}$. So, (15.34)–(15.38) imply the equality

$$\psi_1(\tau_*, m^*) = \max_{m_*}\Big[\, \Delta\psi_1(\tau_*, m^*) + \varphi_1(\tau^*, m_*) \,\Big] \tag{15.39}$$

for $m_* \in G^{(g+1)}$ and $m^* \in G^{(g)}$ (15.37).

Since the set $G^{(g+1)} \subset G^{(g)}$, $m^{*0}_{\tau^*} \in G^{(g+1)}$ and according to (15.34),

(15.36), (15.39), we obtain the inequality

$$\varphi_1(\tau_*, m_{\tau^*}^{*0}) \geq (\tau^* - \tau_*) \max_{v\in Q} \min_{u\in P} m_{\tau^*}^{*0T} \widetilde{f}(\tau_*, u, v) + \varphi_1(\tau^*, m_{*\tau^*}^{0}).$$

After that, again taking into account (15.23)–(15.25), we obtain the condition (15.3) in the case (15.4) also for $\tau^* = t^{[g]}$.

So we have completed the proof of the condition (15.3) of u-stability $3_{*(1)}^{0}u$ in the case (15.4).

Now we consider the case (15.6) in which we need to prove the condition (15.5). Let the partition Δ_* (15.7) have a sufficiently small step so that

$$\tau_* = t^{[g]}, \qquad \tau^* < t^{[g+1]}. \tag{15.40}$$

In this case according to (15.10), (13.34), (13.40)–(13.42) we have the following equality:

$$e_{*(1)}(\tau_*, w_*, \Delta_*) = \max_{m_{\tau_*}^*,\ m_{*\tau_*}} \left[\, m_{\tau_*}^{*T} \widetilde{w}_{\tau_*} + \varphi_1(\tau_*, m_{*\tau_*}) \,\right] \tag{15.41}$$

where

$$m_{\tau_*}^* \in G^{(g)}, \qquad m_{\tau_*}^* = m_{*\tau_*} + \widehat{m}, \qquad m_{*\tau_*} \in G^{(g+1)},$$
$$\widehat{m} = X^T(\tau_*, \vartheta) D^{[g]T} l, \qquad \mu^{[g]*}(l) \leq 1\,. \tag{15.42}$$

We see from (13.33), (13.39) (where we substitute τ_*, w_*, Δ_* and $G^{(g)}$ by τ^*, w^*, Δ^* and $G^{(g+1)}$, respectively) that the value $e_{*(1)}(\tau^*, w^*, \Delta^*)$, has a form similar to (15.12). The only difference is that the argument τ_* is replaced by τ^*. Consider the difference $\Delta e_{*(1)} = e_{*(1)}(\tau^*, w^*, \Delta^*) - e_{*(1)}(\tau_*, w_*, \Delta_*)$. According to (15.12) and (15.41) we have the equality

$$\Delta e_{*(1)} = [m_{\tau^*}^{*0T} \widetilde{w}_{\tau^*} + \varphi_1(\tau^*, m_{\tau^*}^{*0})] - [m_{\tau_*}^{*0T} \widetilde{w}_{\tau_*} + \varphi_1(\tau_*, m_{*\tau_*}^{0})]\,. \tag{15.43}$$

Now we substitute in (15.43) the vector $m_{\tau_*}^{*0}$ by the vector $m_{\tau^*}^{*0} + \widehat{m}$ with some arbitrary $\widehat{m}$ that satisfies the condition

$$\widehat{m} = X^T(t^{[g]}, \vartheta) D^{[g]T} l, \qquad \mu^{[g]*}(l) \leq 1 . \tag{15.44}$$

In (15.43) we also substitute the vector $m^0_{*\tau_*} \in G^{(g+1)}$ by the vector $m^{*0}_{\tau^*} \in G^{(g+1)}$.

Generally speaking, the vectors $m^{*0}_{\tau^*} + \widehat{m}$ and $m^{*0}_{\tau^*}$ are not the maximizing vectors $m^{*0}_{\tau_*}$ and $m^0_{*\tau_*}$ in (15.43). So the inequality

$$\Delta e_{*(1)} \leq m^{*0T}_{\tau^*} \widetilde{w}_{\tau^*} - (m^{*0T}_{\tau^*} + \widehat{m}^T) \widetilde{w}_{\tau_*} +$$

$$+ \varphi_1(\tau^*, m^{*0}_{\tau^*}) - \varphi_1(\tau_*, m^{*0}_{\tau^*}) \tag{15.45}$$

is valid.

Let us choose the vector $\widehat{m}^0$ (15.44) such that

$$m^{0T} \widetilde{w}_{\tau_*} = \max_{\widehat{m}} \widehat{m}^T \widetilde{w}_{\tau_*} \tag{15.46}$$

under the conditions (15.44).

Then, according to (15.44) and to properties of the norm $\mu^{[i]}(D^{[i]} w[t^{[i]}])$ in (11.9), the equality

$$\max_{\widehat{m}} \widehat{m}^T \widetilde{w}_{\tau_*} = \mu^{[g]}(D^{[g]} w_*) \tag{15.47}$$

holds. Here $w_* = w[\tau_*] = w[t^{[g]}]$.

Using (15.43)–(15.47) we now obtain the following estimation:

$$\Delta e_{*(1)} \leq m^{*0T}_{\tau^*} \widetilde{w}_{\tau^*} + \varphi_1(\tau^*, m^{*0}_{\tau^*}) - m^{*0T}_{\tau^*} \widetilde{w}_{\tau_*} -$$

$$- \varphi_1(\tau_*, m^{*0}_{\tau^*}) - \mu^{[g]}(D^{[g]} w_*) . \tag{15.48}$$

Similarly to the first case (15.4), (15.16)–(15.25), choosing for the given $v[\tau_*]$ a suitable pair $\{m^{*0}_{\tau^*},\ u^0[\tau_*]\}$ we come to the inequality

$$m^{*0T}_{\tau^*} \widetilde{w}_{\tau^*} + \varphi_1(\tau^*, m^{*0}_{\tau^*}) \leq$$

$$\leq m_{\tau^*}^{*0T}\widetilde{w}_{\tau_*} + \varphi_1(\tau_*, m_{\tau^*}^{*0}) + \varepsilon(\tau^* - \tau_*) \tag{15.49}$$

where the state $\widetilde{w}_{\tau^*} = X(\vartheta, \tau^*)w[\tau^*]$ is generated from the initial position $\{\tau_*, w_*\}$ by the pair of realizations of actions $u^0[\tau]$, $v[\tau]$, $\tau_* \leq \tau < \tau^*$.

Now using (15.48), (15.49) we obtain the required condition (15.5).

Thus the property of *u-stability* $3^0_{*(1)}u$ of the program extremum $e_{*(1)}(\tau_*, w_*, \Delta)$ is completely proved.

If the condition of the convexity of the set $\{f(t,u,v) : u \in P\}$ is not valid, then instead of the motion of w-model (12.1) we use the w-model (15.9). This enables us to apply again the fixed-point theorem (see p.). Now this theorem is to be used not for the pair $\{m^*_{\tau^*}, u[\tau_*]\}$ (see (15.25)), but for the pair $\{m^*_{\tau^*}, w[\tau^*]\}$. Indeed, the set of all possible vectors $w[\tau^*]$ for the w-model (15.9) for the given actions $v[\tau] = v[\tau_*]$, $\tau_* \leq \tau < \tau^*$ is again convex.

16. Condition of v-stability for the program extremum $e_{*(1)}(\cdot)$

Let us show now that the quantity $e_{*(1)}(\tau_*, w_*, \Delta)$ (13.5), (15.10) satisfies also the following condition of *v-stability*.

$4^0_{*(1)}v$. Suppose there are given a domain G^* (see (7.5)), a position $\{\tau_*, w_*\} \in G^*$, a value $\varepsilon > 0$. Then there exists a value $\delta(\varepsilon) > 0$ such that the following conditions are valid. It is important to remark that for a given bounded domain G^* the value $\delta(\varepsilon) > 0$ can be chosen independent of the initial position $\{\tau_*, w_*\} \in G^*$. Let us choose a partition $\Delta_* = \Delta_\delta = \Delta_\delta\{\tau_i\}$ (12.2), $\tau_{j+1} - \tau_j < \delta$, where $\delta \leq \delta(\varepsilon)$, moments $\tau_* \in \Delta_\delta$ and $\tau^* = \tau_2 \in \Delta_\delta$ and control actions $u[\tau_*[\cdot]\tau^*) = \{u[\tau] = u[\tau_*] \in P,\ \tau_* \leq \tau < \tau^*\}$. Then there exist disturbances $v[\tau_*[\cdot]\tau^*) = \{v[\tau] = v[\tau_*] \in Q,\ \tau_* \leq \tau < \tau^*\}$ which together with $u[\tau_*[\cdot]\tau^*)$ generate from the initial position $\{\tau_*, w_*\}$ a motion $w[\tau_*[\cdot]\tau^*]$ of w-model (12.1) that reaches the state $w[\tau^*] = w^*$ such that

$$e_{*(1)}(\tau^*, w^*, \Delta^*) \geq e_{*(1)}(\tau_*, w_*, \Delta_*) - \varepsilon(\tau^* - \tau_*) \tag{16.1}$$

in the case (15.3), i.e., $\tau_* < \tau^* \leq t^{[g]}$, or

$$\mu^{[g]}(D^{[g]}w[t^{[g]}]) + e_{*(1)}(\tau^*, w^*, \Delta^*) \geq$$

$$\geq e_{*(1)}(\tau_*, w_*, \Delta_*) - \varepsilon(\tau^* - \tau_*) \tag{16.2}$$

in the case (15.6), i.e., $\tau_* = t^{[g]}$.

Here Δ_* and Δ^* are the partitions (15.7) and (15.8); g is the number (12.11); $\mu^{[g]}(\cdot)$ is the norm in (11.9).

As we can see, the property of *v-stability* $4^0_{*(1)}v$ is formulated similarly to the property of *u-stability* $3^0_{*(1)}u$ in Section 15. The difference is that we change in $3^0_{*(1)}u$ mutually the symbols u and v and substitute in (15.3), (15.5) the symbols $\leq$ and $+$ by the symbols $\geq$ and $-$, respectively.

It should be noted also that condition $4^0_{*(1)}v$ is a convenient variant of the general condition 4^0_u from Section 8 for the particular case of the positional functional $\gamma_{*(1)}$ (11.9) and the corresponding value $e_{*(1)}(\tau_*, w_*, \Delta)$ considered in this chapter.

We can prove the property $4^0_{*(1)}v$ by making only a few changes to $3^0_{*(1)}u$ in Section 15.

Proof. At first we consider the case

$$\tau_* < \tau^* < t^{[g]} \tag{16.3}$$

where g is the number (12.11). In this case the values $e_{*(1)}(\tau_*, w_*, \Delta_*)$ and $e_{*(1)}(\tau^*, w^*, \Delta^*)$ have the forms (15.12) and (15.13), respectively. In (15.12), (15.13) the vectors $m^*_{\tau_*}$, $m^{*0}_{\tau_*}$, $m^*_{\tau^*}$, $m^{*0}_{\tau^*}$ satisfy the conditions (15.14), (15.15). Then we consider the difference $\Delta e_{*(1)} = e_{*(1)}(\tau^*, w^*, \Delta^*) - e_{*(1)}(\tau_*, w_*, \Delta_*)$ using the expressions for $e_{*(1)}(\tau^*, \cdot)$ (15.13) and $e_{*(1)}(\tau_*, \cdot)$ (15.12). This difference again has the form (15.16). (In this section we use some expressions from Section 15 because they are similar to the ones we need here in the corresponding cases.)

However, now we substitute in (15.16) not the vector $m^{*0}_{\tau_*}$ by $m^{*0}_{\tau^*}$, but conversely we substitute the vector $m^{*0}_{\tau^*}$ in the expression (15.13) by the vector $m^{*0}_{\tau_*}$ from (15.12). Then we obtain the following estimation:

$$\Delta e_{*(1)} \geq m^{*0T}_{\tau_*}(\widetilde{w}_{\tau^*} - \widetilde{w}_{\tau_*}) + \varphi_1(\tau^*, m^{*0}_{\tau_*}) - \varphi_1(\tau_*, m^{*0}_{\tau_*}) \tag{16.4}$$

where $m^{*0}_{\tau_*} \in G^{(g)}$. Here $G^{(g)}$ is the set (13.35).

Now taking into account the constructions from Section 13 (see (13.33)–(13.42)) we come to the following equalities:

$$\varphi_1(\tau_*, m^*) = \overline{\psi}_1(\tau_*, m^*), \qquad m^* \in G^{(g)} \tag{16.5}$$

where

$$\psi_1(\tau_*, m^*) = \Delta\psi_1(\tau_*, m^*) + \varphi_2(\tau_*, m^*)\,. \tag{16.6}$$

Besides that, the function $\varphi(\tau_*, m^*)$ (16.5), which is the upper convex hull (see Section 13), has the property 4, which is described by the conditions (13.14), (13.15). Therefore, this function due to the Carathéodory theorem can be represented in the form

$$\varphi_1(\tau_*, m^*) = \sum_{s=1}^{n+1} \mu(m^{(s)} \mid m^*) \psi_1(\tau_*, m^{(s)}) =$$

$$= \sum_{s=1}^{n+1} \mu(m^{(s)} \mid m^*)(\Delta\psi_1(\tau_*, m^{(s)}) + \varphi_2(\tau_*, m^{(s)})) \qquad (16.7)$$

where

$$m^* = \sum_{s=1}^{n+1} \mu(m^{(s)} \mid m^*) m^{(s)}, \quad m^{(s)} = m^{(s)}(m^*) \in G^{(g)}. \qquad (16.8)$$

Let us introduce the quantity

$$\widetilde{e}_{*(1)}(\tau_*, w_*, \Delta_*, m^*) = \sum_{s=1}^{n+1} \mu(m^{(s)} \mid m^*) \Big[m^{(s)T} \widetilde{w}_{\tau_*} +$$

$$+ \Delta\psi_1(\tau_*, m^{(s)}) + \varphi_2(\tau_*, m^{(s)}) \Big]. \qquad (16.9)$$

One can see that the equality

$$e_{*(1)}(\tau_*, w_*, \Delta_*) = \max_{m^* \in G^{(g)}} \widetilde{e}_{*(1)}(\tau_*, w_*, \Delta_*, m^*) \qquad (16.10)$$

is valid.

Let $m^{*0}_{\tau_*}$ be the maximizing vector for (16.10), i.e.,

$$e_{*(1)}(\tau_*, w_*, \Delta_*) = \widetilde{e}_{*(1)}(\tau_*, w_*, \Delta_*, m^{*0}_{\tau_*}) =$$

$$= \sum_{s=1}^{n+1} \mu(m^{0(s)} \mid m^{*0}_{\tau_*}) \Big[\, m^{0(s)T} \widetilde{w}_{\tau_*} + \Delta\psi_1(\tau_*, m^{0(s)}) + \varphi_2(\tau_*, m^{0(s)}) \,\Big] \qquad (16.11)$$

where $m^{0(s)} = m^{(s)}(m^{*0}_{\tau_*})$.

Let $s = s^0$ be a number for which the sum

$$m^{0(s^0)T} \widetilde{w}_{\tau_*} + \Delta\psi_1(\tau_*, m^{0(s^0)}) + \varphi_2(\tau_*, m^{0(s^0)}) \qquad (16.12)$$

is maximal. Then for the vector $m^{0(s^0)}$ we obtain the following condition:

$$m^{0(s^0)T} \widetilde{w}_{\tau_*} + \Delta\psi_1(\tau_*, m^{0(s^0)}) + \varphi_2(\tau_*, m^{0(s^0)}) \geq$$

$$\geq \widetilde{e}_{*(1)}(\tau_*, w_*, \Delta_*, m^{*0}_{\tau_*}) = e_{*(1)}(\tau_*, w_*, \Delta_*)\,. \qquad (16.13)$$

According to (16.9), (16.10) in (16.13) equality is attained. So the vector $m^{0(s^0)}$ is the maximizing vector in (16.10) and we may assume

$$m^{0(s^0)} = m^{*0}_{\tau_*}\,. \qquad (16.14)$$

Thus, (16.5)–(16.14) imply the equality

$$e_{*(1)}(\tau_*, w_*, \Delta_*) = m^{*0T}_{\tau_*} \widetilde{w}_{\tau_*} + \Delta\psi_1(\tau_*, m^{*0T}_{\tau_*}) + \varphi_2(\tau_*, m^{*0T}_{\tau_*}), \qquad (16.15)$$

i.e., the following equality

$$\varphi_1(\tau_*, m^{*0}_{\tau_*}) = \Delta\psi_1(\tau_*, m^{*0}_{\tau_*}) + \varphi_2(\tau_*, m^{*0}_{\tau_*}) \qquad (16.16)$$

holds.

From the equality $\varphi_2(\tau_*, m^*) = \varphi_1(\tau^*, m^*)$ (see Section 13) and from (16.4), (16.16) we derive the following estimation:

$$\Delta e_{*(1)} \geq m^{*0T}_{\tau_*} (\widetilde{w}_{\tau^*} - \widetilde{w}_{\tau_*}) + \Delta\psi_1(\tau_*, m^{*0}_{\tau_*}). \qquad (16.17)$$

Again as in Section 15, using the Cauchy formula (15.18) and the expression (15.36) for $\Delta\psi(\tau_*, m^{*0}_{\tau_*})$, we come to the following statement:

For any $\varepsilon > 0$ there exists a number $\delta_\Delta(\varepsilon) > 0$ such that the inequality

$$\Delta e_{*(1)} \geq (\tau^* - \tau_*)\Big[\, m^{*0T}_{\tau_*}\widetilde{f}(\tau_*, u[\tau_*], v[\tau_*]) -$$

$$- \max_{v\in Q}\min_{u\in P}\; m^{*0T}_{\tau_*}\widetilde{f}(\tau_*, u, v)\,\Big] - \varepsilon(\tau^* - \tau_*) \tag{16.18}$$

is valid for $\tau^* - \tau_* \leq \delta_\Delta(\varepsilon)$.

Taking into account (10.2) we can write the equality

$$\max_{v\in Q}\min_{u\in P}\; m^{*0T}_{\tau_*}\widetilde{f}(\tau_*, u, v) \leq \max_{v\in Q}\; m^{*0T}_{\tau_*}\widetilde{f}(\tau_*, u[\tau_*], v) \tag{16.19}$$

for every $u[\tau_*] \in P$.

Now we choose a vector $v[\tau_*] \in Q$ which satisfies the condition

$$m^{*0T}_{\tau_*}\widetilde{f}(\tau_*, u[\tau_*], v[\tau_*]) = \max_{v\in Q}\; m^{*0T}_{\tau_*}\widetilde{f}(\tau_*, u[\tau_*], v)\,. \tag{16.20}$$

According to (16.18)–(16.20) the condition (16.1) is valid. This proves the property $4^0_{*(1)}v$ in the case (16.3).

Let us consider the case

$$\tau^* = t^{[g]} \tag{16.21}$$

and prove that the condition (16.1) is valid again.

In the case (16.21), according to the constructions in Section 13, for the value $e_{*(1)}(\tau^*, w^*, \Delta^*)$ we use the equality (15.26). Then the difference $\Delta e_{*(1)}$ takes the form (15.30).

Substituting in (15.30) the vector $m^{*0}_{\tau^*} \in G^{(g)}$ by the vector $m^{*0}_{\tau_*} \in G^{(g)}$ and the vector $m^0_{*\tau^*}$ by an arbitrary vector $m^*_{*\tau_*} \in G^{(g+1)}$ we obtain the following inequality:

$$\Delta e_{*(1)} \geq \Big[\, m^{*0T}_{\tau_*}\widetilde{w}_{\tau^*} + \varphi_1(\tau^*, m^*_{*\tau_*})\,\Big] - \Big[\, m^{*0T}_{\tau_*}\widetilde{w}_{\tau_*} + \varphi_1(\tau_*, m^{*0}_{\tau_*})\,\Big] \tag{16.22}$$

where the vector $m^*_{*\tau_*}$ satisfies the conditions

$$m^*_{*\tau_*} \in G^{(g+1)}, \qquad m^{*0}_{\tau_*} = m^*_{*\tau_*} + \widehat{m} \tag{16.23}$$

and $\widehat{m}$ is the vector (15.28).

According to the constructions in Section 13 of the quantity $\varphi_1(\tau_*, m)$ we have the following condition:

$$\varphi_1(\tau_*, m^*) = \overline{\psi}_1(\tau_*, m^*), \qquad m^* \in G^{(g)} \tag{16.24}$$

where

$$\psi_1(\tau_*, m^*) = \max_{m_* \in G^{(g+1)}} [\, \Delta\psi_1(\tau_*, m^*) + \varphi_2(\tau_*, m_*)\,] \tag{16.25}$$

and $m^* = m_* + \widehat{m}$, $m_* \in G^{(g+1)}$, the vector $\widehat{m}$ is given by (15.28).

Let for a fixed $m^* \in G^{(g)}$ the vectors $m_*(m^*)$ and $\widehat{m}(m^*) = X^T(t^{[g]}, \vartheta)l$, $\mu^{[g]*}(l) \leq 1$ be the maximizing ones for the maximum problem (16.25). Then due to the property 4 for the function $\varphi(\tau, m)$ (see Section 13, in particular, conditions (13.14), (13.15)) we have the following equality:

$$\varphi_1(\tau_*, m^*) = \sum_{s=1}^{n+1} \mu(m^{(s)} \mid m^*)\, \psi_1(\tau_*, m^{(s)}) =$$

$$= \sum_{s=1}^{n+1} \mu(m^{(s)} \mid m^*)[\, \Delta\psi_1(\tau_*, m^{(s)}) + \varphi_2(\tau_*, m_*(m^{(s)}))\,], \qquad m^{(s)} = m^{(s)}(m^*)\,. \tag{16.26}$$

Here the vector m^* satisfies the equality (16.8). Let $m^{*0}_{\tau_*} \in G^{(g)}$ be the maximizing vector for $e_{*(1)}(\tau_*, w_*, \Delta_*)$, i.e.,

$$e_{*(1)}(\tau_*, w_*, \Delta_*) = \sum_{s=1}^{n+1} \mu(m^{0(s)} \mid m^{*0}_{\tau_*})[\, m^{0(s)T} \widetilde{w}_{\tau_*} +$$

$$+\Delta\psi_1(\tau_*, m^{0(s)}) + \varphi_2(\tau_*, m_*(m^{0(s)}))\ ], \qquad m^{0(s)} = m^{(s)}(m^{*0}_{\tau_*}). \tag{16.27}$$

Assume s^0 is a number that satisfies the condition

$$m^{0(s^0)T}\widetilde{w}_{\tau_*} + \Delta\psi_1(\tau_*, m^{0(s^0)}) + \varphi_2(\tau_*, m_*(m^{0(s^0)})) =$$

$$= \max_s \left[\ m^{0(s)T}\widetilde{w}_{\tau_*} + \Delta\psi_1(\tau_*, m^{0(s)}) + \varphi_2(\tau_*, m_*(m^{0(s)}))\ \right]. \tag{16.28}$$

Then we conclude that the following conditions

$$m^{0(s^0)T}\widetilde{w}_{\tau_*} + \Delta\psi_1(\tau_*, m^{0(s^0)}) + \varphi_2(\tau_*, m_*(m^{0(s^0)})) \geq$$

$$\geq \widetilde{e}_{*(1)}(\tau_*, w_*, \Delta_*, m^{*0}_{\tau_*}) = e_{*(1)}(\tau_*, w_*, \Delta_*) \tag{16.29}$$

are valid, i.e., we may assume that the equality

$$m^{*0}_{\tau_*} = m^{0(s^0)} \tag{16.30}$$

holds. Then (16.22) takes the form

$$\Delta e_{*(1)} \geq m^{*0T}_{\tau_*}\widetilde{w}_{\tau^*} + \varphi_1(\tau^*, m_*(m^{*0}_{\tau_*})) -$$

$$- m^{*0T}_{\tau_*}\widetilde{w}_{\tau_*} - \Delta\psi_1(\tau_*, m^{*0}_{\tau_*}) - \varphi_2(\tau_*, m_*(m^{*0}_{\tau_*})). \tag{16.31}$$

Now according to the equality

$$\varphi_2(\tau_*, m_*) = \varphi_1(\tau^*, m_*) \tag{16.32}$$

which follows from the constructions of the functions $\varphi_j(\tau_*, \cdot)$ and $\varphi_j(\tau^*, \cdot)$ (see Section 13) we obtain the estimation (16.17).

Further, using (16.18)–(16.20) we prove the condition (16.1) in the case (16.21).

The condition $4^0_{*(1)}v$ is proved in the case $\tau_* < \tau^* \leq t^{[g]}$.

Now we consider the case

$$\tau_* = t^{[g]} \tag{16.33}$$

and shall prove the condition (16.2).

In the case (16.33) the value $e_{*(1)}(\tau_*, w_*, \Delta_*)$ has the form (15.41).

We can again assume that the inequality $\tau^* < t^{[g+1]}$ holds.

In this case the value $e_{*(1)}(\tau^*, w^*, \Delta^*)$ has the form (15.13) and for the corresponding difference $\Delta e_{*(1)}$ the equality (15.43) holds. Let us represent the vector $m^{*0}_{\tau_*}$ as the sum

$$m^{*0}_{\tau_*} = m^0_{*\tau_*} + \widehat{m}^0_{\tau_*} \tag{16.34}$$

where

$$\widehat{m}^0_{\tau_*} = X^T(t^{[g]}, \vartheta) D^{[g]T} l, \qquad \mu^{[g]*}(l) \leq 1. \tag{16.35}$$

Then the inequality

$$\Delta e_{*(1)} \geq m^{0T}_{*\tau_*}(\ \widetilde{w}_{\tau^*} - \widetilde{w}_{\tau_*}\ ) - \widehat{m}^{0T}_{\tau_*} \widetilde{w}_{\tau_*} +$$

$$+ \varphi_1(\tau^*, m^0_{*\tau_*}) - \varphi_1(\tau_*, m^0_{*\tau_*}) \tag{16.36}$$

is valid.

Now we proceed as above in the case (16.3) and taking into account the equality

$$\widehat{m}^{0T}_{\tau_*} \widetilde{w}_{\tau_*} = \mu^{[g]}(D^{[g]} w_*) \tag{16.37}$$

(see the corresponding expressions (15.46), (15.47) in Section 15) we obtain the estimation

$$\Delta e_{*(1)} \geq m^{0T}_{*\tau_*}(\ \widetilde{w}_{\tau^*} - \widetilde{w}_{\tau_*}\ ) - \Delta\psi_1(m^0_{*\tau_*}) - \mu^{[g]}(D^{[g]} w_*) \tag{16.38}$$

where $w_* = w[\tau_*] = w[t^{[g]}]$. Further, using (16.18)–(16.20) we come to the required inequality (16.2).

Thus the property of *v-stability* $4^0_{*(1)}v$ for the function $e_{*(1)}(\tau_*, w_*, \Delta)$ is completely proved.

Let us remark again that for a given bounded domain G^* one can choose the value $\delta(\varepsilon) > 0$ independent of the initial position $\{t_*, w_*\} \in G^*$.

17. Approximating solution of the differential game for $\gamma_{(1)}$

It will be shown in this section that the quantity $e_{*(1)}(\tau_*, w_*, \Delta)$ (13.5), (15.10), which was constructed above in Sections 12 and 13 for the functional $\gamma_{*(1)}$ (11.9), approximates the value $\rho^0_{*(1)}(t,x)$ of the differential game for the system (10.1), (10.2) with the quality index $\gamma_{*(1)}$ (11.9). Moreover, it will be shown that this value $e_{*(1)}(\tau_*, w_*, \Delta)$ also approximates the value $\rho^0_{(1)}(t,x)$ of the primary differential game for the system (10.1), (10.2) with the quality index $\gamma_{(1)}$ (10.3).

Let us show at first that the quantity $e_{*(1)}(\tau_*, w_*, \Delta)$ approximates the value of the game $\rho^0_{*(1)}(t,x)$. We have the following:

Assertion 17.1. *For any number $\eta > 0$ there exists a number $\delta(\eta) > 0$ such that for any initial position $\{t_*, w_*\} \in G^*$ the inequality*

$$\rho^0_{*(1)}(t_*, w_*) \le e_{*(1)}(t_*, w_*, \Delta) + \eta \tag{17.1}$$

is valid for the partition $\Delta = \Delta_\delta\{\tau_i\}$ (12.2) that satisfies the condition

$$\delta_\Delta = \max_i\, (\tau_{i+1} - \tau_i) \le \delta(\eta). \tag{17.2}$$

Proof. The quantity $e_{*(1)}(\tau_*, w_*, \Delta_\delta)$ constructed in Sections 12 and 13 satisfies the conditions (15.1), (15.2) and $3^0_{*(1)}u$ (Section 15). One can check that these conditions 1^0, 2^0 and $3^0_{*(1)}u$ imply that the quantity $e_{*(1)}(\cdot)$ satisfies the conditions 1^0, 2^0 and 3^0u in Section 8 for the case of the functional $\gamma_{*(1)}$ (11.9), provided the instants τ_* and τ^* in 3^0u coincide with some instants $\tau_j \in \Delta_\delta$ from $e_{*(1)}(\cdot, \Delta_\delta)$ and $\delta > 0$ is sufficiently small. However, we give here the proof of Assertion 17.1 that does not use this connection between $3^0_{*(1)}u$ and 3^0u. The quality index $\gamma_{*(1)}$ (11.9) is *positional* (4.9), (11.12). So, according to Theorem 9.2, there *exist* the value $\rho^0_{*(1)}(t,x)$ and the saddle point $\{u^0_{*(1)}(\cdot) = u^0_{*(1)}(t,x,\varepsilon),\ v^0_{*(1)}(\cdot) = v^0_{*(1)}(t,x,\varepsilon)\}$ of the game with functional $\gamma_{*(1)}$.

Let us form the motion

$$w_{V^0_{*\eta},u}[t_*[\cdot]\vartheta] = \left\{\ w_{V^0_{*\eta},u}[t],\quad t_* \le t \le \vartheta,\quad w_{V^0_{*\eta},u}[t_*] = w_*\ \right\} \tag{17.3}$$

of w-model (12.1), generated from the initial position $\{t_*, w_*\} \in G^*$ by the control law

$$V^0_{*\eta} = \left\{\ v^0_{*(1)}(\cdot);\ \varepsilon_{v_\eta};\ \Delta_{\delta_\eta}\{t_i\}\ \right\} \tag{17.4}$$

which due to the *definition* (see (6.13)–(6.15)) of the value of the game $\rho^0_{*(1)}(t,x)$

guarantees the inequality

$$\gamma_{*(1)}(\ w_{V^0_{*\eta},u}[t_*[\cdot]\vartheta]\ ) \geq \rho^0_{*(1)}(t_*, w_*) - \frac{\eta}{2} \tag{17.5}$$

for any $\eta > 0$ and for every admissible control actions $u[t_*[\cdot]\vartheta)$ provided

$$\varepsilon_{v_\eta} \leq \varepsilon(\eta), \qquad \delta_\eta \leq \delta(\varepsilon_{v_\eta}, \eta),$$

$$\Delta_{\delta_\eta}\{t_i\}:\ \max_i\ (t_{i+1} - t_i) \leq \delta_\eta\,. \tag{17.6}$$

Let in (17.4) the partition $\Delta_{\delta_\eta}\{t_i\}$ (5.14), (17.6) coincide with the partition $\Delta_{\delta_\eta}\{\tau_j\}$ (12.2) such that $t_i = \tau_j$, $i = j$, $i = 1, \ldots, k_\eta$ and $\delta_\eta \leq \delta(\varepsilon)$, where $\delta(\varepsilon)$ is the quantity in $3^0_{*(1)}u$. Let for the motion $w_{V^0_{*\eta},u}[\cdot]$ (17.3) the control actions $u[t_*[\cdot]\vartheta)$ be formed in the following way. We take

$$u[t_*[\cdot]\vartheta) = \{u[t_i[\cdot]t_{i+1}),\ \ i = 1, \ldots, k_\eta\} \tag{17.7}$$

where

$$u[t_i[\cdot]t_{i+1}) = \{u[t] = u[t_i] \in P,\ \ t_i \leq t < t_{i+1}\} \tag{17.8}$$

are the control actions, which on every time step $t_i \leq t < t_{i+1}$ ($t_i \in \Delta_{\delta_\eta}\{t_i\}$ (17.4)) are formed according to the condition $3^0_{*(1)}u$ of *u-stability* (see Section 15). Here in $3^0_{*(1)}u$ we assume that $\tau_* = t_i$, $\tau^* = t_{i+1}$ and $v[\tau_*[\cdot]\tau^*) = v^0[t_i[\cdot]t_{i+1})$, where $v^0[t_i[\cdot]t_{i+1})$ are the disturbances generated by the law $V^0_{*\eta}$ (17.4).

Then, proceeding as in Chapter I (see Section 9) we prove the inequality

$$\gamma_{*(1)}(\ w_{V^0_{*\eta},u}[t_*[\cdot]\vartheta]\ ) \leq e_{*(1)}(t_*, w_*, \Delta) + \varepsilon \sum_{j=1}^{k}(\tau_{j+1} - \tau_j) \leq$$

$$\leq e_{*(1)}(t_*, w_*, \Delta) + \frac{\eta}{2} \tag{17.9}$$

for every given beforehand $\eta > 0$ provided $\delta_\Delta \le \delta_\eta$, $\varepsilon \le \varepsilon(\eta)$ and

$$\sum_{j=1}^{k(\eta)} (\tau_{j+1}^{(\eta)} - \tau_j^{(\eta)})\varepsilon(\eta) < \frac{\eta}{2}. \tag{17.10}$$

Relations (17.5)–(17.10) imply the required inequality (17.1). That proves the assertion 17.1.

The following assertion also holds.

Assertion 17.2. *For any number $\eta > 0$ there exists a number $\delta(\eta) > 0$ such that for any initial position $\{t_*, w_*\} \in G^*$ the inequality*

$$\rho^0_{*(1)}(t_*, w_*) \ge e_{*(1)}(t_*, w_*, \Delta) - \eta \tag{17.11}$$

is valid if the partition Δ satisfies the condition (17.2).

Assertion 17.2 follows from the existence of the optimal strategy $u^0_{*(1)}(\cdot) = u^0_{*(1)}(t, x, \varepsilon)$ and from the property of *v-stability* $4^0_{*(1)}v$ (Section 16) for the function $e_{*(1)}(\cdot)$. Here we omit the proof of Assertion 17.2 since it is similar to the proof of Assertion 17.1 with some inessential changes.

It follows from Assertions 17.1 and 17.2 that for any number $\eta > 0$ there exists a number $\delta(\eta) > 0$ such that for any initial position $\{t_*, w_*\} \in G^*$ the inequality

$$| \rho^0_{*(1)}(t_*, w_*) - e_{*(1)}(t_*, w_*, \Delta) | \le \eta \tag{17.12}$$

is valid under the conditions (17.2).

Thus, the quantity $e_{*(1)}(\tau_*, w_*, \Delta)$ (13.5), (13.6), (15.10) approximates the value $\rho^0_{*(1)}(t, x)$ of the differential game for the system (10.1), (10.2) with the quality index $\gamma_{*(1)}$ (11.9).

Besides that, it was established in Section 11 (see Lemma 11.1) that the value of the game $\rho^0_{*(1)}(t, x)$ approximates the value of the game $\rho^0_{(1)}(t, x)$. So, for any initial position $\{t_*, x_*\} \in G$ the inequality (11.15) holds under the conditions (11.16). Hence, the quantity $e_{*(1)}(t_*, x_*, \Delta)$ approximates the value $\rho^0_{(1)}(t_*, x_*)$ of the original differential game for the system (10.1), (10.2) with $\gamma_{(1)}$ (10.3)). Therefore the following theorem holds.

Theorem 17.1. *For any number $\eta_* > 0$ there exists a number $\delta(\eta_*) > 0$ such that for any initial position $\{t_*, x_*\} \in G$ (3.12) the inequality*

$$| \rho^0_{(1)}(t_*, x_*) - e_{*(1)}(t_*, x_*, \Delta) | \le \eta_* \tag{17.13}$$

is valid if the partition $\Delta = \Delta\{t_i\}$ (12.2) *satisfies the conditions* (17.2) *and* (11.16) *with suitable* ζ, η.

Here $e_{*(1)}(t_*, x_*, \Delta)$ is the value of program extremum (13.5), (13.10), (13.36), (13.38), (15.10) for the position $\{\tau_*, w_*\} = \{t_*, x_*\}$.

18. Construction of optimal approximating strategies

In this section it will be shown that the function $e_{*(1)}(\tau_*, w_*, \Delta)$, which is *u-stable* (Section 15) and *v-stable* (Section 16) and satisfies the conditions 1^0 and 2^0 in Section 8 (see (15.1), (15.2)), provides the basis for construction of some approximations $u^a_{*(1)}(\cdot)$ and $v^a_{*(1)}(\cdot)$ to the optimal strategies $u^0_{*(1)}(\cdot) = u^0_{*(1)}(t, x, \varepsilon)$ and $v^0_{*(1)}(\cdot) = v^0_{*(1)}(t, x, \varepsilon)$ in the differential game with $\gamma_{*(1)}$ (11.9).

According to the procedure in Sections 8 and 9 the desired *optimal* strategies $u^0_{(1)}(\cdot) = u^0_{(1)}(t, x, \varepsilon)$ and $v^0_{(1)}(\cdot) = v^0_{(1)}(t, x, \varepsilon)$, which form the saddle point $\{u^0_{(1)}(\cdot), v^0_{(1)}(\cdot)\}$ in the differential game for the system (10.1), (10.2) with the quality index $\gamma_{(1)}$ (10.3), can be formed as the *extremal* control vectors $u_e(t, x, \varepsilon) \in P$ and disturbances $v_e(t, x, \varepsilon) \in Q$ determined by the conditions (8.11), (8.12) and (8.16), (8.17), respectively. Here the *accompanying points* $w^{[u]0}(t, x, \varepsilon_u)$ and $w^{[v]0}(t, x, \varepsilon_v)$ satisfy the conditions (8.10), (8.15) with respect to function $\rho(t, w) = \rho^0(t, w)$, which is *u-stable* and *v-stable*, i.e., satisfies the conditions 3^0_u and 4^0_v and also the conditions 1^0 and 2^0 in Section 8.

As was shown in Sections 15 to 17 we can approximate the values $\rho^0_{*(1)}(t, x)$ and $\rho^0_{(1)}(t, x)$ by the function $e_{*(1)}(t, x, \Delta)$ (13.5), (13.10), (15.10) which is *u*-stable (Section 5) and *v*-stable (Section 16) and satisfies the conditions 1^0 (15.1) and 2^0 (15.2). This enables us to construct some approximating procedure for optimal control. This procedure is also based on the method of extremal shift described in Section 8. But here the function $e_{*(1)}(t, x, \Delta)$ is used instead of the function $\rho^0_{*(1)}(t, x)$. Thus, we obtain control laws U^0_a and V^0_a that approximate the control laws generated by the approximating optimal strategies $u^a_{*(1)}(\cdot)$, $v^a_{*(1)}(\cdot)$ for the original differential game with the quality index $\gamma_{*(1)}$. The actions $u^a[t]$ and $v^a[t]$ that are provided by the approximating laws U^0_a and V^0_a will be called, for simplicity, *approximating optimal actions* (based on the function $e_{*(1)}(\cdot)$).

The constructions of these actions are specified here in the following concrete form.

First, let us consider the optimal approximating actions $u^e_{*(1)}[\cdot]$. Let the initial position $\{t_*, x_*\}$ be given and parameter $\varepsilon_u > 0$ and partition $\Delta\{\tau_i\}$ for the interval $[\tau_*, \vartheta]$ be chosen.

According to the method of extremal shift we use again the accompanying point $w^{[u]0}(t, x, \varepsilon_u)$ as in Section 8. But we transform here the conditions (8.7)–(8.10), which describe this point as follows

$$e_{*(1)}(t, w^{[u]0}(t, x, \varepsilon_u), \Delta) =$$

$$= \min_{\{t,w\}\in K^{[u]}(\varepsilon_u,t,x)} e_{*(1)}(t,w,\Delta) =$$

$$= \min_{\{t,w\}\in K^{[u]}(\varepsilon_u,t,x)} \max_{m^*,m_*} [\, \langle\, m^*,\ X(\vartheta,t)w\, \rangle + \varphi_1(t,m_*)\,] \tag{18.1}$$

where m^* and m_* satisfy the conditions (13.36)–(13.42) for $\tau_* = t = \tau_j \in \Delta$.

According to (8.7), (8.8) we can write the condition $w \in K_t^{[u]}(\varepsilon_u, t, x)$ for the vectors $s = x - w$ in the form

$$|\, s\, |=|\, x - w\, |\leq R^{[u]}(\varepsilon_u, t) \tag{18.2}$$

where $R^{[u]}(\varepsilon_u, t)$ is the number (8.8). Now in (8.8) for the system (10.1), (10.2) we have

$$\lambda = \sup_{t_0\leq t\leq\vartheta} |\, A(t)\, | \tag{18.3}$$

where $|\, A(t)\, |= \max_{|x|\leq 1} |\, A(t)x\, |$.

Besides, the concavity of the function $\varphi_1(\cdot, m_*)$ in m_* makes it possible to transpose the minimum and maximum operations in (18.1). So we have the following equalities:

$$e_{*(1)}(t, w^{[u]0}(t,x,\varepsilon_u),\Delta) =$$

$$= \max_{m^*,m_*} \min_{|s|\leq R^{[u]}(\varepsilon_u,t)} \Big[\, \langle\, m^*,\ X(\vartheta,t)x\, \rangle -$$

$$- \langle\, m^*,\ X(\vartheta,t)s\rangle + \varphi_1(t,m_*)\, \Big]. \tag{18.4}$$

Equality (18.4) allows us to obtain a sufficiently constructive expression for the optimal control action $u^a_{*(1)}(t_i, x[t_i], \varepsilon)$. At first we describe the corresponding calculations omitting some special cases. These peculiar cases we discuss after consideration of the scheme outlining the ordinary circumstances.

Let us introduce the minimizing vector $s^{[u]a}(t, m^*, \varepsilon_u)$ which solves the problem (18.4). In accordance with (18.1)–(18.4) we assume

$$s^{[u]a}(t,x,\varepsilon_u) = x - w^{[u]0}(t,x,\varepsilon_u) = s^{[u]a}(t, m_u^{*0}, \varepsilon_u) \tag{18.5}$$

where m_u^{*0} is the solution of the problem (18.4).

It follows from (18.2)–(18.5) that the minimizing vector $s^{[u]a}(t,x,\varepsilon_u)$ (18.5) in (18.4) is given by the equality

$$s^{[u]a}(t,x,\varepsilon_u) = R^{[u]}(\varepsilon_u,t)\frac{X^T(\vartheta,t)m_u^{*0}}{|\, X^T(\vartheta,t)m_u^{*0}\, |} \tag{18.6}$$

where the vector m_u^{*0} is a solution of the following problem:

$$\langle\, m_u^{*0},\ X(\vartheta,t)x\,\rangle + \varphi_1(t,m_{*u}^0) - R^{[u]}(\varepsilon_u,t)\mid X^T(\vartheta,t)m_u^{*0}\mid =$$

$$= \max_{m^*,m_*}\Big[\,\langle\, m^*,\ X(\vartheta,t)x\,\rangle + \varphi_1(t,m_*) - R^{[u]}(\varepsilon_u,t)\mid X^T(\vartheta,t)m^*\mid\Big] \quad (18.7)$$

under the conditions (13.39)–(13.42) for $\tau_* = t$. If $\mid m_u^{*0}\mid= 0$, then we can take $s^{[u]a}(\cdot) = 0$.

Thus, the function $u^a_{*(1)}(t,x,\varepsilon_u)$ which gives the desired *approximating optimal actions* $u^a[t] = u^a_{*(1)}(t_i,x[t_i],\varepsilon_u),\ t_i \le t < t_{i+1},\ i = 1,\dots,k$, is determined by the condition

$$\max_{v\in Q}\Big[\,m_u^{*0T}X(\vartheta,t)f(t,u^a_{*(1)}(t,x,\varepsilon_u),v)\,\Big] =$$

$$= \min_{u\in P}\max_{v\in Q}\Big[\,m_u^{*0T}X(\vartheta,t)f(t,u,v)\,\Big] \quad (18.8)$$

where $f(t,u,v)$ is the function in (10.1), (10.2). We will call this function $u^a_{*(1)}(\cdot)$, for short, an *approximating optimal strategy* (that is based on the function $e_{*(1)}(\cdot)$).

Now we consider the special cases mentioned above. We used equality (18.4), which makes it possibile to transpose operations of max and min. If operation maxmin achieves these values on the unique pair $(s^a,m^0)_{\text{maxmin}}$ then

$$(s^a,m^0)_{\text{minmax}} = (s^a,m^0)_{\text{maxmin}}$$

and all relations (18.6)–(18.8) are valid. However, let us assume that the maxmin operation in (18.4) does not have the unique solution m_u^{*0}, s_u^{*a}, where for instance m_u^{*0} is not a unique solution of the problem (18.7). Then we cannot assert that every solution m^{*0} given by (18.7) provides the desired solution of the primary minmax problem. Only some solutions m^{*0} of (18.4) deliver the required pair (s_u^{*a},m_u^{*0}). It may occur if the concave function $\varphi_1(t,m)$ turns to be not strictly concave in m. In order to avoid this inconvenience we can slightly distort the expression for $e_{*1}(\cdot)$, (15.10). Namely, we can add in this expression the small value $(-\varepsilon_1\mid m\mid^2)$, $\varepsilon_1 > 0$. Then the maximized function in (18.7) will be strictly concave in m. This guaranties the unique solution m_u^{*0} of the problem (18.7). Consequently it gives the unique $s^{[u]0}$ according to (18.6) if $\mid m^{*0}\mid > 0$. If after this distortion we still have $\mid m^{*0}\mid = 0$ then we can distort the primary problem (18.4) in some other more appropriate way (see Sections 24 and 26). Similar circumstances we may meet below; however, as a rule we omit special consideration while keeping them in mind.

Let us now consider the construction of the *approximating optimal strategy* $v^a_{*(1)}(\cdot)$ based on $e_{*(1)}(\cdot)$. The construction of strategy $v^a_{*(1)}(\cdot) = v^a_{*(1)}(t, x, \varepsilon_v)$ is similar to the construction of $u^a_{*(1)}(\cdot)$ in the following way.

It should be noted that in the case of the disturbance $v^a_{*(1)}$ we do not need to deal with special cases similar to those in the case of $u^a_{*(1)}$ considered just above. Because here there is no minmax problem.

According to the method of extremal shift we use the accompanying point $w^{[v]0}(t, x, \varepsilon_v)$ described in Section 8. Now the conditions (8.13)–(8.15), which determine this point, take the form

$$e_{*(1)}(t, w^{[v]0}(t, x, \varepsilon_v), \Delta) = \max_{\{t,w\}\in K^{[v]}_t(\varepsilon_v,t,x)} e_{*(1)}(t, w, \Delta) =$$

$$= \max_{\{t,w\}\in K^{[v]}(\varepsilon_v,t,x)} \max_{m^*,m_*} \Big[\langle\, m^*,\ X(\vartheta, t)w \,\rangle + \varphi_1(t, m_*) \Big] \tag{18.9}$$

under the conditions (13.39), (13.40) where $\tau_* = t = t_i \in \Delta$.

Taking into account (8.13) we transform the inclusion $\{t, w\} \in K^{[v]}(\varepsilon_v, t, x)$ in (18.9) into the condition

$$|\, s \,| = |\, w - x \,| \le R^{[v]}(\varepsilon_v, t) \tag{18.10}$$

where

$$R^{[v]}(\varepsilon_v, t) = (\varepsilon_v + \varepsilon_v(t - t_0))^{1/2} \exp\{\lambda(t - t_0)\}\,. \tag{18.11}$$

Here λ is the number (18.3).

Due to (18.9)–(18.11) we have

$$e_{*(1)}(t, w^{[v]0}(t, x, \varepsilon_v), \Delta) =$$

$$= \max_{m^*,m_*} \max_{|s|\le R^{[v]}(\varepsilon_v,t)} \Big[\langle\, m^*,\ X(\vartheta, t)x \,\rangle + $$
$$+ \langle\, m^*,\ X(\vartheta, t)s \,\rangle + \varphi_1(t, m_*) \Big]. \tag{18.12}$$

Here we use also the equality $w = x + s$ that follows from (18.10).

The vector $s^{[v]a}(t, x, \varepsilon_v)$ which solves the problem (18.12) is determined by the equality

$$s^{[v]a}(t, x, \varepsilon_v) = R^{[v]}(\varepsilon_v, t)\frac{X^T(\vartheta, t)m_v^{*0}}{\mid X^T(\vartheta, t)m_v^{*0} \mid} \tag{18.13}$$

where m_v^{*0} is a solution of the following problem:

$$\langle\, m_v^{*0},\ X(\vartheta, t)x \,\rangle + \varphi_1(t, m_{*v}^0) + R^{[v]}(\varepsilon_v, t) \mid X^T(\vartheta, t)m_v^{*0} \mid =$$

$$= \max_{m^*, m_*} \Big[\langle\, m^*,\ X(\vartheta, t)x \,\rangle + \varphi_1(t, m_*) + R^{[v]}(\varepsilon_v, t) \mid X^T(\vartheta, t)m^* \mid \Big] \tag{18.14}$$

under the conditions (13.39), (13.42) for $\tau_* = t$. If $\mid m_v^{*0} \mid = 0$ we can take $s^{[v]a}(\cdot) = 0$.

Thus, the strategy $v^a_{*(1)}(t, x, \varepsilon_v)$ is determined by the condition

$$\min_{u \in P} \Big[m_v^{*0T} X(\vartheta, t) f(t, u, v^a_{*(1)}(t, x, \varepsilon_v)) \Big] = \max_{v \in Q} \min_{u \in P} \Big[m_v^{*0T} X(\vartheta, t) f(t, u, v) \Big]. \tag{18.15}$$

The following assertion holds.

Assertion 18.1. *The strategies* $u^a_{*(1)}(\cdot) = u^a_{*(1)}(t, x, \varepsilon)$ (18.8) *and* $v^a_{*(1)}(\cdot) = v^a_{*(1)}(t, x, \varepsilon)$ (18.15) *provide an approximating solution of the differential game for the system* (10.1), (10.2) *with the quality index* $\gamma_{*(1)}$ (11.9).

We omit the proof of this assertion because it is similar to the proofs of Theorems 9.1 and 9.2. The only difference is that here the properties of u-stability 3_{*u} and v-stability 4_{*v} are used instead of the properties of u-stability 3^0 and v-stability 4^0.

This concludes approximating of the solution of the problem in the case of the quality index $\gamma_{(1)}$ (11.9).

The results of this section imply the following. For every $\eta > 0$ we can name $\varepsilon(\eta) > 0$ and $\delta(\varepsilon, \eta) > 0$ such that the control laws $U^0_{*(1)}(u^a_{*(1)}(\cdot), \varepsilon, \Delta_\delta)$ and $V^0_{*(1)}(v^a_{*(1)}(\cdot), \varepsilon, \Delta_\delta)$ guarantee the inequalities

$$\gamma_{(1)} \leq \rho^0(t_*, x_*) + \eta,$$

$$\gamma_{(1)} \geq \rho^0(t_*, x_*) - \eta$$

respectively, provided the strategies $u^a_{*(1)}(\cdot)$ (18.8), $v^a_{*(1)}(\cdot)$ (18.15) are based on the value $e_{*(1)}(t, w, \Delta_\delta)$, where Δ_δ is the same as in $U^0_{*(1)}$, $V^0_{*(1)}$ and $\delta < \delta(\varepsilon, \eta)$.

We underline that the optimal approximating strategies $u^a_{*(1)}(\cdot)$ and $v^a_{*(1)}(\cdot)$ determine the actions $u^a[t] = u^a_{*(1)}(t_i, x[t_i], \varepsilon_u)$ or $v^a[t] = v^a_{*(1)}(t_i, x[t_i], \varepsilon_v)$, $t_i \leq t < t_{i+1}$ only for the moments t_i, $i = 1, \dots, k$ each of which coincides with one of the moments $\tau_j \in \Delta_\delta$ from $e_{*(1)}(\tau_*, w_*, \Delta_\delta)$.

19. Conditions of u-stability and v-stability for the program extremum $e_{*(2)}(\cdot)$

In the following two sections we shall consider the case of the positional functional $\gamma_{(2)}$ (10.4). Namely, we construct and justify an approximating solution of the differential game for the system (10.1), (10.2) with the quality index $\gamma_{(2)}$ (10.4). Our considerations here are similar to those given in Sections 15 to 18 for the functional $\gamma_{(1)}$ (10.3). Now we use the approximating functional $\gamma_{*(2)}$ (11.10) and the corresponding value $e_{*(2)}(\tau_*, w_*, \Delta)$ (14.25). Naturally, a different form of the quality index γ_* and the value e_* implies some changes.

The functional $\gamma_{*(2)}$ (11.10) is positional, i.e., it is represented in the form (4.9), (11.13). It follows from Theorem 9.2 that function $\rho_{*(2)}(t, w)$ is the value $\rho^0_{*(2)}(t, x)$ of the game with the quality index $\gamma_{*(2)}$ (11.10) if this function satisfies the conditions 1^0, 2^0, 3^0u and 4^0v in Section 8 for the functional $\gamma_{*(2)}$. Let us consider the function $e_{*(2)}(\tau_*, w_*, \Delta)$ constructed in Sections 12 and 14.

It is obvious that the function $e_{*(2)}(\tau_*, w_*, \Delta)$ constructed in Section 14 satisfies the conditions 1^0 and 2^0 in Section 8. In the case of functional $\gamma_{*(2)}$ the formulation of these conditions is similar to that of conditions 1^0 and 2^0 in Section 15. In the corresponding conditions (15.1) and (15.2) we have only to change the lower indices $*(1)$ by the indices $*(2)$.

Now we assume again that the vector function $f(t, u, v)$ in (10.1) besides the restrictions (3.5), (3.6) and condition (10.2) satisfies the condition of convexity of the set $\{f(t, u, v) : u \in P\}$ (see Section 15, p. 97). As above, this condition is just technical. It simplifies again some definitions and proofs given below. When we use this condition we can indicate the way to do without it. It can be done just as above (see Section 5, (15.9)). Similar remarks about the assumption of the convexity of the set $\{f(t, u, v) : u \in P\}$ and remarks about the way to avoid this assumption will also be useful to keep in mind when we consider the property of u-stability in the other cases below.

Now let us show that the function $e_{*(2)}(\tau_*, w_*, \Delta)$ satisfies some conditions that correspond to 3^0_u and 4^0_v in an appropriate approximating sense. First we show that the function $e_{*(2)}(\tau_*, w_*, \Delta)$ (14.25) satisfies the following condition of *u-stability* that corresponds to 3^0_u.

$3^0_{*(2)}u$. Suppose we know a domain G^* and a position $\{\tau_*, w_*\} \in G^*$, and a value $\varepsilon > 0$. Then we can indicate the value $\delta(\varepsilon) > 0$ such that the following conditions hold. We remark that for a given bounded domain G^* the value $\delta(\varepsilon) > 0$ can be chosen independent of the initial position $\{\tau_*, w_*\} \in G^*$. Assume we are given a partition $\Delta_\delta = \Delta_\delta\{\tau_j\}$ (12.2) : $\tau_{j+1} - \tau_j \le \delta$, where $\delta \le \delta(\varepsilon)$, moments $\tau_* = \tau_1 \in \Delta_\delta$ and $\tau^* = \tau_2 \in \Delta_\delta$ and disturbances $v[\tau_*[\cdot]\tau^*) = \{v[\tau] = v[\tau_*] \in Q,\ \tau_* \le \tau < \tau^*\}$. Then there exist control actions $u[\tau_*[\cdot]\tau^*) = \{u[\tau] = u[\tau_*] \in P,\ \tau_* \le \tau < \tau^*\}$ which together with $v[\tau_*[\cdot]\tau^*)$ generate from the initial position $\{\tau_*, w_*\}$ the motion $w[\tau_*[\cdot]\tau^*]$ of w-model (12.1) that comes to the state $w[\tau^*] = w^*$ such that

$$e_{*(2)}(\tau^*, w^*, \Delta^*) \leq e_{*(2)}(\tau_*, w_*, \Delta_*) + \varepsilon(\tau^* - \tau_*) \tag{19.1}$$

in the case (15.4): $\tau_* < \tau^* \leq t^{[g]}$, or in the case (15.6): $\tau_* = t^{[g]}$ the inequality

$$\max\Big[\, \mu^{[g]}(D^{[g]}w[t^{[g]}]),\ e_{*(2)}(\tau^*, w^*, \Delta^*) \,\Big] \leq$$

$$\leq e_{(2)*}(\tau_*, w_*, \Delta_*) + \varepsilon(\tau^* - \tau_*) \tag{19.2}$$

holds. Here g is the number (12.11), $\mu^{[g]}(\cdot)$ is the norm in (11.10), and Δ_* and Δ^* are the partitions (15.7) and (15.8).

We note that the condition $3^0_{*(2)}u$, as well as the condition $3^0_{*(1)}u$ in Section 15, is a convenient variant of the general condition of u-stability 3^0_u from Section 8 for the particular case considered here. The most important difference between the conditions $3^0_{*(1)}u$ and $3^0_{*(2)}u$ is the difference between the inequalities (15.5) and (19.2) which follows from the form of the functionals $\gamma_{*(1)}$ (11.9) and $\gamma_{*(2)}$ (11.10). Thus, we can prove the condition $3^0_{*(2)}u$ similarly to $3^0_{*(1)}u$ with the corresponding changes.

If the condition of the convexity of the set $\{f(t, u, v) : \ u \in P\}$ does not hold, then the differential equation (12.1) is substituted by the differential inclusion (15.9). In accordance with this substitution of (12.1) by (15.9) in the property of u-stability we make the following changes. As above (see Section 15, p. 98) instead of the existence of control actions $u[\tau_*[\cdot]\tau^*)$ we speak about the existence of a motion $w[\tau_*[\cdot]\tau^*]$ of the w-model (15.9) generated from the initial position $\{\tau_*, w_*\}$ by actions $v[\tau_*[\cdot]\tau^*)$ so that this motion comes to the state $w[\tau^*] = w^*$ such that the conditions (19.1), (19.2) hold.

Let us prove the condition $3^0_{*(2)}u$.

Proof. At first we consider the particular case $\tau^* < t^{[g]}$ in (15.4). Then we have to prove that the inequality (19.1) is true. According to (14.25), (14.26) we take the difference

$$\Delta e_{*(2)} = e_{*(2)}(\tau^*, w^*, \Delta^*) - e_{*(2)}(\tau_*, w_*, \Delta_*) =$$

$$= \max_{m^*_{\tau^*}, \nu_{\tau^*}} \Big[\, m^*_{\tau^*} \widetilde{w}_{\tau^*} + \varphi_1(\tau^*, m^*_{\tau^*}, \nu_{\tau^*}) \,\Big] -$$

$$- \max_{m^*_{\tau_*}, \nu_{\tau_*}} \Big[m^*_{\tau_*} \widetilde{w}_{\tau_*} + \varphi_1(\tau_*, m^*_{\tau_*}, \nu_{\tau_*}) \Big] =$$

$$= m^{*0}_{\tau^*} \widetilde{w}_{\tau^*} + \varphi_1(\tau^*, m^{*0}_{\tau^*}, \nu^0_{\tau^*}) - (m^{*0}_{\tau_*} \widetilde{w}_{\tau_*} + \varphi_1(\tau_*, m^{*0}_{\tau_*}, \nu^0_{\tau_*})) \,. \qquad (19.3)$$

Here due to (14.26) we have $m^{*0}_{\tau^*} \in G_{1,\nu^0_{\tau^*}}(\tau^*)$, $m^{*0}_{\tau_*} \in G_{1,\nu^0_{\tau_*}}(\tau_*)$ and $\nu^0_{\tau^*} = \nu^0_{\tau_*} = 1, G_{1,\nu^0_{\tau^*}}(\tau^*) = G_{1,\nu^0_{\tau_*}}(\tau_*)$. Again as in Section 15 the symbols τ_* and τ^* emphasize that the extrema $e_{*(2)}(\tau_*, \cdot)$, $e_{*(2)}(\tau^*, \cdot)$, and the corresponding qualities m^*, $\widetilde{w}$ and ν defined in Sections 12 to 14 are calculated for the initial time moments τ_* and τ^*.

We substitute in (19.3) the quantities $m^{*0}_{\tau_*}$ and $\nu^0_{\tau_*}$ that maximize $e_{*(2)}(\tau_*, \cdot)$ by the quantities $m^{*0}_{\tau^*}$ and $\nu^0_{\tau^*}$ that maximize $e_{*(2)}(\tau^*, \cdot)$. Then the following estimation

$$\Delta e_{*(2)} \leq m^{*0T}_{\tau^*}(\widetilde{w}_{\tau^*} - \widetilde{w}_{\tau_*}) + \varphi_1(\tau^*, m^{*0}_{\tau^*}, \nu^0_{\tau^*}) - \varphi_1(\tau_*, m^{*0}_{\tau^*}, \nu^0_{\tau^*}) \qquad (19.4)$$

is true.

The summand $m^{*0T}_{\tau^*}(\widetilde{w}_{\tau^*} - \widetilde{w}_{\tau_*})$ in (19.4) is the same as in (15.17). Then according to (15.18)–(15.20) and taking into account (19.4) we obtain the inequality

$$\Delta e_{*(2)} \leq m^{*0T}_{\tau^*} \widetilde{f}(\tau_*, u[\tau_*], v[\tau_*])(\tau^* - \tau_*) +$$

$$+ \varphi_1(\tau^*, m^{*0}_{\tau^*}, \nu^0_{\tau^*}) - \varphi_1(\tau_*, m^{*0}_{\tau^*}, \nu^0_{\tau^*}) + \varepsilon(\tau^* - \tau_*) \qquad (19.5)$$

provided $\tau^* - \tau_* \leq \delta_\Delta(\varepsilon)$.

From the calculations of the quantity $\varphi_1(\cdot)$ in Section 14 we see that the inequality

$$\varphi_1(\tau_*, m^{*0}_{\tau^*}, \nu^0_{\tau^*}) \geq \varphi_1(\tau^*, m^{*0}_{\tau^*}, \nu^0_{\tau^*}) +$$

$$+ (\tau^* - \tau_*) \max_{v \in Q} \min_{u \in P} m^{*0T}_{\tau^*} \widetilde{f}(\tau_*, u, v) \qquad (19.6)$$

holds.

Then due to (19.5), (19.6) and according to the similar considerations in Section 15 (see (15.23)–(15.25)) we obtain the required inequality (19.1) when $\tau^* < t^{[g]}$.

In the other case in (15.4) when $\tau^* = t^{[g]}$ we prove the condition (19.1) similarly with some changes that follow from the transformation of the constructions of the sets $G_{j,\nu}(\tau^*)$ and $G_{j,\nu}(\tau_*)$ (see (14.25)–(14.27)). Namely, in the case $\tau^* = t^{[g]}$ in accordance with (14.25)–(14.27) we have the following equalities (by substituting in (14.27) τ_* by τ^*)

$$e_{*(2)}(\tau^*, w^*, \Delta^*) = \max_{m^*_{\tau^*},\, m_{*\tau^*},\, \nu_{\tau^*}} \Big[m^{*T}_{\tau^*} \widetilde{w}_{\tau^*} + \varphi_1(\tau^*, m_{*\tau^*}, \nu_{\tau^*}) \Big] =$$

$$= m_{\tau^*}^{*0T} \widetilde{w}_{\tau^*} + \varphi_1(\tau^*, m_{*\tau^*}^0, \nu_{\tau^*}^0) \tag{19.7}$$

where the quantities $m_{\tau^*}^*$, $m_{*\tau^*}$ and ν_{τ^*} satisfy the conditions

$$m_{\tau^*}^* = m_{*\tau^*} + \widehat{m}, \qquad m_{*\tau^*} \in G_{1,\nu_{\tau^*}}(\tau^*), \tag{19.8}$$

$$\widehat{m} = X^T(\tau^*, \vartheta) D^{[g]T} l, \qquad \mu^{[g]*}(l) \leq 1 - \nu_{\tau^*}, \tag{19.9}$$

$$G_{1,\nu}(\tau^*) = G_{2,\nu}(\tau_*)\,. \tag{19.10}$$

In the considered case $\tau^* = t^{[g]}$ we have again $\tau_* < t^{[g]}$ and so, according to (14.25), (14.26), the quantity $e_{*(2)}(\tau_*, w_*, \Delta_*)$ is defined by the equality

$$e_{*(2)}(\tau_*, w_*, \Delta_*) = \max_{m_{\tau_*}^*, \nu_{\tau_*}} \left[m_{\tau_*}^{*T} \widetilde{w}_{\tau_*} + \varphi_1(\tau_*, m_{\tau_*}^*, \nu_{\tau_*}) \right] =$$

$$= m_{\tau_*}^{*0T} \widetilde{w}_{\tau_*} + \varphi_1(\tau_*, m_{\tau_*}^{*0}, \nu_{\tau_*}^0) \tag{19.11}$$

where $m_{\tau_*}^* \in G_{1,\nu_{\tau_*}}(\tau_*)$, $\nu_{\tau_*}^0 = 1$.

Taking into account the constructions in Section 14 we have the following equality

$$\varphi_1(\tau_*, m_{\tau^*}^*, 1) = \overline{\psi}_1(\tau_*, m_{\tau^*}^*, 1), \tag{19.12}$$

where according to (19.7)–(19.10)

$$\psi_1(\tau_*, m_{\tau^*}^*, 1) = J(\tau_1, \tau_2, m_{\tau^*}^*) + \max_{m_{*\tau^*}, \nu} \varphi_2(\tau_*, m_{*\tau^*}, \nu) =$$

$$= J(\tau_1, \tau_2, m_{\tau^*}^*) + \max_{m_{*\tau^*}, \nu} \varphi_1(\tau^*, m_{*\tau^*}, \nu), \qquad 0 \leq \nu \leq 1. \tag{19.13}$$

Here $J(\cdot)$ is the quantity (14.8).

Hence according to (19.12) the inequality

$$\varphi_1(\tau_*, m_{\tau^*}^*, 1) \geq J(\tau_1, \tau_2, m_{\tau^*}^*) + \varphi_1(\tau^*, m_{*\tau^*}, \nu) \tag{19.14}$$

for arbitrary $m_{\tau^*}^*$, $m_{*\tau^*}$, $\nu = \nu_{\tau^*}$ from (19.8)–(19.10) and the inequality

$$e_{*(2)}(\tau_*, w_*, \Delta_*) \geq m_{\tau^*}^{*0T} \widetilde{w}_{\tau_*} + J(\tau_1, \tau_2, m_{\tau^*}^{*0}) + \varphi_1(\tau^*, m_{*\tau^*}^0, \nu_{\tau^*}^0) \tag{19.15}$$

are valid. After that we employ (19.3)–(19.6) and (15.22)–(15.25). This proves the condition (19.1) in the case $\tau^* = t^{[g]}$.

Thus, we have proved that the condition (19.1) of u-stability $3^0_{*(2)}u$ is valid in the case $\tau_* < \tau^* \leq t^{[g]}$.

Now we consider the case $\tau_* = t^{[g]}$, in which we have to prove the condition (19.2). Let us assume again that the partition $\Delta_* = \{\tau_1 = \tau_*,\ \tau_2 = \tau^*, \ldots, \tau_{k+1} = \vartheta\}$ has a sufficiently small step so that

$$\tau_* = t^{[g]}, \qquad \tau^* < t^{[g+1]}. \tag{19.16}$$

In this case according to (14.25)–(14.27) we have the following equalities

$$e_{*(2)}(\tau^*, w^*, \Delta^*) = m^{*0T}_{\tau^*}\widetilde{w}_{\tau^*} + \varphi_1(\tau^*, m^{*0}_{\tau^*}, \nu^0_{\tau^*}), \qquad \nu^0_{\tau^*} = 1, \tag{19.17}$$

$$e_{*(2)}(\tau_*, w_*, \Delta_*) =$$
$$= \max_{m^*_{\tau_*}, m_{*\tau_*}, \nu} \Big[m^{*T}_{\tau_*}\widetilde{w}_{\tau_*} + \varphi_1(\tau_*, m_{*\tau_*}, \nu) \Big] \tag{19.18}$$

where $m_{*\tau_*} \in G_{1,\nu}(\tau_*)$, $0 \leq \nu \leq 1$ and

$$m^*_{\tau_*} = m_{*\tau_*} + \widehat{m}, \qquad \widehat{m} = X^T(\tau_*, \vartheta)D^{[g]T}l, \qquad \mu^{[g]*}(l) \leq 1 - \nu\,. \tag{19.19}$$

According to the definitions of the domains $G_{\gamma,\nu}(\tau_*)$, $G_{\lambda,\nu}(\tau^*)$ and the functions $\varphi_j(\tau_*, m, \nu)$, $\varphi_j(\tau^*, m, \nu)$ and to (19.18), (19.19) we can check that the equality

$$e_{*(2)}(\tau_*, w_*, \Delta_*) = \max_{m_{*\tau_*}, \nu} [(1-\nu)\mu^{[g]}(D^{[g]}w[\tau_*]) + m^T_{*\tau_*}\widetilde{w}_{\tau_*} +$$

$$+\varphi_1(\tau_*, m_{*\tau_*}, \nu)] = (1-\nu^0)\mu^{[g]}(D^{[g]}w_*) + m^{0T}_{*\tau_*}\widetilde{w}_{\tau_*} + \varphi_1(\tau_*, m^0_{*\tau_*}, \nu^0) \geq$$
$$\geq \max[\mu^{[g]}(D^{[g]}w_*),\ (\widetilde{m}^{0T}_{*\tau_*}\widetilde{w}_{\tau_*} + \varphi_1(\tau_*, \widetilde{m}^0_{*\tau_*}, 1)] \tag{19.20}$$

holds. Here

$$m_{*\tau_*} = \nu\widetilde{m}_{*\tau_*}, \qquad \widetilde{m}_{*\tau_*} \in G_{1,1}(\tau_*) = G_{1,1}(\tau^*). \tag{19.21}$$

Using (19.17)–(19.21) and substituting in (19.20), (19.21) $\widetilde{m}^0_{*\tau_*}$ by the maximizing vector $m^{*0}_{\tau^*} \in G_{1,1}(\tau^*)$ for $e_{*(2)}(\tau^*, w^*, \Delta^*)$ we obtain the following estimation:

$$e_{*(2)}(\tau_*, w_*, \Delta_*) \geq \max[\mu^{[(g)]}(D^{[g]}w_*), m_{\tau_*}^{*0T}\widetilde{w}_{\tau_*} + \varphi_1(\tau_*, m_{\tau^*}^{*0}, 1)]\,. \qquad (19.22)$$

Again, as in the first case, choosing a suitable pair $\{m_{\tau^*}^{*0}, u^0[\tau_*]\}$ for the given $v[\tau_*]$ we obtain the estimation

$$e_{*(2)}(\tau^*, w^*, \Delta^*) =$$

$$= m_{\tau^*}^{*0T}\widetilde{w}_{\tau^*} + \varphi_1(\tau^*, m_{\tau^*}^{*0}, 1) \leq m_{\tau^*}^{*0T}\widetilde{w}_{\tau_*} + \varphi_1(\tau_*, m_{\tau^*}^{*0}, 1) + \varepsilon(\tau^* - \tau_*) \quad (19.23)$$

where the state $\widetilde{w}_{\tau^*} = X(\vartheta, \tau^*)w[\tau^*]$ and $w[\tau^*]$ are generated from the initial position $\{\tau_*, w_*\}$ by the pair of realizations of the actions $u^0[\tau] = u^0[\tau_*]$, $v[\tau] = v[\tau_*]$, $\tau_* \leq \tau < \tau^*$.

Let at first the condition

$$\widetilde{m}_{*\tau_*}^{0T}\widetilde{w}_* + \varphi_1(\tau_*, \widetilde{m}_{*\tau_*}^{0}, 1) > \mu^{[g]}(D^{[g]}w_*) \qquad (19.24)$$

be valid. Then according to (19.20) and according to (19.22)–(19.24) we have the inequality

$$e_{*(2)}(\tau^*, w^*, \Delta^*) \leq e_{*(2)}(\tau_*, w_*, \Delta_*) + \varepsilon(\tau^* - \tau_*). \qquad (19.25)$$

Due to (19.20), (19.24) and (19.25) we obtain the inequality

$$\max\, [\, \mu^{[g]}(D^{[g]}w_*),\ e_{*(2)}(\tau^*, w^*, \Delta^*)\,] \leq$$

$$\leq \max\, [\, \mu^{[g]}(D^{[g]}w_*),\ e_{*(2)}(\tau_*, w_*, \Delta_*)\,] + \varepsilon(\tau^* - \tau_*) \leq$$
$$\leq e_{*(2)}(\tau_*, w_*, \Delta_*)\ + \varepsilon(\tau^* - \tau_*) \qquad (19.26)$$

which coincides with the condition (19.2) for $3^0_{*(2)}u$ in the considered case (19.24).

Assume now that the condition

$$\mu^{[g]}(D^{[g]}w_*) \geq \widetilde{m}^{0T}_{*\tau_*}\widetilde{w}_* + \varphi_1(\tau_*, \widetilde{m}^0_{*\tau_*}, 1) \tag{19.27}$$

is valid. Then according to (19.20) we have

$$e_{*(2)}(\tau_*, w_*, \Delta_*) \geq \mu^{[g]}(D^{[g]}w_*)\,. \tag{19.28}$$

On the other hand we have the inequality

$$\widetilde{m}^{0T}_{*\tau_*}\widetilde{w}_* + \varphi_1(\tau_*, \widetilde{m}^0_{*\tau_*}, 1) \geq m^{*0T}_{\tau^*}\widetilde{w}_* + \varphi_1(\tau_*, m^{*0}_{\tau^*}, 1)\,. \tag{19.29}$$

And according to (19.23) we obtain again the inequality (19.2).

Thus, the property of u-stability $3^0_{*(2)}u$ is completely proved.

As above in Section 15, p. 98 if the convexity of the set $\{f(t,u,v):\ u \in P\}$ is not valid, then instead of the motion of w-model (12.1) we use again the w-model (15.9). This enables us to apply again the fixed-point theorem, but now for the pair $\{m^*_{\tau^*}, w[\tau^*]\}$ (see p. 101).

Similarly to the property $3^0_{*(2)}u$ we can formulate the particular case of *v-stability* property for the function $e_{*(2)}(\tau_*, w_*, \Delta_*)$ (14.23). The formulation of this property can be obtained if we change mutually the symbols u and v, and substitute the symbols $\leq$ and $+$ by the symbols $\geq$ and $-$, respectively.

Namely, the property of v-stability for the program extremum $e_{*(2)}(\cdot)$ (Section 14) is formulated in the following form.

$4^0_{*(2)}v$. Suppose we are given a domain G^*, a value $\varepsilon > 0$ and a position $\{\tau_*, w_*\} \in G^*$. Then we can indicate the value $\delta(\varepsilon) > 0$ such that the following conditions hold. We remark that the value $\delta(\varepsilon) > 0$ can be chosen independent of the initial position $\{\tau_*, w_*\} \in G^*$. Assume we are given a partition $\Delta_\delta = \Delta\{\tau_j\}$, $\tau_{j+1} - \tau_j < \delta$, $\delta \leq \delta(\varepsilon)$, moments $\tau_* = \tau_1$, $\tau^* = \tau_2 \in \Delta_\delta$ and control actions $u[\tau_*[\cdot]\tau^*) = \{u[\tau] = u[\tau_*] \in P,\ \tau_* \leq \tau < \tau^*\}$. Then there exist control actions $v[\tau_*[\cdot]\tau^*) = \{v[\tau] = v[\tau_*] \in Q,\ \tau_* \leq \tau < \tau^*\}$ which together with $u[\tau_*[\cdot]\tau^*)$ generate from the initial position $\{\tau_*, w_*\}$ the motion $w[\tau_*[\cdot]\tau^*]$ of w-model (12.1) that comes to the state $w[\tau^*] = w^*$ such that

$$e_{*(2)}(\tau^*, w^*, \Delta^*) \geq e_{*(2)}(\tau_*, w_*, \Delta_*) - \varepsilon(\tau^* - \tau_*) \tag{19.30}$$

in the case $\tau_* < \tau^* \leq t^{[g]}$, or

$$\max [\ \mu^{[g]}(D^{[g]}w[t^{[g]}]),\ e_{*(2)}(\tau^*, w^*, \Delta^*)\ ] \geq$$

$$\geq e_{*(2)}(\tau_*, w_*, \Delta_*) - \varepsilon(\tau^* - \tau_*) \tag{19.31}$$

in the case $\tau_* = t^{[g]}$.

We can prove the property $4^0_{*(2)}v$ similarly to $3^0_{*(2)}u$ with some slight changes. These changes are the following. At first for the estimation of $\Delta e_{*(2)}$ conversely to (19.3) we use now not the maximizing quantities $m^{*0}_{\tau^*}$ and $\nu^0_{\tau^*}$ but the quantities $m^{*0}_{\tau_*}$ and $\nu^0_{\tau_*}$. Besides, we use now the 4th property of the upper convex hull $\varphi(\cdot)$ (see (13.14) and (13.15) in Section 13). This permits us to estimate the difference $\varphi_1(\tau^*, m^{*0}_{\tau_*}, \nu^0_{\tau_*}) - \varphi_1(\tau_*, m^{*0}_{\tau_*}, \nu^0_{\tau_*})$ in a proper way. We obtain the required estimations, which prove the property $4^0_{*(2)}v$. We omit here the details of the considered proof. (The scheme of the proof of the property $4^0_{*(2)}v$ actually repeats that of the property $4^0_{*(1)}v$ in Section 16. The changes are clear.)

20. Approximating solution of the differential game for $\gamma_{(2)}$

The following assertion holds.

Theorem 20.1. *For any number $\eta > 0$ there exists a number $\delta(\eta) > 0$ such that for any initial position $\{t_*, x_*\} \in G$ (3.12) the inequality*

$$|\ \rho^0_{(2)}(t_*, x_*) - e_{*(2)}(t_*, x_*, \Delta)\ | \leq \eta \tag{20.1}$$

is valid if the partition $\Delta = \Delta\{t_i\}$ (12.2) satisfies the conditions (17.2) and (11.15) with suitable ξ, η.

Here $\rho^0_{(2)}(t_*, x_*)$ is the value of the differential game for the system (10.1), (10.2) with the positional quality index $\gamma_{(2)}$ (10.4); $e_{*(2)}(t_*, x_*, \Delta)$ is the program extremum $e_{*(2)}(\tau_*, w_*, \Delta)$ (where $\tau_* = t_*$, $w_* = x_*$) which was constructed in Section 14.

The proof of Theorem 20.1 is similar to the proof of Theorem 17.1 in Section 17. Here we again use the corresponding conditions 1^0, 2^0, $3^0_{*(2)}u$ and $4^0_{*(2)}v$ for the function $e_{*(2)}(\cdot)$ (14.23)–(14.25).

Further repeating with slight changes the constructions from Section 18 we can determine the approximating optimal strategies $u^a_{*(2)}(\cdot) = u^a_{*(2)}(t, x, \varepsilon)$, $v^a_{*(2)}(\cdot) = v^a_{*(2)}(t, x, \varepsilon)$, which provide an approximating solution for the differential game with the functional $\gamma_{(2)}$. These strategies are similar to the strategies $u^a_{*(1)}(\cdot)$, $v^a_{*(1)}(\cdot)$ in Section 18. They also replace in some approximating sense the optimal strategies $u^0_{*(2)}(\cdot) = u^0_{*(2)}(t, x, \varepsilon)$, $v^0_{*(2)}(\cdot) = v^0_{*(2)}(t, x, \varepsilon)$, which form the saddle point in the differential game for the functionals $\gamma_{*(2)}$. Namely, now we can use the function $e_{*(2)}(\tau_*, w_*, \Delta)$, which satisfies the conditions $1^0 - 4^0_{*(2)}v$ and also the condition (20.1). Thus, we can repeat all the constructions from Section 18 where we need only to substitute the value $e_{*(1)}(\cdot)$ by the value $e_{*(2)}(\cdot)$ and to substitute the symbols $u^a_{*(1)}(\cdot)$ and $v^a_{*1)}(\cdot)$ by the symbols $u^a_{*(2)}(\cdot)$ and $v^a_{*(2)}(\cdot)$, respectively.

This concludes the approximating solution of the problem in the case of the quality index $\gamma_{(2)}$ (10.4).

Chapter III
Pure strategies for quasi-positional functionals

In this chapter we consider the same problem as in Chapters I and II but for functionals $\gamma_{(i)}$ combining typical estimates of the motion of x-system and also the actions u and v. We call these functionals $\gamma_{(i)}$ *quasi-positional* because the optimal strategies $u^0_{(i)}(\cdot) = u^0_{(i)}(y_{(i)}, \varepsilon)$ and $v^0_{(i)}(\cdot) = v^0_{(i)}(y_{(i)}, \varepsilon)$ are based on the *information images* $y_{(i)}$ that can include now not only current position $\{t, x\}$ of the controlled x-system but also other variables. We give some classification of these functionals $\gamma_{(i)}$ and describe an effective computation of the values of the game $\rho^0_{(i)}(y_{(i)})$ and constructions of the pure optimal strategies $u^0_{(i)}(\cdot)$ and $v^0_{(i)}(\cdot)$ which form the saddle point in the corresponding differential games.

21. Quasi-positional functionals

As was shown in Section 4 the functional $\gamma_{(3)}$ which is a sum of two positional functionals $\gamma_{(1)}$ (10.3) and $\gamma_{(2)}$ (10.4), i.e.,

$$\gamma_{(3)} = \gamma_{(1)} + \gamma_{(2)} \tag{21.1}$$

is not a positional functional γ (4.2), (4.9). Therefore, for the functional $\gamma_{(3)}$ (21.1) we cannot provide a proper formalization for the differential game with a saddle point in the class of *universal* strategies $u(t, x, \varepsilon)$ and $v(t, x, \varepsilon)$ whose information image is the current position $\{t, x\}$ of the controlled x-system. But it can be proved that for the functional $\gamma_{(3)}$ (21.1) one can solve the problem in the class of strategies $u(x[t^0_*[\cdot]t], \varepsilon)$ and $v(x[t^0_*[\cdot]t], \varepsilon)$ whose information image is the *history* $x[t^0_*[\cdot]t]$ of the motion.

Moreover, if we consider some functionals $\gamma_{(i)}$ that combine estimates of the motion $x[\cdot]$ with some estimates of the control actions $u[\cdot]$ and disturbances $v[\cdot]$ then the *sufficient* information image for the strategies $u_{(i)}(\cdot)$ and $v_{(i)}(\cdot)$ also ought to include some other characteristics of the evolution of the controlled system.

Let us call $x[t^0_*[\cdot]t]$ the *history of the motion* up to the time moment t. An admissible history of the motion has to satisfy the condition $\{\tau, x[\tau]\} \in G$, $t^0_* \le \tau \le t$, where G is the set (3.12). And also we call the triple

$$\{x[t^0_*[\cdot]t], u[t^0_*[\cdot]t), v[t^0_*[\cdot]t)\} \tag{21.2}$$

the *history of the process* up to the time moment t (or from t^0_* to t, to avoid ambiguity). An admissible history of the process ought to satisfy $\{\tau, x[\tau]\} \in G$, $u[\tau] \in P$, $v[\tau] \in Q$, $t^0_* \le \tau < t$.

corresponding group for each of the functionals $\gamma_{(i)}$, $i = 3, \ldots, 8$ depending on a sufficient information image $y_{(i)}[t]$ corresponding to the functional $\gamma_{(i)}$. When we speak about the *sufficient information image* $y_{(i)}[t]$ for this or that functional $\gamma_{(i)}$ we imply that the game with this functional $\gamma_{(i)}$ has a saddle point $\{u^0_{(i)}(\cdot), v^0_{(i)}(\cdot)\}$ in the class of strategies $u_{(i)}(\cdot) = u_{(i)}(y_{(i)}[t], \varepsilon)$ and $v_{(i)}(\cdot) = v_{(i)}(y_{(i)}[t], \varepsilon)$ with this information image $y_{(i)}[t]$. However, the value of the corresponding functional $\gamma_{(i)}$ generated by the optimal strategies on the whole time interval $[t^0_*,\ \vartheta]$ does not coincide, generally speaking, with the value $\rho^0_{(i)}(y_{(i)}[t])$, which is used by us to characterize the value of the game in some way. Nevertheless, a corresponding suitable function $\rho^0_{(i)}(y_{(i)}[t_*])$ will below be called the *value* of the game *for the future* $t_* \le t \le \vartheta$ for the given *initial* value $y_{(i)}[t_*]$ of the information image. We call it so because it characterizes the results of *feedback* control on the basis of optimal strategies $u^0_{(i)}(\cdot)$ and $v^0_{(i)}(\cdot)$ on the *future* time interval $[t_*,\ \vartheta] \subset [t^0_*,\ \vartheta]$. But for the estimation of the functional $\gamma_{(i)}$ on the whole time interval $[t^0_*,\ \vartheta]$ it is also necessary to use some additional operations, for instance, the operations of sum or maximum. These operations involve also together, with the value $\rho^0_{(i)}(y_{(i)}[t_*])$, the terms that are defined by the history $\{x[t^0_*[\cdot]t_*],\ u[t^0_*[\cdot]t_*),\ v[t^0_*[\cdot]t_*)\}$ of the process on the time interval $[t^0_*,\ t_*]$. This will be explained in detail when we consider concrete types of the functionals $\gamma_{(i)}$ $i = 4, \ldots, 8$ (see Sections 22 to 27).

We shall consider six quality indices $\gamma_{(i)}$ (21.1), (21.4)–(21.8) representing different combinations of the estimates of the motion $x[t^0_*[\cdot]\vartheta]$ and integral estimates of the control actions $u[t^0_*[\cdot]\vartheta)$ and disturbances $v[t^0_*[\cdot]\vartheta)$.

We assume that the saddle point condition in a small game similar to (10.2) holds. Now this condition has the following form

$$\min_{u \in P} \max_{v \in Q} [\langle\ l,\ \ f(t,u,v)\ \rangle + q\sigma(t,u,v)] =$$

$$= \max_{v \in Q} \min_{u \in P} [\langle\ l,\ \ f(t,u,v)\ \rangle + q\sigma(t,u,v)] \tag{21.9}$$

where $f(\cdot)$ and $\sigma(\cdot)$ are the functions in (10.1) and (21.4)–(21.8), respectively; l is an arbitrary n-dimensional vector; q is an arbitrary scalar.

Under the condition (21.9) we shall solve the problem of feedback control for the minimax or for the maximin of the functionals $\gamma_{(i)}$ in the class of *pure* strategies $u_{(i)}(\cdot) = u_{(i)}(y_{(i)}, \varepsilon)$ and $v_{(i)}(\cdot) = v_{(i)}(y_{(i)}, \varepsilon)$, where $y_{(i)}$ is the corresponding information image.

22. The differential game for the functional $\gamma_{(4)}$

Let us consider the functional $\gamma_{(4)}$ (21.4). It will be proved in this section that the functional $\gamma_{(4)}$ belongs to the *first* group of the functionals see Section 21.

So for the quality index $\gamma_{(4)}$ the problem is solved in the class of pure *positional* strategies $u(\cdot) = u(t, x, \varepsilon)$ and $v(\cdot) = v(t, x, \varepsilon)$. In this case the strategies $u(\cdot)$ and $v(\cdot)$, ensured results $\rho_u(\cdot)$ and $\rho_v(\cdot)$ (for the future) and the optimal strategies $u^0(\cdot)$ and $v^0(\cdot)$ are understood as in Chapter I (see Sections 5 to 7). So according to what was said in Section 21 we have

$$y_{(4)} = \{t, x\}, \tag{22.1}$$

i.e., the sufficient information image $y_{(4)}$ in the case of the functional $\gamma_{(4)}$ is the current position of the system (10.1), (21.9).

The following result holds.

Theorem 22.1. *The differential game for the system* (10.1) *under the condition* (21.9) *for the quality index* $\gamma_{(4)}$ (21.4) *has a value* $\rho^0_{(4)}(t, x)$ *(for the future) and a saddle point* $\{u^0_{(4)}(\cdot), v^0_{(4)}(\cdot)\}$*, which is a pair of pure optimal strategies* $u^0_{(4)}(\cdot) = u^0_{(4)}(t, x, \varepsilon)$ *and* $v^0_{(4)}(\cdot) = v^0_{(4)}(t, x, \varepsilon)$.

According to the remark in Section 21 (see p. 134), the value $\rho^0_{(4)}(\cdot) = \rho^0_{(4)}(t, x)$ and the saddle point $\{u^0_{(4)}(\cdot), v^0_{(4)}(\cdot)\} = \{u^0_{(4)}(t, x, \varepsilon), v^0_{(4)}(t, x, \varepsilon)\}$ are understood in Theorem 22.1 as follows. Let the initial moment $t_* \in [t^0_*, \vartheta)$ be given and the position $\{t_*, x_*\} \in G$ together with the history of the process from t^0_* up to t_* be realized. Then for the value of the functional $\gamma_{(4)}$ over the time interval $[t^0_*, \vartheta]$ (21.4) we consider the following inequalities

$$\rho^0_{(4)}(t_*, x_*) - \eta \le \gamma_{(4)}(x[t^0_*[\cdot]\vartheta], u[t^0_*[\cdot]\vartheta), v[t^0_*[\cdot]\vartheta)) -$$

$$- \int_{t^0_*}^{t_*} \mu(t, D_*(t)x[t])dt - \sum_{i=g^0_*}^{g_*-1} \mu^{[i]}(D^{[i]}_* x[t^{[i]}_*]) -$$
$$- \int_{t^0_*}^{t_*} \sigma(t, u[t], v[t])dt \le \rho^0_{(4)}(t_*, x_*) + \eta\,. \tag{22.2}$$

In (22.2) $x_* = x[t_*]$, $u[t] = u[t^0_*[t]t_*)$, $v[t] = v[t^0_*[t]t_*)$, $x[t] = x[t^0_*[t]t_*]$, $x[t^{[i]}_*] = x[t^0_*[t^{[i]}_*]t_*]$; g^0_* and g^* are the numbers:

$$g^0_* = \min_{t^{[i]}_* \ge t^0_*} i, \qquad g_* = \min_{t^{[i]}_* \ge t_*} i. \tag{22.3}$$

The right-hand inequality in (22.2) is guaranteed if for $t \ge t_*$ the control law

$U^0 = \{u^0_{(4)}(\cdot);\ \varepsilon_u;\ \Delta_\delta\{t_i^{(u)}\}\}$ is chosen and $\varepsilon_u > 0,\ \delta > 0$ are sufficiently small. In this case we have $\gamma_{(4)} = \gamma_{(4)}(x[t_*^0[\cdot]t_*),\ x_{U^0,v}[t_*[\cdot]\vartheta],\ u[t_*^0[\cdot]t_*),\ u^0[t_*[\cdot]\vartheta),\ v[t_*^0[\cdot]t_*),\ v[t_*[\cdot]\vartheta))$. The left-hand inequality is guaranteed if for $t \geq t_*$ the control law $V^0 = \{v^0(\cdot);\ \varepsilon_v;\ \Delta_\delta\{t_i^{(v)}\}\}$ is chosen and $\varepsilon_v > 0,\ \delta > 0$ are sufficiently small. Thus, in that case $\gamma_{(4)} = \gamma_{(4)}(x[t_*^0[\cdot]t_*),\ x_{V^0,u}[t_*[\cdot]\vartheta],\ u[t_*^0[\cdot]t_*),\ u[t_*[\cdot]\vartheta),\ v[t_*^0[\cdot]t_*),\ v^0[t_*[\cdot]\vartheta))$. Both inequalities in (22.2) are true if the laws U^0 and V^0 are chosen in pairs, i.e., if $\gamma_{(4)} = \gamma_{(4)}(x[t_*^0[\cdot]t_*),\ x_{V^0,U^0}[t_*[\cdot]\vartheta],\ u[t_*^0[\cdot]t_*),\ u^0[t_*[\cdot]\vartheta),\ v[t_*^0[\cdot]t_*),\ v^0[t_*[\cdot]\vartheta))$.

Theorem 22.1 is given in [58],[66],[77]. Here we omit the details of the proof and only briefly outline it. It repeats the proof of Theorem 9.2 from Chapter I with the following changes.

According to the well known transformation we introduce $(n+1)$-dimensional vector variable $\widehat{x} = \{x, \widehat{x}_{n+1}\}$. Here x is the n-dimensional phase vector x of the original controlled system (10.1) and $\widehat{x}_{n+1}$ is an additional scalar variable. This variable is the value of the integral, which involves the control actions and disturbances that are realized on the time interval that ends at the current time moment. So we can reduce the functional $\gamma_{(4)}$ (21.4) to the positional functional $\widehat{\gamma}_{(4)}(\widehat{x}[t_*[\cdot]\vartheta])$. This functional turns out to be positional with respect to the *extended position* $\{t, \widehat{x}\}$ and *extended motion* $\widehat{x}[t_*^0[\cdot]\vartheta]$, respectively. Namely, for all $t_* \in [t_0,\ \vartheta)$, $t^* \in (t_*,\ \vartheta]$, $x[t_*] = x_* \in G$, $\widehat{x}_{n+1}[t_*] = \widehat{x}_{n+1*}$ we have

$$\widehat{\gamma}_{(4)}(\widehat{x}[t_*[\cdot]\vartheta]) = \beta_{(4)}(\widehat{x}[t_*[\cdot]t^*), \alpha_{(4)}) =$$

$$= \int_{t_*}^{t^*} \mu(t, D_*(t)x[t])dt + \sum_{i=g_*}^{g^*-1} \mu^{[i]}(D_*^{[i]}x[t_*^{[i]}]) + \alpha_4 \qquad (22.4)$$

where

$$\alpha_{(4)} = \widehat{\gamma}_{(4)}(\widehat{x}[t^*[\cdot]\vartheta]) =$$

$$= \int_{t^*}^{\vartheta} \mu(t, D_*(t)x[t])dt + \sum_{i=g^*}^{N_*} \mu^{[i]}(D_*^{[i]}x[t_*^{[i]}]) + \widehat{x}_{n+1}[\vartheta] \qquad (22.5)$$

and

$$\widehat{x}_{n+1}[t] = \widehat{x}_{n+1}[t_*] + \int_{t_*}^{t} \sigma(\tau, u[\tau], v[\tau])d\tau\,. \qquad (22.6)$$

And thus, $\widehat{x}_{n+1}[t]$ satisfies the differential equation

$$\frac{d}{dt}\widehat{x}_{n+1} = \sigma(t, u, v). \qquad (22.7)$$

Since the quality index $\widehat{\gamma}_{(4)} = \widehat{\gamma}_{(4)}(\widehat{x}[t_*[\cdot]\vartheta])$ (22.4) is positional with respect to the position $\{t, \widehat{x}\} = \{t, x, \widehat{x}_{n+1}\}$, we conclude that the value of the game $\widehat{\rho}^0_{(4)}(t, \widehat{x})$ *exists* according to Theorem 9.2. It was proved in [66], [77] that in the case of the quality index $\widehat{\gamma}_{(4)}(\widehat{x}[t_*[\cdot]\vartheta])$ (22.4) the following equality

$$\widehat{\rho}^0_{(4)}(t_*, \widehat{x}_*) = \widehat{\rho}^0_{(4)}(t_*, x_*) + \widehat{x}_{*n+1} \tag{22.8}$$

is valid.

It follows from (10.1), (22.7) and (21.9) that using the method of *extremal shift to the accompanying points* (see Section 8) for the game with the quality index $\widehat{\gamma}_{(4)}(\widehat{x}[t_*[\cdot]\vartheta])$ and, consequently, with the quality index $\gamma_{(4)}$ (21.4) (where t^0_* is substituted by t_*) we can construct the strategies $u^0_{(4)}(\cdot) = u^0_{(4)}(t, \widehat{x}, \varepsilon)$, $v^0_{(4)}(\cdot) = v^0_{(4)}(t, \widehat{x}, \varepsilon)$ as extremal ones $u_e(\cdot)$ and $v_e(\cdot)$. More then that, (22.8) implies that we can construct these strategies as functions $u^0_{(4)}(\cdot) = u^0_{(4)}(t, x, \varepsilon)$, $v^0_{(4)}(\cdot) = v^0_{(4)}(t, x, \varepsilon)$ which do not depend on the additional variable $\widehat{x}_{n+1}$. Let us show this.

Now we use the extended position $\{t, \widehat{x}\}$ of the controlled x-object. In accordance with that we employ here the *extended $\widehat{w}$-model*, which is described by $(n+1)$-dimensional phase vector $\widehat{w}^{[u]} = \{w^{[u]}, \widehat{w}^{[u]}_{n+1}\}$. The motion of the extended $\widehat{w}$-model satisfies the differential equations

$$\dot{w}^{[u]} = A(t)w^{[u]} + f(t, u, v), \tag{22.9}$$

$$\frac{d}{dt}\widehat{w}^{[u]}_{n+1} = \sigma(t, u, v) \tag{22.10}$$

with the corresponding initial conditions

$$w^{[u]}[t_*] = w^{[u]}_*, \qquad \widehat{w}^{[u]}_{n+1}[t_*] = \widehat{w}^{[u]}_{*n+1}. \tag{22.11}$$

According to the definition of the accompanying point (see (8.7)–(8.10)) in the case of the quality index $\widehat{\gamma}_{(4)}(\widehat{x}[t_*[\cdot]\vartheta])$ (22.4) we have

$$\widehat{\rho}^0(t, \widehat{w}^{[u]0}(t, \widehat{x}, \varepsilon_u)) = \min_{\widehat{w}^{[u]} \in K^{[u]}_t(\varepsilon_u, t, \widehat{x})} \widehat{\rho}^0(t, \widehat{w}^{[u]}) \tag{22.12}$$

where $\widehat{w}^{[u]0}(t,\widehat{x},\varepsilon_u) = \{w^{[u]0}(t,\widehat{x},\varepsilon_u), \widehat{w}^{[u]0}_{n+1}(t,\widehat{x},\varepsilon_u)\}$ and $K^{[u]}_t(\varepsilon_u,t,\widehat{x})$ is the set:

$$K^{[u]}_t = \Big\{\widehat{w}^{[u]} : |\,\widehat{x}-\widehat{w}^{[u]}\,|^2 = \sum_{i=1}^{n}(x_i - w^{[u]}_i)^2 +$$

$$+ (\widehat{x}_{n+1} - \widehat{w}^{[u]}_{n+1})^2 \leq (R^{[u]}(\varepsilon_u,t))^2, \quad \{t,\widehat{w}^{[u]}\} \in \widehat{G}^*\Big\} \tag{22.13}$$

where $R^{[u]}(\varepsilon_u,t)$ is the quantity (8.8). Here $\widehat{G}^*$ is the *extended region* which satisfies the conditions with respect to the extended motions $\widehat{w}[t]$ similar to the conditions on the region G^* with respect to the motions $w[t]$.

According to (22.8), (22.12)–(22.13) we obtain the minimization problem

$$\widehat{\rho}^0_{(4)}(t,\widehat{w}^{[u]0}(t,\widehat{x},\varepsilon_u)) = \min_{w^{[u]},\widehat{w}^{[u]}_{n+1}} \Big[\,\widehat{\rho}^0_{(4)}(t,w^{[u]}) + \widehat{w}^{[u]}_{n+1}\,\Big] \tag{22.14}$$

under the condition

$$\Big\{w^{[u]},\widehat{w}^{[u]}_{n+1}\Big\} : \ \Big(\sum_{i=1}^{n}(x_i - w^{[u]}_i)^2 + (\widehat{x}_{n+1} - \widehat{w}^{[u]}_{n+1})^2\Big) \leq$$

$$\leq (R^{[u]}(\varepsilon_u,t))^2. \tag{22.15}$$

Therefore, we have to solve the minimum problem (22.14) under the condition (22.15).

If we now introduce the vector $s^{[u]} = x - w^{[u]}$ and the quantity $\widehat{s}^{[u]}_{n+1} = \widehat{x}_{n+1} - \widehat{w}^{[u]}_{n+1}$ and take into account (22.14), (22.15), then we come to the following problem

$$\widehat{\rho}^0_{(4)}(t,\{x - s^{[u]0}(t,x,\widehat{x}_{n+1},\varepsilon_u), \widehat{x}_{n+1} - \widehat{s}^{[u]0}_{n+1}(t,x,\widehat{x}_{n+1},\varepsilon_u)\}) =$$

$$= \min_{s^{[u]},\ \widehat{s}^{[u]}_{n+1}} \Big[\,\rho^0_{(4)}(t,x-s^{[u]}) - \widehat{s}^{[u]}_{n+1}\,\Big] + \widehat{x}_{n+1} \tag{22.16}$$

under the condition

$$\Big\{s^{[u]},\widehat{s}^{[u]}_{n+1}\Big\} : |\,s^{[u]}\,|^2 + (\widehat{s}^{[u]}_{n+1})^2 \leq (R^{[u]}(\varepsilon_u,t))^2. \tag{22.17}$$

Solving this conditional minimum problem we find the quantities $s^{[u]0} = s^{[u]0}(t, x, \varepsilon_u)$ and $\widehat{s}^{[u]0}_{n+1} = \widehat{s}^{[u]0}_{n+1}(t, x, \varepsilon_u)$. Only these quantities are used in the expressions that define the optimal strategy $u^0_{(4)}(t, x, \varepsilon) = u_e(t, x, \varepsilon)$ (see (8.11), (8.12)). Here the corresponding conditions have the form

$$\max_{v \in Q}\Big[\langle\, s^{[u]0}(t, x, \varepsilon_u),\ f(t, x, u_e(t, x, \varepsilon_u), v)\,\rangle + \widehat{s}^{[u]0}_{n+1}(t, x, \varepsilon_u)\sigma(t, u_e(t, x, \varepsilon_u), v)\Big] =$$

$$= \min_{u \in P}\max_{v \in Q}\Big[\langle\, s^{[u]0}(t, x, \varepsilon_u),\ f(t, x, u, v)\,\rangle + \widehat{s}^{[u]0}_{n+1}(t, x, \varepsilon_u)\sigma(t, u, v)\Big]. \quad (22.18)$$

These equalities confirm that in the construction of the strategy $u^0_{(4)}(\cdot) = u^0_{(4)}(t, x, \varepsilon)$ the coordinate $\widehat{x}_{(n+1)}$ is not used. Likewise, one can prove the corresponding assertion for the strategy $v^0_{(4)}(\cdot) = v^0_{(4)}(t, x, \varepsilon)$. Namely, the minimization problem (22.16) in the case of strategy $v_{(4)}(\cdot)$ changes to the maximization problem

$$\widehat{\rho}^0_{(4)}(t, x + s^{[v]0}(t, x, \widehat{x}_{n+1}, \varepsilon_v)), \widehat{x}_{(n+1)} + \widehat{s}^{[v]0}_{n+1}(t, x, \widehat{x}_{n+1}, \varepsilon_v)) =$$

$$= \max_{s^{[v]},\ \widehat{s}^{[v]}_{n+1}} \Big[\rho^0_{(4)}(t, x - s^{[v]}) - \widehat{s}^{[v]}_{n+1}\Big] + \widehat{x}_{n+1} \quad (22.19)$$

under the condition

$$\Big\{s^{[v]}, \widehat{s}^{[v]}_{n+1}\Big\} : \mid s^{[v]} \mid^2 + (\widehat{s}^{[v]}_{n+1})^2 \leq (R^{[v]}(t, \varepsilon_v))^2. \quad (22.20)$$

Here

$$s^{[v]} = w^{[v]} - x, \qquad \widehat{s}^{[v]}_{n+1} = \widehat{w}^{[v]}_{n+1} - \widehat{x}_{n+1}. \quad (22.21)$$

Solving this conditional problem we find the quantities $s^{[v]0} = s^{[v]0}(t, x, \varepsilon_v)$ and $\widehat{s}^{[v]0}_{n+1} = \widehat{s}^{[v]0}(t, x, \varepsilon_v)$. The conditions that determine the strategy $v^0_{(4)}(\cdot) = v_e(\cdot)$ have the form

$$\min_{u \in P}\Big[\langle\, s^{[v]0}(t, x, \varepsilon_v),\ f(t, x, u, v_e(t, x, \varepsilon_v))\,\rangle + \widehat{s}^{[v]0}_{n+1}(t, x, \varepsilon_v)\sigma(t, u, v_e(t, x, \varepsilon_v))\Big] =$$

$$= \max_{v \in Q}\min_{u \in P}\Big[\langle\, s^{[v]0}(t, x, \varepsilon_v),\ f(t, x, u, v)\,\rangle + \widehat{s}^{[v]0}_{n+1}(t, x, \varepsilon_v)\sigma(t, u, v)\Big]. \quad (22.22)$$

Now using the extremal conditions for the strategies $u^0_{(4)}(\cdot) = u_e(\cdot)$ (22.18) and $v^0_{(4)}(\cdot) = v_e(\cdot)$ (22.22), similarly to the considerations for the functional $\gamma_{(1)}$ (see Sections 9 and 10), the inequalities (22.2) can be proved for the functional $\gamma_{(4)}$. In this way we complete the proof of Theorem 22.1.

We see again that the optimal universal strategies $u^0_{(4)}(\cdot) = u^0_{(4)}(t, x, \varepsilon)$ and $v^0_{(4)}(\cdot) = v^0_{(4)}(t, x, \varepsilon)$ can be obtained by the extremal shift method using the value of the game $\rho^0_{(4)}(t, x) = \widehat{\rho}^0_{(4)}(t, x)$, which is the value of the game for the future.

In the following section we shall describe an effective procedure for calculating the approximating value of the game $\rho^a_{(4)}(\cdot)$ and constructing the approximating optimal strategies $u^a_{(4)}(\cdot)$ and $v^a_{(4)}(\cdot)$.

23. Calculation of the approximating value of the game for $\gamma_{(4)}$

Calculation of the approximating value $\rho^a_{(4)}(t, x)$ of the differential game for the system (10.1), (10.2), (21.9) with quality index $\gamma_{(4)}$ (21.4) repeats with some slight changes the procedure that was described and justified in Sections 10 to 13 for the quality index $\gamma_{(1)}$ (10.3). Below we discuss these changes.

Let us introduce the scalar variable

$$\widehat{w}^{[c]}_{n+1}[t] = \int_{t_*}^{t} \sigma(\tau, u[\tau], v[\tau])d\tau + c \tag{23.1}$$

where $\sigma(t, u, v)$ is a function in (21.4)–(21.7); c is a constant. As a rule we choose for c a sufficiently large positive number such that for all possible $t \in [t_0,\ \vartheta]$, $\tau \in [t_0,\ \vartheta]$, $u[\tau] \in P$, $v[\tau] \in Q$ the quantity $\widehat{w}^{[c]}_{n+1}[t]$ (23.1) is nonnegative. We use here the notation employed in Sections 10 to 13. Besides the quality index $\gamma_{(4)}$ (22.4), introduced for the x-object (10.1), (10.2), (21.9), let us consider also a functional $\gamma^{[c]}_{(4)}(\widehat{w}^{[c]}[t_*[\cdot]\vartheta])$ for a $\widehat{w}$-model. This $\widehat{w}$-model is described by the $(n+1)$-dimensional phase vector $\widehat{w}^{[c]}[t] = \{w^{[c]}[t], \widehat{w}^{[c]}_{n+1}[t]\}$ that satisfies the differential equations

$$\dot{w}^{[c]} = A(t)w^{[c]} + f(t, u, v), \tag{23.2}$$

$$\frac{d}{dt}\widehat{w}^{[c]}_{n+1} = \sigma(t, u, v) \tag{23.3}$$

and the initial conditions

$$w^{[c]}[t_*] = w_*, \tag{23.4}$$

$$\widehat{w}_{n+1}^{[c]}[t_*] = c. \tag{23.5}$$

In accordance with (22.4) we introduce the functional $\gamma_{(4)}^{[c]}$ which has the form

$$\gamma_{(4)}^{[c]}(\widehat{w}^{[c]}[t_*[\cdot]\vartheta]) =$$

$$= \gamma_{(4)}^{[c]}(\widehat{w}([t_*[\cdot]\vartheta]) = \widehat{\gamma}_{(4)}(\widehat{w}[t_*[\cdot]\vartheta]) + c = \int_{t_*}^{\vartheta} \mu(t, D_*(t)w[t])dt +$$

$$+ \sum_{i=g_*}^{N_*} \mu^{[i]}(D_*^{[i]} w[t_*^{[i]}]) + \widehat{w}_{n+1}^{[c]}[\vartheta]. \tag{23.6}$$

We introduce also a certain *approximating* functional $\gamma_{*(4)}^{[c]}$ for the $\widehat{w}$-model

$$\gamma_{*(4)}^{[c]}(\widehat{w}^{[c]}[t_*[\cdot]\vartheta]) - \sum_{i=g}^{N} \mu^{[i]}(D^{[i]} w^{[c]}[t^{[i]}]) + \widehat{w}_{n+1}^{[c]}[\vartheta]. \tag{23.7}$$

As above, $t^{[g]} = \min t^{[i]}$, $t^{[i]} \geq t_*$, where $t_* \in [t_*^0,\ \vartheta)$ is the initial time moment.

We observe that the functional $\gamma_{*(4)}^{[c]}$ (23.7) is connected with the approximating quality index $\gamma_{*(4)}$ to the quality index $\gamma_{(4)}$ (21.4) by the following equality:

$$\gamma_{*(4)}^{[c]}(\widehat{w}^{[c]}[t_*[\cdot]\vartheta]) = \gamma_{*(4)}(w[t_*^0[\cdot]\vartheta], u[t_*^0[\cdot]\vartheta), v[t_*^0[\cdot]\vartheta)) +$$

$$+ c - \int_{t_*^0}^{t_*} \sigma(t, u[t], v[t])dt - \sum_{j=g_*^0}^{j=g-1} \mu^{[j]}(D^{[j]} w^{[c]}[t^{[j]}]), \tag{23.8}$$

$$\widehat{w}_{n+1}^{[c]}[\vartheta] = \int_{t_*^0}^{\vartheta} \sigma(\tau, u[\tau], v[\tau])d\tau + c - \int_{t_*^0}^{t_*} \sigma(\tau, u[\tau], v[\tau])d\tau. \tag{23.9}$$

Similarly to $\rho_{*(1)}(\tau_*, w_*, \Delta)$ (12.10) we introduce the stochastic program maximin $\rho_{*(4)}^{[c]}(\cdot)$ in the following way:

$$\rho_{*(4)}^{[c]}(\tau_*, \widehat{w}_*, \Delta) = \max_{v[\cdot]} \min_{u[\cdot]} M\{\gamma_{*(4)}^{[c]}(\widehat{w}[\cdot, \omega])\} =$$

$$= \max_{v[\cdot]} \min_{u[\cdot]} M\Big\{\sum_{i=g}^{N} \mu^{[i]}(D^{[i]}w^{[c]}[t^{[i]},\omega]) + \widehat{w}_{n+1}^{[c]}[\vartheta,\omega]\Big\} \tag{23.10}$$

where the number g_*^0 is given by (10.5) for $t_* = t_*^0$. The numbers g and N are the same as in (12.11).

We employ here the same probability constructions as in Section 12.

Now we use again the $\nu^{[i]}$-dimensional random variables $l_{(1)}^{(i)}(\omega)$, $i = 1,\ldots,N$ that are defined on the probability space $\{\Omega, \mathcal{B}, \mathbf{P}\}$. Here $\nu^{[i]} \in [1,\ n]$ are determined by the dimensions of the matrices $D^{[i]}$. We consider these variables $l^{(i)}(\omega)$ as components of the random variable $l_{(4)}(\cdot)$, for which we set the norm

$$\| \, l_{(4)}(\cdot) \, \|^* = \operatorname*{ess\,sup}_{\omega} \max_{i} \mu^{[i]*}(l_{(4)}^{(i)}(\omega)) \tag{23.11}$$

where $\mu^{[i]*}(l)$ is the norm adjoint to $\mu^{[i]}(l)$.

Further, as in Chapter II (see Section 12), we introduce the approximating program extremum

$$e_{*(4)}^{[c]}(\tau_*, w_*, \Delta) =$$

$$\max_{0\le l\le 1,\ \|l_{(4)}(\cdot)\|^*\le 1} \Big[\sum_{i=g}^{N} M\Big\{l_{(4)}^{(i)T}(\omega)\Big\} D^{[i]} X(t^{[i]},\tau_*) w_* + l\,c +$$

$$+M\Big\{\sum_{j=1}^{k}(\tau_{j+1}-\tau_j)\max_{v\in Q}\min_{u\in P}\Big[M\Big\{\sum_{i=d(j)}^{k} l_{(4)}^{(i)T}(\omega) D^{[i]} X(t^{[i]},\tau_j) *$$

$$* f(\tau_j,u,v) \mid \xi_1,\ldots,\xi_j\Big\} + l\sigma(\tau_j,u,v)\Big]\Big\}\Big] \tag{23.12}$$

where $d(j) = \min(i)$ for $i:\ t^{[i]} \ge \tau_{j+1}$.

Constructing the quantity $e_{*(4)}(\cdot)$ (23.12) we use the following additional circumstance. From the construction of the quantity $\widehat{w}_{n+1}^{[c]}[t]$ (23.1), where c is sufficiently large, it follows that the inequality $\widehat{w}_{n+1}[\vartheta] > 0$ is valid, and so the following equality

$$\widehat{w}_{n+1}[\vartheta] = \mid \widehat{w}_{n+1}[\vartheta] \mid \tag{23.13}$$

is true.

Besides, in this construction instead of the restriction on the value $l = l_{n+1}(\omega)$ of the form $\mid l \mid \le 1$ we use the condition $0 \le l \le 1$. We can do so due to the definition (23.1) where the positive constant c in (23.1) is sufficiently large. Hence, the corresponding maximum in l for the corresponding summands in (23.12) is achieved for $l \equiv 1$. Thus, the value $e_{*(4)}^{[c]}(\cdot)$ (23.12) has the form

$$e_{*(4)}^{[c]}(\tau_*, w_*, \Delta) =$$

$$\max_{\|l_{(4)}(\cdot)\|^*\le 1}\Big[\sum_{i=g}^{N} M\Big\{l_{(4)}^{(i)T}(\omega)\Big\}D^{[i]}X(t^{[i]},\tau_*)w_*+$$

$$+M\Big\{\sum_{j=1}^{k}(\tau_{j+1}-\tau_j)\max_{v\in Q}\min_{u\in P}\Big[M\Big\{\sum_{i=d(j)}^{N} l_{(4)}^{(i)T}(\omega)D^{[i]}X(t^{[i]},\tau_j)*$$

$$*\, f(\tau_j,u,v)\mid \xi_1,\ldots,\xi_j\Big\}+\sigma(\tau_j,u,v)\Big]\Big\}+c\Big]\,. \qquad (23.14)$$

It can be proved that the value $e_{*(4)}^{[c]}(\tau_*,w_*,\Delta)$ approximates the value $\rho_{*(4)}^{[c]}(\tau_*,\widehat{w}_*,\Delta)$ (23.10) in the same sense as it was established in Section 12 for the corresponding values $\rho_{*(1)}(\cdot)$ and $e_{*(1)}(\cdot)$.

Now, employing the notation similar to (13.1)–(13.4) in Chapter II, and using the deterministic vectors $m^{[i]}$, $l^{[i]}$, we obtain that the value $e_{*(4)}^{[c]}(\tau_*,w_*,\Delta)$ (23.14) can be represented in the form

$$e_{*(4)}^{[c]}(\tau_*,w_*,\Delta)=\max_{m^*}\Big[m^{*T}\widetilde{w}_*+\kappa_{(4)}(\tau_*,m_*,\Delta)\Big]+c \qquad (23.15)$$

where

$$\kappa_{(4)}(\tau_*,m_*,\Delta)=$$

$$=\max_{\|l_{(4)}(\cdot)\|^*\le 1} M\Big\{\sum_{j=1}^{k}(\tau_{j+1}-\tau_j)\max_{v\in Q}\min_{u\in P}\Big[\Big(\sum_{i=d(j)}^{N} m_j^{(i)T}\Big)*$$

$$*\,\widetilde{f}(\tau_j,u,v)+\sigma(\tau_j,u,v)\Big]\Big\}. \qquad (23.16)$$

Here $m_*=\sum_{i=h}^{N}m^{[i]}$ is the component of m^*, where $h=g$ if $\tau_*<t^{[g]}$, otherwise $h=g+1$.

A procedure for constructing the quantity $\kappa_{(4)}(\tau_*,m_*,\Delta)$ is presented below. The justification of this procedure is similar to that of the procedure of constructing $\kappa_{(1)}(\tau_*,m_*,\Delta)$ in Section 13 for the quality index $\gamma_{*(1)}$ (11.9). This justification is omitted here. The procedure of the recurrent construction of the value $\kappa_{(4)}(\cdot)$ (23.16) mainly coincides with the procedure for $\kappa_{(1)}(\cdot)$ (13.10), (13.36), (13.38). The only difference is that we have here some additional terms that include the quantity $\sigma(\tau_j,u,v)$.

Thus, the procedure of constructing $\kappa_{(4)}(\cdot)$ in (23.15) is the following. We set

$$\varphi_{k+1}(\tau_*,m)=0,\qquad \Delta\psi_k(\tau_*,m)=J(\tau_k,\tau_{k+1},m)=$$

$$=(\tau_{k+1}-\tau_k)\max_{v\in Q}\min_{u\in P}\Big[m^T\widetilde{f}(\tau_k,u,v)+\sigma(\tau_k,u,v)\Big]\,,$$

$$\psi_k(\tau_*,m)=\Delta\psi_k(\tau_*,m),\qquad \varphi_k(\tau_*,m)=\overline{\psi}_k(\tau_*,m) \qquad (23.17)$$

for $m \in G_k(\tau_*)$. Here, as above, $\overline{\psi}$ is the upper convex hull of the function ψ in the corresponding region G. The set $G_k(\tau_*)$ is $G_k(\tau_*) = \{m : \mu^{[N]*}(l^{[N]}) \leq 1\}$ where $m = D^{[N]T}l^{[N]}$. Assume that the set $G_{j+1}(\tau_*)$ and the function $\varphi_{j+1}(\tau_*, m)$ are already constructed, where $j \geq 1$. Assume at first that $\tau_{j+1} < t^{[i]}$, $i = d(j)$. Then we take $G_j(\tau_*) = G_{j+1}(\tau_*)$ and

$$\Delta\psi_j(m) = J(\tau_j, \tau_{j+1}, m), \quad \psi_j(\tau_*, m) = \Delta\psi_j(\tau_*, m) + \varphi_{j+1}(\tau_*, m),$$

$$\varphi_j(\tau_*, m) = \overline{\psi}_j(\tau_*, m), \quad m \in G_j(\tau_*). \tag{23.18}$$

If $\tau_{j+1} = t^{[i]}$, $i = d(j)$ then $G_j(\tau_*)$ is the set of the vectors $m = m_* + \widehat{m}$, $m_* \in G_{j+1}(\tau_*)$, $\widehat{m} = X^T(t^{[i]}, \vartheta)D^{[i]T}l$, $\mu^{[i]*}(l) \leq 1$ and

$$\psi_j(\tau_*, m) = \max_{m_*}\left[\Delta\psi_j(\tau_*, m) + \varphi_{j+1}(\tau_*, m_*)\right],$$

$$\varphi_j(\tau_*, m) = \overline{\psi}_j(\tau_*, m), \quad m \in G_j(\tau_*). \tag{23.19}$$

Induction in j is continued up to $\tau_1 = \tau_*$. The following cases are possible: 1) $\tau_* < t^{[g]}$ and 2) $\tau_* = t^{[g]}$. In case 1) we obtain a function $\varphi_1(\tau_*, m)$ such that

$$\kappa_{(4)}(\tau_*, m_*, \Delta) = \varphi_1(\tau_*, m_*), \quad m_* \in G_1(\tau_*) = G^{(g)}. \tag{23.20}$$

In case 2), we have

$$\kappa_{(4)}(\tau_*, m_*, \Delta) = \varphi_1(\tau_*, m_*), \quad m_* \in G_1(\tau_*) = G^{(g+1)}. \tag{23.21}$$

The quantity $\kappa_{(4)}(\cdot)$ determines $e_{*(4)}(\cdot)$ according to (23.15). In (23.15) in case 1) we have $m^* = m_* \in G^{(g)}$, and in case 2) we have $m^* = m_* + \widehat{m}$, $m_* \in G^{(g+1)}$, $\widehat{m} = X^T(\tau_*, \vartheta)D^{[g]T}l$, $\mu^{[g]*}(l) \leq 1$.

This concludes the construction of the value $e^{[c]}_{*(4)}(\cdot)$ (23.15) for the functional $\gamma^{[c]}_{*(4)}$ (23.7).

Similarly to Sections 15 and 16 one can establish the corresponding conditions of u and v-stability for the quantity $e^{[c]}_{*(4)}(\cdot)$ (23.15). And likewise in Sections 7 and 18 we obtain the *approximating* solution of the differential game for the system (10.1), (21.9) with the quality index $\gamma_{(4)}$ (21.4) in the class of pure positional approximating strategies $u^a_{*(4)}(t, x, \varepsilon)$ and $v^a_{*(4)}(t, x, \varepsilon)$. Actually all the calculations from Sections 15 to 18 are repeated here without essential changes.

Thus, the value $e^{[c]}_{*(4)}(\tau_*, w_*, \Delta)$ (23.15) approximates the value of the game $\rho^0_{*(4)}(t,x)$ for the quality index $\gamma_{*(4)}$ (23.8) (where $c = 0$). Namely,

$$\rho^0_{*(4)}(t,x) = \lim_{\delta\to 0} e^{[c]}_{*(4)}(t,x,\Delta_\delta) - c \tag{23.22}$$

where $\delta = \max_j(\tau_{j+1} - \tau_j)$, $\tau_j \in \Delta_\delta\{\tau_j\}$ (12.2).

In their turn, according to Lemma 11.1 in Section 11, the quality index $\gamma_{*(4)}$ (23.7) (where $c = 0$) and the value $\rho^0_{*(4)}(t,x)$ approximate the quality index $\gamma_{(4)}$ (21.4) and the value of the game $\rho^0_{(4)}(t,x)$ (for the future). This concludes the approximating solution of the problem in the case of quality index $\gamma_{(4)}$ (21.4).

After calculating the value $\kappa_{(4)}(\tau_*, m_*, \Delta)$ (23.20), (23.21) one can calculate the approximating optimal actions $u^a(\cdot)$ and $v^a(\cdot)$ according to the procedures (22.16)–(22.22) and (18.1)–(18.15) described in Sections 18 and 22. Due to the rule of extremal shift towards the accompanying point, these actions can be found from conditions similar to conditions (22.18) and (22.22). However, in these conditions the vectors $\widehat{s}^{[u]0}$ from (22.16) and $\widehat{s}^{[v]0}$ from (22.19) are substituted by the analogous vectors $\widehat{s}^{[u]a}$ and $\widehat{s}^{[v]a}$ which are calculated on the basis of the value $e^{[c]}_{*(4)}(t,w,\Delta)$ which replaces $\rho^0_{(4)}(t,w)$. Thus, according to (22.16), (22.19) the vectors $\widehat{s}^{[u]a}(t,x,\varepsilon_u)$ and $\widehat{s}^{[v]a}(t,x,\varepsilon_v)$ are given by the conditions

$$e^{[c]}_{*(4)}(t, x - s^{[u]a}, \Delta) - \widehat{s}^{[u]a}_{n+1} + c =$$

$$= \max_{m^*,m_*}\Big[\min_{s,\widehat{s}_{n+1}}(m^{*T}X(\vartheta,t)(x-s) - \widehat{s}_{n+1} + \varphi_1(t,m_*)\Big] + c, \tag{23.23}$$

$$e^{[c]}_{*(4)}(t, x + s^{[v]a}, \Delta) + \widehat{s}^{[v]a}_{n+1} + c =$$

$$= \max_{m^*,m_*}\Big[\max_{s,\widehat{s}_{n+1}}(m^{*T}X(\vartheta,t)(x+s) + \widehat{s}_{n+1} + \varphi_1(t,m_*)\Big] + c. \tag{23.24}$$

In (23.23) we have employed the fact that the operations min and max can be transposed (see Section 18). In (23.23) and (23.24) the vectors $\widehat{s} = \{s, \widehat{s}_{(n+1)}\}$ are subject to the condition

$$|\, s\,|^2 + \widehat{s}^2_{n+1} \le R^2(\varepsilon,t) \tag{23.25}$$

where one can put, for example,

$$R^2(t,\varepsilon) = \Big[\,\varepsilon + \varepsilon(t-t_0)\Big]exp(2\lambda(t-t_0))\,. \tag{23.26}$$

In the general case we have

$$m = m^* \in G^{(g)}, \quad m_* = G^{(h)} \tag{23.27}$$

where $h = g$ or $h = g + 1$. Then the minimizing vector $\widehat{s}^{[u]a}$ in (23.23) and the maximizing vector $\widehat{s}^{[v]a}$ in (23.24) are determined by the equalities

$$s^{[u]a}(t, x, \varepsilon_u) = \frac{X^T(\vartheta, t)m_u^0(t, x, \varepsilon_u)}{\sqrt{1+ \mid X^T(\vartheta, t)m_u^0(t, x, \varepsilon_u) \mid^2}} R(\varepsilon_u, t), \tag{23.28}$$

$$\begin{aligned}
\widehat{s}_{n+1}^{[u]a}(t, x, \varepsilon_u) &= \frac{R(\varepsilon_u, t)}{\sqrt{1+ \mid X^T(\vartheta, t)m_u^0(t, x, \varepsilon_u) \mid^2}}, \\
s^{[v]a}(t, x, \varepsilon_v) &= \frac{X^T(\vartheta, t)m_v^0(t, x, \varepsilon_v)}{\sqrt{1+ \mid X^T(\vartheta, t)m_v^0(t, x, \varepsilon_v) \mid^2}} R(\varepsilon_v, t), \\
\widehat{s}_{n+1}^{[v]a}(t, x, \varepsilon_v) &= \frac{R(\varepsilon_v, t)}{\sqrt{1+ \mid X^T(\vartheta, t)m_v^0(t, x, \varepsilon_v) \mid^2}}.
\end{aligned} \tag{23.29}$$

Here the vectors m_u^0 and m_v^0 are, respectively, the solutions of the following maximizing problems:

$$m_u^{0T} X(\vartheta, t)x - \sqrt{1+ \mid X^T(\vartheta, t)m_u^0 \mid^2} R(\varepsilon_u, t) + \varphi_1(t, m_{*u}^0) =$$

$$= \max_{m, m_*} \Big(m^T X(\vartheta, t)x - \sqrt{1+ \mid X^T(\vartheta, t)m \mid^2} R(\varepsilon_u, t) + \varphi_1(t, m_*) \Big), \tag{23.30}$$

$$m_v^{0T} X(\vartheta, t)x + \sqrt{1+ \mid X^T(\vartheta, t)m_v^0 \mid^2} R(\varepsilon_v, t) + \varphi_1(t, m, m_{*v}^0) =$$

$$= \max_{m, m_*} \Big(m^T X(\vartheta, t)x + \sqrt{1+ \mid X^T(\vartheta, t)m \mid^2} R(\varepsilon_v, t) + \varphi_1(t, m_*) \Big). \tag{23.31}$$

Consequently, in the considered case the approximating control actions $u^a(t_i, x[t_i], \varepsilon_u)$ and $v^a(t_i, x[t_i], \varepsilon_v)$ are given by the conditions

$$\max_{v \in Q} \Big[\langle s^{[u]a}(t_i, x[t_i], \varepsilon_u), \ f(t_i, u^a(t_i, x[t_i], \varepsilon_u)) \rangle$$
$$+ \widehat{s}_{n+1}^{[u]a}(t_i, x[t_i], \varepsilon_u) \sigma(t_i, u^a(t_i, x[t_i], \varepsilon_u), v) \Big]$$

$$= \min_{u \in P} \max_{v \in Q} \Big[\langle s^{[u]a}(t_i, x[t_i], \varepsilon_u), \ f(t_i, u, v) \rangle + \widehat{s}_{n+1}^{[u]a} \sigma(t_i, u, v) \Big], \tag{23.32}$$

$$\min_{u\in P}\Big[\ \langle s^{[v]a}(t_i,x[t_i],\varepsilon_v),\ f(t_i,u,v^{[v]a}(t_i,x[t_i],\varepsilon_v))\rangle$$
$$+s^{[v]a}_{n+1}(t_i,x[t_i],\varepsilon_v)\sigma(t_i,u,v^a(t_i,x[t_i],\varepsilon_v))\ \Big]$$

$$=\max_{v\in Q}\min_{u\in P}\Big[\ \langle s^{[v]a}(t_i,x[t_i],\varepsilon_v),\ f(t_i,u,v)\rangle+\widehat{s}^{[v]a}_{n+1}(t_i,x[t_i],\varepsilon_v)\sigma(t_i,u,v)\ \Big]. \tag{23.33}$$

Let us note the following. The summand $-\sqrt{1+\mid X^T(\vartheta,t)m\mid^2}\,R(t,\varepsilon)$ in the maximized expression (23.30), generally speaking, stabilizes the control algorithm. Really, the function in the right-hand side of (23.30) is strictly concave in m. Thus, the maximizing value m turns out to be unique and, for sufficiently small step δ, to be stable enough. In the case of the functional $\gamma_{(4)}(\cdot)$ the term $-\sqrt{1+\mid X^T(\vartheta,t)m\mid^2}$ appears naturally according to the character of the problem and due to the additional variable $\widehat{w}_{n+1}$. However, in the case of the other functionals $\gamma_{(i)}(\cdot)$ one can also try to stabilize the control process by adding fictitious variables $\widehat{x}_{n+1},\widehat{w}_{n+1}$. These variables can be subjected to the equations

$$\dot{\widehat{x}}_{n+1}[t]=0,\qquad \widehat{x}_{n+1}[t^0_*]=c,\qquad \dot{\widehat{w}}_{n+1}[\tau]=0,\qquad \widehat{w}_{n+1}[t^0_*]=c \tag{23.34}$$

and we can add the summand $\widehat{x}_{n+1}[\vartheta]$ to the functional $\gamma_{(i)}(\cdot)$ so that we obtain a new functional

$$\widehat{\gamma}_{(i)}(\cdot)=\gamma_{(i)}(\cdot)+\widehat{x}_{n+1}[\vartheta]. \tag{23.35}$$

The new minimax problem for $\widehat{\gamma}_{(i)}(\cdot)$ is actually equivalent to the initial minimax problem for $\gamma_{(i)}(\cdot)$. However, in the corresponding auxiliary problems some stabilizing terms appear. This way of stabilizing the control process sometimes proves useful.

24. Differential games with the functional $\gamma_{(3)}$

Let us consider the functional $\gamma_{(3)}$ (21.1). As was said at the beginning of Section 21 this functional is a sum of two positional functionals $\gamma_{(1)}$ and $\gamma_{(2)}$. However, $\gamma_{(3)}$ is not a positional functional (4.2)–(4.9). One can check that the functional $\gamma_{(3)}$ belongs to the *second* group of quasi-positional functionals introduced in Section 21. That is, the differential game with $\gamma_{(3)}$ has a saddle point in the class of strategies

$$u_{(3)}(\cdot)=u_{(3)}(x[t^0_*[\cdot]t],\varepsilon) \tag{24.1}$$

and

$$v_{(3)}(\cdot) = v_{(3)}(x[t_*^0[\cdot]t], \varepsilon). \tag{24.2}$$

Thus, the history $x[t_*^0[\cdot]t]$ (21.3) of the motion of x-object (10.1) realized up to the current time moment t provides a sufficient information image $y_{(3)}[t]$ for the strategies $u_{(3)}(\cdot)$ and $v_{(3)}(\cdot)$. This fact was given in [63], [66]. Here we give only a brief sketch of the proof.

First of all let us define the *ensured result* for the strategy $u_{(3)}(\cdot)$ (24.1) and the initial history of the motion $x[t_*^0[\cdot]t_*]$, $t_* \in [t_*^0,\ \vartheta)$ by the equality

$$\rho_{(3)u}(u_{(3)}(\cdot); x[t_*^0[\cdot]t_*]) =$$

$$= \overline{\lim_{\varepsilon \to 0}} \lim_{\delta \to 0} \sup_{\Delta_\delta} \sup_{v[t_*[\cdot]\vartheta)} \gamma_{(3)u}(x_{U,v}[t_*^0[\cdot]\vartheta]) \tag{24.3}$$

where

$$x_{U,v}[t_*^0[\cdot]\vartheta] = \{x[t_*^0[\cdot]t_*), x_{U,v}[t_*[\cdot]\vartheta]\}. \tag{24.4}$$

Here

$$x_{U,v}[t_*[\cdot]\vartheta] = \{x_{U,v}[t], \quad t_* \le t \le \vartheta; \quad x_{U,v}[t_*] = x[t_*^0[t_*]t_*]\} \tag{24.5}$$

is the motion of the object (10.1), which is defined as the solution of the step-by-step differential equation

$$\dot{x}_{U,v}[t] = A(t)x_{U,v}[t] + f(t, u(x[t_*^0[\cdot]t_i], \varepsilon), v[t]),$$

$$t_i \le t < t_{i+1}, \quad i = 1, \dots, k, \quad t_1 = t_*, \quad x[t_1] = x[t_*^0[t_1]t_*]. \tag{24.6}$$

The strategy $u_{(3)}^0(\cdot)$ and ensured result $\rho_{(3)u}^0(\cdot) = \rho_{(3)u}^0(x[t_*^0[\cdot]t_*])$ are *optimal* if the equalities

$$\rho_{(3)u}(u_{(3)}^0(\cdot); x[t_*^0[\cdot]t_*]) = \min_{u_{(3)}(\cdot)} \rho_{(3)u}(u_{(3)}(\cdot); x[t_*^0[\cdot]t_*]) =$$

$$= \rho_{(3)u}^0(x[t_*^0[\cdot]t_*]) \tag{24.7}$$

are valid for every admissible history $x[t_*^0[\cdot]t_*]$. The values of the ensured result $\rho_{(3)v}(v_{(3)}(\cdot); x[t_*^0\ [\cdot]t_*])$, optimal ensured result $\rho_{(3)v}^0(\cdot)$ and optimal strategy

$v^0_{(3)}(\cdot)$ are defined similarly. Namely, the strategy $v^0_{(3)}(\cdot) = v^0_{(3)}(x[t^0_*[\cdot]t], \varepsilon)$ and ensured result $\rho^0_{(3)v}(x[t^0_*[\cdot]t_*])$ are *optimal* if the equalities

$$\rho_{(3)v}(v^0_{(3)}(\cdot); x[t^0_*[\cdot]t_*]) =$$

$$= \max_{v_{(3)}(\cdot)} \rho_{(3)v}(v_{(3)}(\cdot); x[t^0_*[\cdot]t_*]) = \rho^0_{(3)v}(x[t^0_*[\cdot]t_*]) \tag{24.8}$$

hold for every admissible history $x[t^0_*[\cdot]t_*]$.

The value $\rho_{(3)v}(v_{(3)}(\cdot); x[t^0_*[\cdot]t_*])$ is defined by the equality

$$\rho_{(3)v}(v_{(3)}(\cdot); x[t^0_*[\cdot]t_*]) =$$

$$= \underline{\lim_{\varepsilon \to 0}} \liminf_{\delta \to 0} \inf_{\Delta_\delta} \inf_{u[t_*[\cdot]\vartheta)} \gamma_{(3)u}(x_{V,u}[t^0_*[\cdot]\vartheta]) \tag{24.9}$$

where the motion $x_{V,u}[\cdot]$ is determined similarly to (24.4) if in (24.4)–(24.6) we change U for V and v for u.

The following result is valid.

Theorem 24.1. *The differential game for the system* (10.1) *under the condition* (21.9) *for the quality index* $\gamma_{(3)}$ (21.1) *has a value* $\rho^0_{(3)}(x[t^0_*[\cdot]t_*]) = \rho^0_{(3)u}(x[t^0_*[\cdot]t_*]) = \rho^0_{(3)v}(x[t^0_*[\cdot]t_*])$ *and a saddle point* $\{u^0_{(3)}(\cdot), v^0_{(3)}(\cdot)\}$ *which is a pair of pure optimal strategies* $u^0_{(3)}(\cdot) = u^0_{(3)}(x[t^0_*[\cdot]t], \varepsilon)$ (24.1), (24.7) *and* $v^0_{(3)}(\cdot) = v^0_{(3)}(x[t^0_*[\cdot]t], \varepsilon)$ (24.2), (24.8).

Theorem 24.1 is proved similarly to the corresponding Theorem 9.2 in Chapter 1 with the following changes. Here we deal with the *accompanying histories* $w^{[u]0}(\cdot; x[t^0_*[\cdot]t], \varepsilon) = \{w^{[u]0}(\tau; x[t^0_*[\cdot]t], \varepsilon),\ t^0_* \le \tau \le t\}$ and $w^{[v]0}(\cdot; x[t^0_*[\cdot]t], \varepsilon) = \{w^{[v]0}(\tau; x[t^0_*[\cdot]t], \varepsilon),\ t^0_* \le \tau \le t\}$ instead of the *accompanying points* $w^{[u]0}(t, x, \varepsilon)$ and $w^{[v]0}(t, x, \varepsilon)$. Instead of the function $\rho(t, w)$ (8.1)–(8.4) we can construct the functional $\rho(w[t^0_*[\cdot]t])$ which satisfies the following conditions

$$\begin{aligned} &| \rho(w^{(1)}[t^0_*[\cdot]t]) - \rho(w^{(2)}[t^0_*[\cdot]t]) | \le \\ &\le L^* \sup_{t^0_* \le \tau \le t} | w^{(1)}[\tau] - w^{(2)}[\tau] |, \end{aligned} \tag{24.10}$$

$$\gamma(w[t^0_*[\cdot]\vartheta]) = \rho(w[t^0_*[\cdot]\vartheta]). \tag{24.11}$$

In the properties of u-stability 3_u^0 and v-stability 4_v^0 the words "the position $\{\tau_*, w_*\}$" and "the position $\{\tau^*, w^*\}$" are substituted by the expressions "the history $w[t_*^0[\cdot]\tau_*]$" and "the history $w[t_*^0[\cdot]\tau^*]$". The inequalities (8.3) and (8.4) are replaced by the inequalities

$$\rho(w[t_*^0[\cdot]\tau^*]) \leq \rho(w[t_*^0[\cdot]\tau_*]) + \varepsilon \tag{24.12}$$

and

$$\rho(w[t_*^0[\cdot]\tau^*]) \geq \rho(w[t_*^0[\cdot]\tau_*]) - \varepsilon \tag{24.13}$$

respectively.

As in the case of the function $\rho(t, w)$ in Section 8 the existence of the functional $\rho(w[t_*^0[\cdot]t])$ (24.10)–(24.13) is proved with the help of appropriate Q-procedures. Here we omit this proof. The accompanying histories $w^{[u]0}(\cdot)$ and $w^{[v]0}(\cdot)$ are determined by the following conditions:

$$\rho(w^{[u]0}(\cdot; x[t_*^0[\cdot]t], \varepsilon)) = \min_{w[\cdot]} \rho(w[t_*^0[\cdot]t]), \tag{24.14}$$

$$\rho(w^{[v]0}(\cdot; x[t_*^0[\cdot]t], \varepsilon)) = \max_{w[\cdot]} \rho(w[t_*^0[\cdot]t]) \tag{24.15}$$

under the restrictions

$$| x[\tau] - w[\tau] | \leq R(\varepsilon, \tau), \quad t_*^0 \leq \tau \leq t. \tag{24.16}$$

Here we can choose, for instance,

$$R(\varepsilon, \tau) = [\varepsilon + \varepsilon(\tau - t_0)]^{1/2} \exp(\lambda(\tau - t_0)) \tag{24.17}$$

(see (8.13)).

We introduce the vector-functions

$$s^{[u]0}(\cdot; x[t_*^0[\cdot]t], \varepsilon) = \{x[\tau] - w^{[u]0}[\tau], \quad t_*^0 \leq \tau \leq t\}, \tag{24.18}$$

$$s^{[v]0}(\cdot; x[t^0_*[\cdot]t], \varepsilon) = \{w^{[v]0}[\tau] - x[\tau], \quad t^0_* \le \tau \le t\}. \tag{24.19}$$

These vector-functions substitute here the vectors $s^{[u]0}(\cdot)$ (8.12) and $s^{[v]0}(\cdot)$ (8.17).

The strategies $u_e(\cdot) = u_e(x[t^0_*[\cdot]t], \varepsilon_u)$ and $v_e(\cdot) = v_e(x[t^0_*[\cdot]t], \varepsilon_v)$ extremal to the functional $\rho(w[t^0_*[\cdot]t])$ are defined by the conditions

$$\max_v \langle\, s^{[u]0}(t; x[t^0_*[\cdot]t], \varepsilon_u),\; f(t, u_e(x[t^0_*[\cdot]t], \varepsilon_u), v)\,\rangle =$$

$$= \min_u \max_v \langle\, s^{[u]0}(t; x[t^0_*[\cdot]t], \varepsilon_u),\; f(t, u, v)\,\rangle, \tag{24.20}$$

$$\min_u \langle\, s^{[v]0}(t; x[t^0_*[\cdot]t], \varepsilon_v),\; f(t, u, v_e(x[t^0_*[\cdot]t], \varepsilon_v))\,\rangle =$$

$$= \max_v \min_u \langle\, s^{[v]0}(t; x[t^0_*[\cdot]t], \varepsilon_v),\; f(t, u, v)\,\rangle. \tag{24.21}$$

These conditions correspond here to the conditions (8.11), (8.16).

The conditions (24.20), (24.21) ensure the closeness of the motions $x[t^0_*[\cdot]\vartheta]$ generated by the extremal strategies $u_e(\cdot)$ or $v_e(\cdot)$ to the auxiliary motions $w[t^0_*[\cdot]\vartheta]$. These motions $w[t^0_*[\cdot]\vartheta]$ are constructed here similarly to the auxiliary motions $w[\cdot]$ in Section 9. The closeness between the corresponding motions $x[t^0_*[\cdot]\vartheta]$ and $w[t^0_*[\cdot]\vartheta]$ together with the conditions of u-stability (24.12) or v-stability (24.13) justify the assertion that the strategies $u_e(\cdot)$ and $v_e(\cdot)$ are optimal.

Thus, we see that the functional $\rho(\cdot)$ (24.10)–(24.13) gives the value of the game

$$\rho^0_{(3)}(x[t^0_*[\cdot]t]) = \rho(x[t^0_*[\cdot]t]) \tag{24.22}$$

and the extremal strategies $u_e(\cdot)$ and $v_e(\cdot)$ provide the saddle point $\{u^0_{(3)}(\cdot), v^0_{(3)}(\cdot)\}$, $u^0_{(3)}(\cdot) = u_e(\cdot)$, $v^0_{(3)}(\cdot) = v_e(\cdot)$.

25. Constructions of approximating optimal strategies for $\gamma_{(3)}$

In this section we describe the construction of the approximating optimal strategies for the functional $\gamma_{(3)}(\cdot)$ (21.1) on the basis of the program extremum $e_{*(3)}(\cdot)$. This value $e_{*(3)}(\cdot)$ approximates the value of the game $\rho^0_{(3)}(\cdot)$ (24.22).

First of all we describe the main stages of the construction of the program extremum $e_{*(3)}(\cdot)$. For $\gamma_{(3)}$ the *approximating* functional $\gamma_{*(3)}$ has the form

$$\gamma_{*(3)}(x[t^0_*[\cdot]\vartheta]) = \sum_{i=g^0_{*(1)}}^{N_{(1)}} \mu^{[i]}_{(1)}\Big(D^{[i]}_{(1)}\, x[t^{[i]}_{(1)}]\Big) +$$

$$+ \max_{i=g^0_{*(2)},\ldots,N_{(2)}} \mu^{[i]}_{(2)}\Big(D^{[i]}_{(2)}\, x[t^{[i]}_{(2)}]\Big). \tag{25.1}$$

Here the quantities $\mu^{[i]}(\cdot)$, $D^{[i]}$, $t^{[i]}$, g, N are defined similarly to the corresponding quantities in (11.9) and (11.10). In particular,

$$N_{(s)} = \max i, \quad t_0 \le t^{[i]}_{(s)} \le \vartheta,$$

$$g^0_{*(s)} = \min i, \quad t^0_* \le t^{[i]}_{(s)} \le \vartheta, \quad s = 1, 2. \tag{25.2}$$

The lower indices (1) and (2) indicate the corresponding functional $\gamma_{(1)}$ (10.3) or $\gamma_{(2)}$ (10.4).

The definition of the value $\rho^0_{*(3)}(x[t^0_*[\cdot]t_*])$ for the functional $\gamma_{*(3)}(\cdot)$ is obtained from the definition of the value $\rho^0_{(3)}(x[t^0_*[\cdot]t_*])$ for the functional $\gamma_{(3)}(\cdot)$ as a particular case.

Due to this definition of the value $\rho^0_{*(3)}(x[t^0_*[\cdot]t_*])$ and according to what was said above in Sections 12 and 23 the stochastic program maximin for $\gamma_{*(3)}$ (25.1) has the form

$$\rho_*(\tau_*, w[t^0_*[\cdot]\tau_*], \Delta) = \max_{v[\cdot]} \min_{u[\cdot]} M\Big\{ \sum_{i=g^0_{*(1)}}^{g_{(1)}} \mu^{[i]}_{(1)}\Big(D^{[i]}_{(1)}\, w[t^{[i]}_{(1)}]\Big) +$$

$$+ \sum_{i=g_{*(1)}}^{N_{(1)}} \mu^{[i]}_{(1)}\Big(D^{[i]}_{(1)}\, w[t^{[i]}_{(1)}, \omega]\Big) +$$

$$+ \max\Big[\max_{g^0_{*(2)} \le i \le g_{(2)}} \mu^{[i]}_{(2)}\Big(D^{[i]}_{(2)}\, w[t^{[i]}_{(2)}]\Big),$$

$$\max_{g_{*(2)} \le i \le N_{(2)}} \mu^{[i]}_{(2)}\Big(D^{[i]}_{(2)}\, w[t^{[i]}_{(2)}, \omega]\Big)\Big]\Big\}. \tag{25.3}$$

Here

$$g_{(s)} = \max_{t^0_* \le t^{[i]}_{(s)} < \tau_*} i, \quad g_{*(s)} = \min_{\tau_* \le t^{[i]}_{(s)} \le \vartheta} i, \quad s = 1, 2. \tag{25.4}$$

An approximating stochastic program extremum is represented in the form

$$e_{*(3)}(\tau_*, w[t^0_*[\cdot]\tau_*], \Delta) =$$

$$= \max_{\|l_{(3)}(\cdot)\|^* \le 1} \Big[\sum_{p=g^0_{*(2)}}^{g_{(2)}} l^{[p]T}_{(2)} D^{[p]}_{(2)} w[t^{[p]}_{(2)}] +$$

$$+ \sum_{q=g^0_{*(1)}}^{g_{(1)}} l^{[q]T}_{(1)} D^{[q]}_{(1)} w[t^{[q]}_{(1)}] + (m^{*T}_{(2)} + m^{*T}_{(1)})\widetilde{w}_* +$$

$$+ M\Big\{ \sum_{j=1}^{k} (\tau_{j+1} - \tau_j) \max_{v \in Q} \min_{u \in P} \Big[\Big(\sum_{i=d(j)_{(2)}}^{N_{(2)}} m^{(i)T}_{j(2)} +$$

$$+ \sum_{i=d(j)_{(1)}}^{N_{(1)}} m^{(i)T}_{j(1)} \Big) \widetilde{f}(\tau_j, u, v) \Big] \Big\} \Big] \tag{25.5}$$

where

$$d_{(j)_{(s)}} = \min_{\tau_j < t^{[i]}_{(s)} \le \vartheta} i, \quad s = 1, 2. \tag{25.6}$$

The quantities employed here are defined as above (see (13.1)–(13.4)). In (25.5) $l_{(3)}(\cdot)$ is a random vector variable whose components are random vectors $l^{(i)}_{(2)}(\omega)$, $i = g_{*(2)}, \dots, N_{(2)}$, random vectors $l^{(i)}_{(1)}(\omega)$, $i = g_{*(1)}, \dots, N_{(1)}$, constant vectors $l^{[i]}_{(2)}$, $i = g^0_{*(2)}, \dots, g_{(2)}$ and $l^{[i]}_{(1)}$, $i = g^0_{*(1)}, \dots, g_{(1)}$.

The norm $\| l_{(3)}(\cdot) \|^*$ is defined by the equality

$$\| l_{(3)}(\cdot) \|^* = \max \Big[\Big(\sum_{p=g^0_{*(2)}}^{g_{(2)}} \mu^{[p]*}_{(2)}(l^{[p]}_{(2)}) + \operatorname{ess\,max}_{\omega \in \Omega} \sum_{i=g_{*(2)}}^{N_{(2)}} \mu^{[i]*}_{(2)}(l^{(i)}_{(2)}(\omega)) \Big),$$

$$\max_{g^0_{*(1)} \le q \le g_{(1)}} \Big(\mu^{[q]*}_{(1)}(l^{[q]}_{(1)}) \Big), \max_{g_{*(1)} \le i \le N_{(1)}} \Big(\operatorname{ess\,max}_{\omega \in \Omega} \Big(\mu^{[i]*}_{(1)}(l^{(i)}_{(1)}(\omega)) \Big) \Big) \Big]. \tag{25.7}$$

If this or that set of indices i, j, p or q in (25.1)–(25.7) turns out to be void, then the corresponding term in the expressions (25.1), (25.3), (25.5), (25.7) is absent.

Let us introduce deterministic vectors $m^{[i]}_{(s)}$ and $l^{[i]}_{(s)}$ $(s = 1, 2)$ in accordance with the remark in Section 13 (see (13.6)).

So, the value $e_{*(3)}(\cdot)$ (25.5) can be represented in the form

$$e_{*(3)}(\tau_*, w[t^0_*[\cdot]\tau_*], \Delta) =$$

$$= \max_{\{l\}} \widetilde{e}_{*(3)}(\tau_*, w[t^0_*[\cdot]\tau_*], \Delta; \{l\}) =$$

$$= \max_{\{l\}} \Big[\sum_{p=g^0_{*(2)}}^{g_{(2)}} l^{[p]T}_{(2)} D^{[p]}_{(2)} w[t^{[p]}_{(2)}] +$$

$$+ \sum_{q=g^0_{*(1)}}^{g_{(1)}} l^{[q]T}_{(1)} D^{[q]}_{(1)} w[t^{[q]}_{(1)}] + m^{*T} \widetilde{w}_* + \kappa_{(3)}(\tau_*, m_*, \Delta, \nu) \Big]. \tag{25.8}$$

Here m^* and m_* are given by the following conditions

$$m_* = \sum_{i=g_{*(2)}}^{N_{(2)}} m^{[i]}_{(2)} + \sum_{i=g_{*(1)}}^{N_{(1)}} m^{[i]}_{(1)}, \qquad m^* = m_* \tag{25.9}$$

if $\tau_* < t^{[g_{*(1)}]}_{(1)}$, $\tau_* < t^{[g_{*(2)}]}_{(2)}$, by the conditions

$$m_* = \sum_{i=g_{*(2)}-1}^{N_{(2)}} m^{[i]}_{(2)} + \sum_{i=g_{*(1)}}^{N_{(1)}} m^{[i]}_{(1)}, \qquad i > 0,$$

$$m^* = m^{[g_{*(2)}]}_{(2)} + m_* \tag{25.10}$$

if $\tau_* < t^{[g_{*(1)}]}_{(1)}$, $\tau_* = t^{[g_{*(2)}]}_{(2)}$ and by the conditions

$$m_* = \sum_{i=g_{*(2)}}^{N_{(2)}} m^{[i]}_{(2)} + \sum_{i=g_{*(1)}-1}^{N_{(1)}} m^{[i]}_{(1)}, \qquad i > 0,$$

$$m^* = m^{[g_{*(1)}]}_{(1)} + m_* \tag{25.11}$$

in case $\tau_* = t^{[g_{*(1)}]}_{(1)}$, $\tau_* < t^{[g_{*(2)}]}_{(2)}$. In (25.9)–(25.11)

$$m^{[i]}_{(s)} = X^T(\vartheta, t^{[i]}_{(s)})\, D^{[i]T}_{(s)} l^{[i]}_{(s)}. \tag{25.12}$$

In (25.8) the symbol $\{l\}$ denotes the set of vectors $l^{[q]}_{(1)}$, $l^{[p]}_{(2)}$, $l^{[i]}_{(1)}$ and $l^{[i]}_{(2)}$.

The restrictions for these vectors have the form

$$\sum_{p=g^0_{*(2)}}^{g_{(2)}} \mu^{[p]*}_{(2)}(l^{[p]}_{(2)}) \leq 1-\nu,$$

$$\sum_{i=g_{*(2)}}^{N_{(2)}} \mu^{[i]*}_{(2)}(l^{[i]}_{(2)}) \leq \nu, \quad 0 \leq \nu \leq 1,$$

$$\mu^{[q]*}_{(1)}(l^{[q]}_{(1)}) \leq 1, \quad \mu^{[i]*}_{(1)}(l^{[i]}_{(1)}) \leq 1. \tag{25.13}$$

We remind the reader that all the moments $t^{[i]}_{(s)}$ are assumed to be different.

The value $\kappa_{(3)}(\tau_*, m_*, \Delta, \nu)$ is the function that corresponds to similar functions in the cases of the other functionals considered above (see for instance (13.10) and (14.2)). In order to obtain the expression for $\kappa_{(3)}(\tau_*, m_*, \Delta, \nu)$ we employ again a recurrent procedure of taking upper convex hulls $\varphi_j(\tau_*, m, \nu) = \overline{\psi}(\tau_*, m, \nu)$ for some sequence of appropriate auxiliary functions.

As in the case of the quality indices $\gamma_{(2)}$ we introduce the parameter $\nu \in [0,\ 1]$ that determines the admissible sum of the norms for $l^{[i]}_{(2)}$ for the remaining time interval $[\tau_j,\ \tau_{k+1}]$. The procedure for constructing the quantity $\kappa_{(3)}(\cdot)$ on the basis of the functions $\varphi_j(\cdot)$ repeats the corresponding procedures for the cases of quality indices $\gamma_{(1)}$, $\gamma_{(2)}$ with some changes. Namely, at the instances $\tau_j = t^{[i]}_{(2)}$ we change the values ν and the sets $G_{j,\nu}(\tau_*)$ as in the case of the quality index $\gamma_{(2)}$ (see Section 14). And at the instances $t^{[i]}_{(1)}$ we change the sets $G_j(\tau_*)$ as in the case of the quality index $\gamma_{(1)}$ (see Section 13). A more detailed description of these constructions are omitted here.

We have the expression for the value $e_{*(3)}(\cdot)$ (25.8)–(25.13) where

$$\kappa_{(3)}(\tau_*, m_*, \Delta, \nu) = \varphi_1(\tau_*, m_*, \Delta, \nu). \tag{25.14}$$

One can check that this value $e_{*(3)}(\cdot)$ satisfies the proper conditions of u-stability and v-stability that correspond to the conditions (24.12), (24.13). On this basis we can check that the value $e_{*(3)}(\cdot)$ really approximates the value $\rho^0_{*(3)}(\cdot)$ of the game with functional $\gamma_{*(3)}(\cdot)$. Consequently, the value $e_{*(3)}(\cdot)$ approximates also the value $\rho^0_{(3)}(\cdot)$.

Thus, we can construct the approximating optimal strategies $u^a(x[t^0_*[\cdot]t],\, \varepsilon_u)$ and $v^a(x[t^0_*[\cdot]t],\ \varepsilon_v)$ in the form of extremal strategies to the approximating value $\rho^a_{(3)}(x[t^0_*[\cdot]t_*])$ determined by the program extremum $e_{*(3)}(\cdot)$. The extremal strategies $u^a(\cdot)$ and $v^a(\cdot)$ are determined by the conditions similar to the conditions (24.20), (24.21). Namely,

$$\max_v \langle\ s^{[u]a}(t; x[t_*^0[\cdot]t], \varepsilon_u),\ f(t, u_e^a(x[t_*^0[\cdot]t], \varepsilon_u, v)\ \rangle =$$

$$= \min_u \max_v \langle\ s^{[u]a}(t; x[t_*^0[\cdot]t], \varepsilon_u),\ f(t, u, v)\ \rangle, \tag{25.15}$$

$$\min_u \langle\ s^{[v]a}(t; x[t_*^0[\cdot]t], \varepsilon_v),\ f(t, u, v_e^a(x[t_*^0[\cdot]t], \varepsilon_v))\ \rangle =$$

$$= \max_v \min_u \langle\ s^{[v]a}(t; x[t_*^0[\cdot]t], \varepsilon_v),\ f(t, u, v)\ \rangle. \tag{25.16}$$

But here the vectors $s^{[u]a}(t; x[t_*^0[\cdot]t], \varepsilon_u)$ and $s^{[v]a}(t; x[t_*^0[\cdot]t], \varepsilon_v)$ are determined by the conditions similar to conditions in Section 24 (24.14)–(24.19) where the value of the game $\rho(\cdot)$ is substituted by $e_{*(3)}(\cdot)$. We obtain here the following conditions for the vector-functions $s^{[u]a}(\cdot)$ and $s^{[v]a}(\cdot)$:

$$e_{*(3)}(t, x[t_*^0[\cdot]t] - s^{[u]a}(\cdot; x[t_*^0[\cdot]t], \varepsilon_u), \Delta) =$$

$$= \min_{s[\cdot]} \max_{\{l\}} \tilde{e}(t, x[t_*^0[\cdot]t] - s[\cdot], \Delta; \{l\}),$$

$$|\ s[\tau]\ | \leq R(\varepsilon_u, \tau), \quad t_*^0 \leq \tau \leq t, \tag{25.17}$$

$$e_{*(3)}(t, x[t_*^0[\cdot]t] + s^{[v]a}(\cdot; x[t_*^0[\cdot]t], \varepsilon_v), \Delta) =$$

$$= \max_{s[\cdot]} \max_{\{l\}} \tilde{e}(t, x[t_*^0[\cdot]t] + s[\cdot], \Delta; \{l\}),$$

$$|\ s[\tau]\ | \leq R(\varepsilon_v, \tau), \quad t_*^0 \leq \tau \leq t. \tag{25.18}$$

As above in Sections 18 and 20 we can transpose minimum and maximum operations in (25.17) and transpose the operations maximum and maximum in (25.18). Thus, we come to the following conditions for the maximizing set $\{l^0\}$ of vectors l :

$$e_{*(3)}(t, w^{[u]0}[t_*^0[\cdot]t], \Delta) =$$

$$= \max_{\{l\}} \min_{s[\cdot]} \tilde{e}(t, x[t_*^0[\cdot]t] - s[\cdot], \Delta; \{l\}), \tag{25.19}$$

$$e_{*(3)}(t, w^{[v]0}[t_*^0[\cdot]t], \Delta) =$$

$$= \max_{\{l\}} \max_{s[\cdot]} \tilde{e}(t, x[t_*^0[\cdot]t] + s[\cdot], \Delta; \{l\}). \tag{25.20}$$

This concludes the approximating solution of the considered problem for the quasi-positional functional $\gamma_{(3)}$. Below in Sections 28 to 31 the theory of Chapters II and III will be illustrated by examples of computer simulation of the control processes for certain model systems. These examples involve some concrete *positional* and *quasi-positional* quality indices γ.

26. Differential games for the functionals $\gamma_{(i)}$, $i = 5, \dots, 8$

Let us consider the functional $\gamma_{(5)}$ (21.5). This functional is a sum of the positional functional $\gamma_{(2)}$ (10.4) and an integral term, depending on the control actions and disturbances, i.e.,

$$\gamma_{(5)}(x[t_*^0[\cdot]\vartheta], u[t_*^0[\cdot]\vartheta), v[t_*[\cdot]\vartheta)) = \gamma_{(2)}(x[t_*^0[\cdot]\vartheta])+$$

$$+\int_{t_*^0}^{\vartheta} \sigma(t, u[t], v[t])dt. \tag{26.1}$$

It can be checked that the functional $\gamma_{(5)}$ (26.1) is not positional and does not belong to the *first* group of functionals that was introduced in Section 21. The proof of this fact is similar to the considerations in Section 4 for the case of functional γ (4.3), which is a particular case of the functional $\gamma_{(3)}$ (21.1). The functional $\gamma_{(5)}$ (26.1) (as well as the functionals γ (4.3) and $\gamma_{(3)}$ (21.1)) belongs to the *second* group of quasi-positional functionals (see Section 21). That is, the history $x[t_*^0[\cdot]t]$ (21.3) of the motion of x-object (10.1) realized up to the current time moment t provides a sufficient information image $y_{(5)}[t]$ for the strategies $u_{(5)}(\cdot)$ and $v_{(5)}(\cdot)$. In other words, in the case of quality index $\gamma_{(5)}$ we have

$$y_{(5)}[t] = x[t_*^0[\cdot]t]. \tag{26.2}$$

Here due to the additivity of the integral component in (21.5), (26.1) there is again no need to use the adjugate coordinate $\widehat{x}_{n+1}[t]$ (22.6) as an extra argument of the strategies $u_{(5)}(\cdot)$ and $v_{(5)}(\cdot)$. This fact is proved similarly to the case of quality index $\gamma_{(4)}$ (21.4) (see Section 22). We have here only the following changes. Instead of the current position $\{t, x\}$ we deal here with the history of the motion realized up to the current time moment t (26.2).

The control problem for the quality index $\gamma_{(5)}$ (26.1) will be solved in the class of pure strategies:

$$u_{(5)}(\cdot) = \{u_{(5)}(x[t_*^0[\cdot]t], \varepsilon) \in P, \quad \varepsilon > 0\}, \tag{26.3}$$

$$v_{(5)}(\cdot) = \{v_{(5)}(x[t_*^0[\cdot]t], \varepsilon) \in Q, \quad \varepsilon > 0\}. \tag{26.4}$$

In accordance with (10.4), (21.2), (21.5), (26.1) we introduce the notation

$$\gamma_{(5)}(x[t_*^0[\cdot]\vartheta], u[t_*^0[\cdot]\vartheta), v[t_*^0[\cdot]\vartheta)) =$$

$$= \max\Big[\sup_{t_*^0 \le t \le \vartheta} \mu(t, D_*(t)x[t]),$$

$$\max_{i=g_*^0,\dots,N_*} \mu^{[i]}(D_*^{[i]}x[t_*^{[i]}])\Big] + \int_{t_*^0}^{\vartheta} \sigma(t, u[t], v[t])dt. \tag{26.5}$$

where g_*^0 and N_* are the numbers in (10.5) with $t_* = t_*^0$.

Let us define the *ensured result* (for the future) for the strategy $u_{(5)}(\cdot)$ (26.3) and the initial history of the motion $x[t_*^0[\cdot]t_*]$ (21.3), $t_* \in [t_*^0, \vartheta)$ by the equality

$$\rho_{(5)u}(u_{(5)}(\cdot); x[t_*^0[\cdot]t_*]) =$$

$$= \overline{\lim_{\varepsilon\to 0}} \lim_{\delta\to 0} \sup_{\Delta_\delta} \sup_{v[t_*[\cdot]\vartheta)} \gamma_{(5)}(x_{U,v}[t_*^0[\cdot]\vartheta],$$

$$u[t_*^0[\cdot]\vartheta), v[t_*^0[\cdot]\vartheta)) - \int_{t_*^0}^{t_*} \sigma(t, u[t], v[t])dt \tag{26.6}$$

where

$$x_{U,v}[t_*^0[\cdot]\vartheta] = \{x[t_*^0[\cdot]t_*), x_{U,v}[t_*[\cdot]\vartheta]\} \tag{26.7}$$

and

$$x_{U,v}[t_*[\cdot]\vartheta] = \{x_{U,v}[t], t_* \le t \le \vartheta;$$

$$x_{U,v}[t_*] = x[t_*^0[t_*]t_*]\} \tag{26.8}$$

is the motion of the object (10.1), which is defined as the solution of the step-by-step differential equation

$$\dot{x}_{U,v}[t] = A(t)x_{U,v}[t] + f(t, u(x[t_*^0[\cdot]t_i], \varepsilon), v[t]),$$

$$t_i \le t < t_{i+1}, \quad i = 1, \dots, k, \quad t_1 = t_*, \quad x[t_1] = x[t_*^0[t_*]t_*]. \tag{26.9}$$

The strategy $u_{(5)}^0(\cdot)$ and the ensured result $\rho_{(5)u}^0(\cdot) = \rho_{(5)u}^0(x[t_*^0[\cdot]t_*])$ are *optimal* if the equalities

$$\rho_{(5)u}(u_{(5)}^0(\cdot); x[t_*^0[\cdot]t_*]) =$$

$$= \min_{u_{(5)}(\cdot)} \rho_{(5)u}(u_{(5)}(\cdot); x[t_*^0[\cdot]t_*]) =$$

$$= \rho_{(5)u}^0(x[t_*^0[\cdot]t_*]) \tag{26.10}$$

are valid for every admissible history $x[t_*^0[\cdot]t_*]$. The values of the ensured result $\rho_{(5)v}(v_{(5)}(\cdot); x[t_*^0[\cdot]t_*])$, optimal ensured result $\rho_{(5)v}^0(\cdot)$ and optimal strategy $v_{(5)}^0(\cdot)$ are defined similarly.

The strategy $v^0_{(5)}(\cdot) = v^0_{(5)}(x[t^0_*[\cdot]t], \varepsilon)$ and ensured result $\rho^0_{(5)v}(x[t^0_*[\cdot]t_*])$ are *optimal* if the equalities

$$\rho_{(5)v}(v^0_{(5)}(\cdot); x[t^0_*[\cdot]t_*]) =$$

$$= \max_{v_{(5)}(\cdot)} \rho_{v(5)}(v_{(5)}(\cdot); x[t^0_*[\cdot]t_*]) =$$
$$= \rho^0_{(5)v}(x[t^0_*[\cdot]t_*]) \qquad (26.11)$$

hold for every admissible history $x[t^0_*[\cdot]t_*]$.

The value $\rho_{(5)v}(v_{(5)}(\cdot); x[t^0_*[\cdot]t_*])$ is defined by the equality

$$\rho_{(5)v}(v_{(5)}(\cdot); x[t^0_*[\cdot]t_*]) =$$

$$= \underline{\lim_{\varepsilon \to 0}}\ \liminf_{\delta \to 0}\ \inf_{\Delta_\delta}\ \inf_{u[t_*[\cdot]\vartheta)} \gamma_{(5)}(x_{V,u}[t^0_*[\cdot]\vartheta], u[t^0_*[\cdot]\vartheta), v[t^0_*[\cdot]\vartheta))-$$
$$- \int_{t^0_*}^{t_*} \sigma(t, u[t], v[t])dt \qquad (26.12)$$

where the motion $x_{V,u}[\cdot]$ is determined similarly to (26.7) if in (26.7)–(26.9) we change U for V and v for u.

Now let us consider the differential game for the system (10.1) under condition (21.9) with the quality index $\gamma_{(6)}$ (21.6), i.e.,

$$\gamma_{(6)} = \gamma_{(1)} + \gamma_{(2)} + \int_{t^0_*}^{\vartheta} \sigma(t, u[t], v[t])dt \qquad (26.13)$$

where $\gamma_{(1)}$ and $\gamma_{(2)}$ are the positional functionals (10.3) and (10.4), respectively. Let us remark that the functional $\gamma_{(5)}(\cdot)$ (21.5) is a particular case of the functional $\gamma_{(6)}$.

We can check that the functional $\gamma_{(6)}$ (26.13) belongs also to the *second* group of quasi-positional functionals introduced in Section 21. That is, the differential game with $\gamma_{(6)}$ has a saddle point in the class of strategies $u^0_{(6)}(\cdot) = u^0_{(6)}(x[t^0_*[\cdot]t], \varepsilon)$ and $v^0_{(6)}(\cdot) = v^0_{(6)}(x[t^0_*[\cdot]t], \varepsilon)$. This is based on the fact that the functional $\gamma_{(6)}$ is the sum of functional $\gamma_{(3)} = \gamma_{(1)} + \gamma_{(2)}$ (21.1) and the integral term that depends on the control actions and disturbances. As was shown in Sections 4 and 24 the quality index $\gamma_{(3)}$ (21.1) is a functional from the *second* group of quasi-positional functionals, i.e., the corresponding differential game for $\gamma_{(3)}$ has a solution in the class of strategies $u^0_{(3)}(\cdot) = u^0_{(3)}(x[t^0_*[\cdot]t], \varepsilon)$ and $v^0_{(3)}(\cdot) = v^0_{(3)}(x[t^0_*[\cdot]t], \varepsilon)$. Using the additivity of the extra integral term in the expression $\gamma_{(6)}$ (26.13) we can prove that the auxiliary coordinate $\widehat{x}_{n+1}[t]$ (22.6) is not needed for the construction of the optimal strategies $u^0_{(6)}(\cdot)$ and $v^0_{(6)}(\cdot)$ by the method of extremal shift (see (22.8)–(22.22)). This proof is similar to the considerations in Section 22 for $\gamma_{(4)}$ (21.4) and the above considerations in this section for $\gamma_{(5)}$ (26.1).

The differential game for $\gamma_{(i)}$, $i = 5, 6$ has a solution in the class of strategies $u^0_{(i)}(\cdot)$ and $v^0_{(i)}(\cdot)$ with the information image $y_{(i)}[t] = x[t^0_*[\cdot]t]$, $i = 5, 6$.

Thus, the following result is valid.

Theorem 26.1. *The differential game for the system* (10.1) *under the condition* (21.9) *for the quality indices* $\gamma_{(i)}$, $i = 5, 6$ *has a value* $\rho^0_{(i)}(x[t^0_*[\cdot]t_*]) = \rho^0_{(i)u}(x[t^0_*[\cdot]t_*]) = \rho^0_{(i)v}(x[t^0_*[\cdot]t_*])$ *(for the future) and a saddle point* $\{u^0_{(i)}(\cdot), v^0_{(i)}(\cdot)\}$ *which is a pair of pure optimal strategies* $u^0_{(i)}(\cdot) = u^0_{(i)}(x[t^0_*[\cdot]t], \varepsilon)$ *and* $v^0_{(i)}(\cdot) = v^0_{(i)}(x[t^0_*[\cdot]t], \varepsilon)$.

Theorem 26.1 is published in [66]. Here we discuss it only briefly. Theorem 26.1, according to the definitions of the value $\rho^0_{(i)}(x[t^0_*[\cdot]t_*])$ and the observation in the end of Section 21, means that for the value $\rho^0_{(i)}(x[t^0_*[\cdot]t_*])$, $t_* \in [t^0_*, \vartheta]$ the following inequalities

$$\rho^0_{(i)}(x[t^0_*[\cdot]t_*]) - \eta \le \gamma_{(i)}(x[t^0_*[\cdot]\vartheta], u[t^0_*[\cdot]\vartheta), v[t^0_*[\cdot]\vartheta))-$$

$$-\int_{t^0_*}^{t_*} \sigma(t, u[t], v[t])dt \le \rho^0_{(i)}(x[t^0_*[\cdot]t_*]) + \eta, \qquad i = 5, 6 \qquad (26.14)$$

hold provided $\varepsilon \le \varepsilon(\eta)$ and $\delta \le \delta(\eta, \varepsilon)$. These inequalities are understood in the same sense as the inequalities (22.2) in Section 22 for the quality index $\gamma_{(4)}$.

Theorem 26.1 is proved similarly to the corresponding Theorem 22.1 with the following difference. Here we deal with the *accompanying histories* $w^{[u]0}(\cdot, x[t^0_*[\cdot]t], \varepsilon)$ and $w^{[v]0}(\cdot, x[t^0_*[\cdot]t], \varepsilon)$ instead of the *accompanying points* $w^{[u]0}(t, x, \varepsilon)$ and $w^{[v]0}(t, x, \varepsilon)$. We employ here these histories $w^{[u]0}(\cdot)$ and $w^{[v]0}(\cdot)$ similarly to considerations in Section 24 in the case of functional $\gamma_{(3)}$. The proof of Theorem 26.1 combines the constructions used in the proofs of Theorem 22.1 and Theorem 24.1. The optimal strategies $u^0_{(i)}(x[t^0_*[\cdot]t], \varepsilon_u)$ and $v^0_{(i)}(x[t^0_*[\cdot]t], \varepsilon_v)$, $i = 5, 6$ are constructed similarly to the optimal strategies $u^0_{(4)}(t, x, \varepsilon_u)$ and $v^0_{(4)}(t, x, \varepsilon_v)$ in Section 22. Only instead of the augmented vectors $\widehat{s}^{[u]0}(t, x, \varepsilon_u)$ and $\widehat{s}^{[v]0}(t, x, \varepsilon_v)$ we use here the augmented vectors $\widehat{s}^{[u]0}(t, x[t^0_*[\cdot]t], \varepsilon_u)$ and $\widehat{s}^{[v]0}(t, x[t^0_*[\cdot]t], \varepsilon_v)$. The details of these constructions are omitted here.

Let us consider the quality index $\gamma_{(7)}$ (21.7). Actually the functional $\gamma_{(7)}$ is similar to the positional functional $\gamma_{(2)}$ (10.4), but only for the *extended* motion

$$\widehat{x}[t_*^0[\cdot]\vartheta] = \{x[t_*^0[\cdot]\vartheta], \widehat{x}_{n+1}[t_*^0[\cdot]\vartheta]\} \tag{26.15}$$

where

$$x[t_*^0[\cdot]\vartheta] = \{x[t], \quad t_*^0 \leq t \leq \vartheta\} \tag{26.16}$$

is the motion of x-object (10.1), (21.9). The function

$$\widehat{x}_{n+1}[t_*^0[\cdot]\vartheta] = \{\widehat{x}_{n+1}[t], t_*^0 \leq t \leq \vartheta\} \tag{26.17}$$

is defined by the auxiliary coordinate $\widehat{x}_{n+1}[t]$ (22.6) where $t_* = t_*^0$, $x[t_*^0] = 0$. This coordinate satisfies the equation (22.7).

Thus, the functional $\gamma_{(7)} = \widehat{\gamma}_{(7)}(\widehat{x}[t_*^0[\cdot]\vartheta])$ is similar to the functional $\gamma_{(2)}$. And the functional $\gamma_{(7)}$ is positional, but now with respect to the augmented position $\{t, \widehat{x}[t]\}$, so this position is a sufficient information image for the strategies $u_{(7)}(\cdot)$ and $v_{(7)}(\cdot)$. In the quality index $\gamma_{(7)}$ the integral term containing the control actions and disturbances is not a summand in contrast with $\gamma_{(4)}$. Therefore, to construct the optimal strategies $u^0_{(7)}(\cdot)$ and $v^0_{(7)}(\cdot)$ we cannot get rid of additional scalar coordinate $\widehat{x}_{n+1}[t]$ (22.6) (unlike the case of $\gamma_{(4)}$). According to Section 21 the functional $\gamma_{(7)}$ (21.7) belongs to the *third* group of quasi-positional functionals. Thus, a solution of the problem exists in the class of pure strategies

$$u_{(7)}(\cdot) = \{u_{(7)}(t, \widehat{x}, \varepsilon) \in P, \quad \widehat{x} \in \mathbf{R}^{n+1}, \quad \varepsilon > 0\}, \tag{26.18}$$

$$v_{(7)}(\cdot) = \{v_{(7)}(t, \widehat{x}, \varepsilon) \in Q, \quad \widehat{x} \in \mathbf{R}^{n+1}, \quad \varepsilon > 0\} \tag{26.19}$$

which are functions of the *augmented* position

$$\{t, \widehat{x}\} = \{t, x_1, \ldots, x_n, \widehat{x}_{n+1}\} = y_{(7)} \tag{26.20}$$

of the x-object (10.1), (21.9).

The ensured results (for the future) $\rho_{(7)}(u_{(7)}(\cdot); t_*, \widehat{x}_*)$, $\rho_{(7)}(v_{(7)}(\cdot);\ t_*, \widehat{x}_*)$ for the strategies (26.18), (26.19), the optimal strategies $u^0_{(7)}(\cdot)$, $v^0_{(7)}(\cdot)$ and the optimal ensured results $\rho^0_{u_{(7)}}(t_*, \widehat{x}_*)$ and $\rho^0_{v_{(7)}}(t_*, \widehat{x}_*)$ are defined as in Section 6 if we substitute $\{t, x\}$ by $\{t, \widehat{x}\}$ (26.20).

The following result holds.

Theorem 26.2. *The differential game for the system* (10.1) *under the condition* (21.9) *with the quality index* $\gamma_{(7)}$ (21.7) *has a value* $\rho^0_{(7)}(t,\widehat{x}) = \rho^0_{(7)u}(t,\widehat{x}) = \rho^0_{(7)v}(t,\widehat{x})$ *and a saddle point* $\{u^0_{(7)}(\cdot), v^0_{(7)}(\cdot)\}$ *formed by a pair of pure optimal (uniform in the domain* $\widehat{G} \in [t_0,\ \vartheta] \times \mathbf{R}^{(n+1)}$*) strategies* $u^0_{(7)}(\cdot) = u^0_{(7)}(t,\widehat{x},\varepsilon)$ *and* $v^0_{(7)}(\cdot) = v^0_{(7)}(t,\widehat{x},\varepsilon)$.

According to the remark at the end of Section 21 Theorem 26.2 has the following meaning: We have the inequalities

$$\max\Big[\sup_{t^0_*\le t<t_*} \mu(t, D_*(t)x[t]),$$

$$\max_{i=g^0_*,\dots,g_*-1} \mu(D^{[i]}_* x[t^{[i]}_*]),\ \rho^0_{(7)}(t_*,\widehat{x}_*)\Big] - \eta \le$$

$$\le \widehat{\gamma}_{(7)}(\widehat{x}[t^0_*[\cdot]\vartheta]) \le \max\Big[\sup_{t^0_*\le t<t_*} \mu(t, D_*(t)x[t]),$$

$$\max_{i=g^0_*,\dots,g_*-1} \mu(D^{[i]}_* x[t^{[i]}_*]),\ \rho^0_{(7)}(t_*,\widehat{x}_*)\Big] + \eta. \tag{26.21}$$

Here g^0_* is the number (22.3) that corresponds to the starting moment t^0_* and g_* is the number (22.3) for any initial moment $t_* \in [t^0_*,\ \vartheta)$.

Sufficient conditions for the inequalities (26.21) are similar to those for the inequalities (22.2).

The proof of Theorem 26.2 is based on the fact that the functional $\gamma_{(7)} = \widehat{\gamma}_{(7)}(\widehat{x}[t^0_*[\cdot]\vartheta])$ of the augmented motion (26.15)–(26.17) of the x-object (10.1), (21.9) is actually a *positional functional* γ (4.9) but with respect to the position $\{t,\widehat{x}\}$.

Really, the functional $\widehat{\gamma}_{(7)}(\widehat{x}[t_*[\cdot]\vartheta])$ can be represented in the form

$$\widehat{\gamma}_{(7)}(\widehat{x}[t_*[\cdot]\vartheta]) = \beta_{(7)}(\widehat{x}[t_*[\cdot]t^*), \alpha_{(7)}) =$$

$$= \max\Big[\sup_{t_*\le t<t^*} \mu(t, D_*(t)x[t]),\ \max_{i=g_*,\dots,g^*-1} \mu^{[i]}(D^{[i]}_* x[t^{[i]}_*]);\ \alpha_{(7)}\Big] \tag{26.22}$$

where

$$\alpha_{(7)} = \max\Big[\sup_{t^*\le t\le\vartheta} \mu(t, D_*(t)x[t]),\ \max_{i=g^*,\dots,N} \mu^{[i]}(D^{[i]}_* x[t^{[i]}_*]),\ \widehat{x}_{n+1}[\vartheta]\Big]. \tag{26.23}$$

In (26.22), (26.23) $\widehat{x}_{n+1}[t]$ is the quantity (22.6); $g^* = \min(i)$ under the condition $t_*^{[i]} \geq t^*$.

The optimal strategies $u^0_{(7)}(\cdot) = u^0_{(7)}(t, \widehat{x}, \varepsilon)$ and $v^0_{(7)}(\cdot) = v^0_{(7)}(t, \widehat{x}, \varepsilon)$ are constructed with the help of an extremal shift to the accompanying points $\widehat{w}^{[u]0}(t, \widehat{x}, \varepsilon)$, $\widehat{w}^{[v]0}(t, \widehat{x}, \varepsilon)$. These constructions are omitted here. They repeat with some clear changes the corresponding constructions in Section 8. (In particular, the accompanying point $\widehat{w}^{[u]0}(t_*, \widehat{x}_*, \varepsilon_u) = \{w^{[u]0}(t_*, \widehat{x}_*, \varepsilon_u), \widehat{w}^{[u]0}_{n+1}(t_*, \widehat{x}_*, \varepsilon_u)\}$ is determined by the conditions (22.12), (22.13), and the extremal shift by the condition (22.18) with $s^{[u]0} = x - w^{[u]0}(t, \widehat{x}, \varepsilon_u)$, $\widehat{s}^{[u]0}_{n+1} = \widehat{x}_{n+1} - \widehat{w}^{[u]0}_{n+1}(t, \widehat{x}, \varepsilon_u)$).

This concludes the solution of the problem in the case of the functional $\gamma_{(7)}$ (21.7).

Let us consider the functional $\gamma_{(8)}$ (21.8). It is a sum of the positional functional $\gamma_{(1)}$ (10.3) and the positional functional $\gamma_{(7)}$ (21.7). Actually, it is similar to the functionals $\gamma_{(3)}$(21.1) and $\gamma_{(6)}$ (21.6). However, $\gamma_{(8)}$ contains a nonadditive component that depends on the control actions $u[t]$ and the disturbances $v[t]$, $t^0_* \leq t \leq \vartheta$. For the quality index $\gamma_{(8)}$ the differential game is formalized in the class of pure strategies $u_{(8)}(\cdot)$ and $v_{(8)}(\cdot)$ with the information image

$$y_{(8)}[t] = \{x[t^0_*[\cdot]t], \widehat{x}_{n+1}[t]\} \tag{26.24}$$

where $x[t^0_*[\cdot]t]$ is the history (21.3) of the motion of x-system (10.1), (21.9) and $\widehat{x}_{n+1}[t]$ is the quantity (22.6) where $t_* = t^0_*$ and $x[t^0_*] = 0$. Thus, the functional $\gamma_{(8)}$ (21.8) belongs to the *fourth* group of quasi-positional functionals (see Section 21). That is, for the functional $\gamma_{(8)}$ (21.8) the problem is solved in the class of pure strategies

$$u_{(8)}(\cdot) = \{u_{(8)}(t, x[t^0_*[\cdot]t], \widehat{x}_{n+1}, \varepsilon) \in P,$$

$$\widehat{x}_{n+1} \in \mathbf{R}^1, \quad \varepsilon > 0\}, \tag{26.25}$$

$$v_{(8)}(\cdot) = \{v_{(8)}(t, x[t^0_*[\cdot]t], \widehat{x}_{n+1}, \varepsilon) \in Q,$$

$$\widehat{x}_{n+1} \in \mathbf{R}^1, \ \varepsilon > 0\}. \tag{26.26}$$

The *ensured result* for the strategy $u_{(8)}(\cdot)$ is defined by the equality

$$\rho_{(8)u}(u_{(8)}(\cdot); x[t^0_*[\cdot]t_*], \widehat{x}_{n+1}[t_*]) =$$

$$= \overline{\lim_{\varepsilon \to 0}} \lim_{\delta \to 0} \sup_{\Delta_\delta} \sup_{v[t_*[\cdot]\vartheta)} \gamma_{(8)}(x_{U,v}[t^0_*[\cdot]\vartheta],$$

$$u[t_*^0[\cdot]\vartheta), v[t_*^0[\cdot]\vartheta)). \qquad (26.27)$$

The strategy $u_{(8)}^0(\cdot)$ and ensured result $\rho_{(8)u}^0(\cdot)$ are optimal if

$$\rho_{(8)u}(u_{(8)}^0(\cdot); x[t_*^0[\cdot]t_*], \widehat{x}_{n+1}[t_*]) =$$

$$= \min_{u_{(8)}(\cdot)} \rho_{(8)u}(u_{(8)}(\cdot); x[t_*^0[\cdot]t_*], \widehat{x}_{n+1}[t_*]) =$$

$$= \rho_{(8)u}^0(x[t_*^0[\cdot]t_*], \widehat{x}_{n+1}[t_*]) \qquad (26.28)$$

for every admissible $x[t_*^0[\cdot]t_*]$ (21.3) and $\widehat{x}_{n+1}[t_*]$ (22.6).

Similarly we can define the ensured result $\rho_{(8)u}(v_{(8)}(\cdot); \cdot)$, the optimal strategy $v_{(8)}^0(\cdot)$ and the optimal ensured result $\rho_{(8)v}^0(\cdot)$.

The following theorem is valid.

Theorem 26.3. *The differential game for the system* (10.1) *under the condition* (21.9) *with the quality index* $\gamma_{(8)}$ (21.8) *has a value* $\rho_{(8)}^0(x[t_*^0[\cdot]t_*], \widehat{x}_{n+1}[t_*]) = \rho_{(8)u}^0(x[t_*^0[\cdot]t_*], \widehat{x}_{n+1}[t_*]) = \rho_{(8)v}^0(x[t_*^0[\cdot]t_*], \widehat{x}_{n+1}[t_*])$ *and a saddle point* $\{u_{(8)}^0(\cdot) = u_{(8)}^0(t, x[t_*^0[\cdot]t], \widehat{x}_{n+1}, \varepsilon),\ v_{(8)}^0(\cdot) = v_{(8)}^0(t, x[t_*^0[\cdot]t], \widehat{x}_{n+1}, \varepsilon)\}$.

According to the definition of the ensured results the sense of Theorem 26.3 is that the following inequalities

$$\rho_{(8)}^0(x[t_*^0[\cdot]t_*], \widehat{x}_{n+1}[t_*]) - \eta \le$$

$$\le \gamma_{(8)}(x[t_*^0[\cdot]\vartheta], u[t_*^0[\cdot]\vartheta), v[t_*^0[\cdot]\vartheta)) \le$$

$$\le \rho_{(8)}^0(x[t_*^0[\cdot]t_*], \widehat{x}_{n+1}[t_*]) + \eta \qquad (26.29)$$

are valid under conditions similar to the conditions for the inequalities (22.2).

The proof of Theorem 26.3 is similar to the consideration for the cases of the quality indices $\gamma_{(5)}$ (21.5) and $\gamma_{(6)}$ (21.6). The difference is connected with the fact that instead of the information image $y_{(5)} = y_{(6)} = \{x[t_*^0[\cdot]t]\}$ we use the information image $y_{(8)} = \{t, x[t_*^0[\cdot]t], \widehat{x}_{n+1}[t]\}$.

Besides, if in the case of the quality index $\gamma_{(7)}$ (21.7) we include the history $x[t_*^0[\cdot]t]$ to the information image (as in the case of $\gamma_{(8)}$ (21.8)), then using the principle of the optimization for the ensured result, we do not need to use the full information about the coordinate $\widehat{x}_{n+1}[t]$ (22.6) in the case of both functionals $\gamma_{(7)}$ and $\gamma_{(8)}$. Namely, forming the strategy $u_{(8)}^0(\cdot)$ we can use the information about the history $u[t_*^0[\cdot]t)$ of our own control actions instead of the information about $\widehat{x}_{n+1}[t]$. Then we can input, instead of the actual current values $\widehat{x}_{n+1}[t]$, some imaginary values $\widehat{x}_{n+1}^*[t]$. This value $\widehat{x}_{n+1}^*[t]$ would be obtained if the most unfavourable disturbances $v[t_*^0[\cdot]t)$ were realized. These disturbances ought to be compatible with the history $\{x[t_*^0[\cdot]t], u[t_*^0[\cdot]t)\}$ and also ought to maximize the value $\widehat{x}_{n+1}[t]$. So, when we form the strategy $u_{(8)}^0(\cdot)$ we can supplement the information image, which contains the history $x[t_*^0[\cdot]t]$, with the histories $u[t_*^0[\cdot]t), v[t_*^0[\cdot]t)$.

On the other hand, when we form the strategy $v_{(8)}^0(\cdot)$ we can supplement the considered information image, which includes the history $v[t_*^0[\cdot]t)$ and the most unfavourable history $u[t_*^0[\cdot]t)$. Here $u[t_*^0[\cdot]t)$ should be compatible with the histories $\{x[t_*^0[\cdot]t], v[t_*^0[\cdot]t)\}$ and should also minimize the value $\widehat{x}_{n+1}[t]$.

This concludes the solution of the considered problem in the cases of the quasi-positional functionals $\gamma_{(i)}$, $i = 5, \dots, 8$.

27. Construction of approximating optimal strategies for games with functionals $\gamma_{(i)}$, $i = 5, \dots, 8$

In this section we describe very briefly the main stages of the construction of the program extrema $e_{*(i)}(\cdot)$ which approximate the values of the game $\rho_{(i)}^0(\cdot)$ in the cases of quasi-positional functionals $\gamma_{(i)}$, $i = 5, \dots, 8$, introduced in Section 21 and considered in Section 26. These main stages are similar to those given in Sections 13, 14, 23, and 25 for the functionals $\gamma_{(1)}$, $\gamma_{(2)}$, $\gamma_{(4)}$ and $\gamma_{(3)}$, respectively. Therefore, here we restrict ourselves only to some remarks. These remarks outline only the difference of this or that case from the cases considered in Sections 13, 14, 23, and 25.

First of all let us consider the case of the quasi-positional functional $\gamma_{(6)}$ (21.6). In this case the procedure for constructing $e_{*(6)}(\cdot) = e_{*(6)}(\tau_*, w[t_*^0[\cdot]\tau_*], \Delta, \nu)$ in comparison with the construction for $e_{*(3)}(\cdot)$ (25.5)–(25.14) undergoes the changes connected with the fact that we supplement the corresponding constructions with the terms that contain the quantity $\sigma(\tau_j, u, v)$. These additional constructions are similar to those given in Section 23 for the functional $\gamma_{(4)}$ and connected there also with the additional terms that contain the quantity $\sigma(\tau_j, u, v)$.

The functional $\gamma_{(5)}$ (21.5) is a particular case of the functional $\gamma_{(6)}$. In this particular case we have $\gamma_{(1)} = 0$. Therefore we obtain the procedure for constructing $e_{*(5)}(\cdot) = e_{*(5)}(\tau_*, w[t_*^0[\cdot]\tau_*], \Delta, \nu)$ from the procedure for $e_{*(6)}(\cdot)$ if we eliminate the terms connected with $\gamma_{(1)}$.

In the case of positional functional $\gamma_{(7)}$ (21.7) the construction of the corresponding value $e_{*(7)}(\cdot) = e_{*(7)}(\tau_*, \widehat{w}_*, \Delta)$ repeats the procedure for $e_{*(2)}(\cdot) = e_{*(2)}(\tau_*, w_*, \Delta)$ (see Section 14) with the following changes. In the corresponding constructions we substitute the position $\{\tau_*, w_*\}$ in Section 14 by the augmented position $\{\tau_*, \widehat{w}_*\} = \{\tau_*, w_*, \widehat{w}_{*n+1}\}$. This is connected with changing the information image $y_{(2)} = \{t, x\}$ for $y_{(7)} = \{t, \widehat{x}\}$. For this reason we also have to introduce the augmented vector $m \in \mathbf{R}^{n+1}$ to the corresponding procedure.

Finally, the value $e_{*(8)}(\cdot) = e_{*(8)}(\tau_*, w[t_*^0[\cdot]\tau_*], \widehat{w}_{n+1}[\tau_*], \Delta)$ for the functional $\gamma_{(8)}$ (21.8), i.e., $\gamma_{(8)} = \gamma_{(1)} + \gamma_{(7)}$, is constructed according to the same rule as the value $e_{*(3)}(\cdot) = e_{*(3)}(\tau_*, w[t_*^0[\cdot]\tau_*], \Delta)$ for $\gamma_{(3)} = \gamma_{(1)} + \gamma_{(2)}$. Some understandable changes are connected now with the change of the information image $y_{(3)} = \{t, x[t_*^0[\cdot]t]\}$ corresponding to $\gamma_{(3)}$ for $y_{(8)} = \{t, x[t_*^0[\cdot]t], \widehat{x}_{n+1}[t]\}$ that corresponds to $\gamma_{(8)}$.

After constructing the program extrema $e_{*(i)}(\cdot)$, $i = 5, \ldots, 8$ which approximate the values of the game $\rho^0_{*(i)}(\cdot)$ and, consequently, the values of the game $\rho^0_{(i)}(\cdot)$, we can construct the approximating optimal strategies $u^a_{(i)}(\cdot)$ and $v^a_{(i)}(\cdot)$. These strategies are constructed in the form of extremal strategies with respect to the approximating value of the game $\rho^a_{(i)}(\cdot)$. These approximating values $\rho^a_{(i)}(\cdot)$ are determined by the program extrema $e_{*(i)}(\cdot)$. The constructions of approximating optimal strategies $u^a_{(i)}(\cdot) = u_e(\cdot)$ and $v^a_{(i)}(\cdot) = v_e(\cdot)$ on the basis of $e_{*(i)}(\cdot)$ are similar to those considered in Sections 13, 14, 23, and 25 for functionals $\gamma_{(1)}$, $\gamma_{(2)}$, $\gamma_{(4)}$, $\gamma_{(3)}$ with the understandable changes connected with the changes of the informational images $y_{(i)}$.

This concludes the approximating solution of the considered problem for the quasi-positional functionals $\gamma_{(i)}$, $i = 3, \ldots, 8$ introduced in Section 21.

Below in Sections 28 to 31 the theory of Chapters II and III will be illustrated by examples of computer simulation of the control processes for certain model systems. These examples involve some concrete *positional* and *quasi-positional* quality indices γ.

28. Example of constructing the optimal strategies for $\gamma_{(1)}$

In this section we illustrate the construction of the approximate optimal strategies $\{u^a(\cdot) = u^a(t, x, \varepsilon),\ v^a(\cdot) = v^a(t, x, \varepsilon)\}$ for the example of the model problem in Section 1. Namely, with the help of this example we will analyze the construction of the control actions and the disturbances by the method of the *extremal shift* to the *accompanying points*. These points are computed on the basis of the *program extremum* $e_{*(1)}(\cdot)$ (see Section 12). In addition we also illustrate the auxiliary construction of the upper convex hulls $\varphi_j(\tau_*, m)$ (see Section 13). In the considered concrete case this construction is sufficiently transparent. The obtained algorithms for calculating the approximating optimal actions $u^a[t]$ and $v^a[t]$ can be simulated with the help of available computers.

In Section 1 the material point M with variable mass $m[t]$ is considered. The motion of this point is described by the Mescherskii differential equation (1.7). Here we assume that the motion of the point M in plane $\{r_1, r_2\}$ is described by the differential equation of a somewhat more general form than (1.7). Namely, let the equation have the form

$$\ddot{r} = \alpha(t)\dot{r} + \beta(t)r + k(t)u + n(t)v$$

$$r^T = \{r_1, r_2\} \qquad t_0 \le t < \vartheta\,. \tag{28.1}$$

The restrictions for the control action $u^T = \{u_1, u_2\}$ and for the disturbance $v^T = \{v_1, v_2\}$ have the form

$$u \in P, \quad P = \left\{ u = \begin{bmatrix} u_1 \\ u_2 \end{bmatrix} : \frac{u_1^2}{a_1^2[t]} + \frac{u_2^2}{a_2^2[t]} \leq 1 \right\} \tag{28.2}$$

$$v \in Q, \quad Q = \left\{ v = \begin{bmatrix} v_1 \\ v_2 \end{bmatrix} : \frac{v_1^2}{b_1^2[t]} + \frac{v_2^2}{b_2^2[t]} \leq 1 \right\} \tag{28.3}$$

where

$$a_1[t] > a_2[t] \geq a > 0, \quad t_0 \leq t \leq \vartheta, \quad a - \text{ const}, \tag{28.4}$$

$$b_2[t] > b_1[t] \geq b > 0, \quad t_0 \leq t \leq \vartheta, \quad b - \text{ const}. \tag{28.5}$$

We assume that all the functions considered here of the time $t \in [t_0, \vartheta]$ are piecewise-continuous and continuous from the right.

The quality index is

$$\gamma = (r_1^2[\vartheta] + r_2^2[\vartheta])^{1/2}. \tag{28.6}$$

Denote

$$x_1 = r_1, \quad x_2 = \dot{r}_1, \quad x_3 = r_2, \quad x_4 = \dot{r}_2 \tag{28.7}$$

and reduce the equation (28.1) to the differential system in normal form

$$\begin{aligned} \dot{x}_1 &= x_2 \\ \dot{x}_2 &= \beta(t)x_1 + \alpha(t)x_2 + k(t)u_1 + n(t)v_1 \\ \dot{x}_3 &= x_4 \\ \dot{x}_4 &= \beta(t)x_3 + \alpha(t)x_4 + k(t)u_2 + n(t)v_2 . \end{aligned} \tag{28.8}$$

According to (28.7) the quality index γ (28.6) takes a form

$$\gamma = (x_1^2[\vartheta] + x_3^2[\vartheta])^{1/2}. \tag{28.9}$$

The system (28.8) has the following vector form

$$\dot{x} = A(t)x + B(t)u + C(t)v \tag{28.10}$$

where x is a 4-dimensional vector

$$x = \begin{bmatrix} x_1 \\ x_2 \\ x_3 \\ x_4 \end{bmatrix}, \tag{28.11}$$

u and v are 2-dimensional vectors

$$u = \begin{bmatrix} u_1 \\ u_2 \end{bmatrix} \in P, \qquad v = \begin{bmatrix} v_1 \\ v_2 \end{bmatrix} \in Q$$

with P and Q being the sets (28.2) and (28.3), respectively.

The matrix-valued functions $A(t)$, $B(t)$ and $C(t)$ in (28.10) have the form

$$A(t) = \begin{pmatrix} 0 & 1 & 0 & 0 \\ \beta(t) & \alpha(t) & 0 & 0 \\ 0 & 0 & 0 & 1 \\ 0 & 0 & \beta(t) & \alpha(t) \end{pmatrix}, \tag{28.12}$$

$$B(t) = \begin{bmatrix} 0 & 0 \\ k(t) & 0 \\ 0 & 0 \\ 0 & k(t) \end{bmatrix}, \qquad C(t) = \begin{bmatrix} 0 & 0 \\ n(t) & 0 \\ 0 & 0 \\ 0 & n(t) \end{bmatrix}. \tag{28.13}$$

The quality index γ (28.9) is the seminorm $\mu_*(x[\vartheta])$ of the 4-dimensional vector $x[\vartheta]$. This seminorm $\mu_*(x)$ is the Euclidean norm

$$\mu = | Dx[\vartheta] | \tag{28.14}$$

of the 2-dimensional vector

$$x' = Dx \tag{28.15}$$

where the matrix D is determined by the equality

$$D = \begin{pmatrix} 1 & 0 & 0 & 0 \\ 0 & 0 & 1 & 0 \end{pmatrix}. \tag{28.16}$$

The fundamental matrix $X(t, \vartheta)$ for the equation

$$\dot{x} = A(t)x \tag{28.17}$$

is obtained by backward integration (generally speaking, numerical) of the matrix differential equation

$$\dot{X} = A(t)X \tag{28.18}$$

on the time interval $t_0 \leq t \leq \vartheta$ with the initial condition

$$X(\vartheta, \vartheta) = E = \begin{pmatrix} 1 & 0 & 0 & 0 \\ 0 & 1 & 0 & 0 \\ 0 & 0 & 1 & 0 \\ 0 & 0 & 0 & 1 \end{pmatrix}. \tag{28.19}$$

In the considered case the fundamental matrix $X(t, \vartheta)$ has the following block form:

$$X(t, \vartheta) = \left(\begin{array}{cc|cc} x_{11}(t, \vartheta) & x_{12}(t, \vartheta) & \multicolumn{2}{c}{0} \\ x_{21}(t, \vartheta) & x_{22}(t, \vartheta) & & \\ \hline & & x_{11}(t, \vartheta) & x_{12}(t, \vartheta) \\ \multicolumn{2}{c|}{0} & x_{21}(t, \vartheta) & x_{22}(t, \vartheta) \end{array}\right). \tag{28.20}$$

Carrying out the transformations (12.14), (13.1)–(13.6), (13.33)–(13.36), (13.39) and taking into account the remark about (13.5), (13.6) we obtain the expression for $e_{*(1)}(\tau_*, w_*, \Delta)$ in the following form

$$e_{*(1)}(\tau_*, w_*, \Delta) = \max_{m_*} \widetilde{e}_*(\tau_*, w_*, m_*, \Delta) \tag{28.21}$$

where

$$m_*^T = \{l_1, 0, l_2, 0\}, \qquad | \, l \, | = (l_1^2 + l_2^2)^{1/2} \leq 1, \tag{28.22}$$

$$\widetilde{e}_*(\tau_*, w_*, m_*, \Delta) = l_1[x_{11}(\vartheta, \tau_*)w_{*1} + x_{12}(\vartheta, \tau_*)w_{*2}] +$$
$$+ l_2[x_{11}(\vartheta, \tau_*)w_{*3} + x_{12}(\vartheta, \tau_*)w_{*4}] + \kappa(\tau_*, m_*, \Delta). \tag{28.23}$$

Here $l^T = \{l_1, l_2\}$ is the vector-parameter employed in the condition (13.16). Below in this section instead of the 4-dimensional vector m_* in (28.22) it is convenient to deal with a new 2-dimensional vector $m^T = \{m_1, m_2\}$ taking $m_1 = l_1$, $m_2 = l_2$. Thus, we will construct the functions $\varphi_j(\tau_*, m)$ as the functions of the new vector m in the domain $| \, m \, | \leq 1$.

The quantities $\Delta\psi_j(\tau_*, m)$ that determine the construction of the recurrent sequence for $\varphi_j(\tau_*, m)$ (see (13.16)–(13.36)) in the considered case have the form

$$\Delta\psi_j(\tau_*, m) = (\tau_{j+1} - \tau_j)\Big(\sqrt{\eta_1[\tau_j]m_1^2 + \eta_2[\tau_j]m_2^2} -$$
$$- \sqrt{\xi_1[\tau_j]m_1^2 + \xi_2[\tau_j]m_2^2}\Big), \qquad m^T = \{m_1, m_2\}. \tag{28.24}$$

Here the quantities $\eta_i[\tau_j]$, $\xi_i[\tau_j]$, $j = 1, \ldots, k$; $i = 1, 2$ can be obtained by direct calculation using the elements of the fundamental matrix $X(t, \vartheta)$ (28.20), the elements of the matrix $B(t)$, $C(t)$ (28.13) and the parameters $a_i[t]$, $b_i[t]$, $i = 1, 2$ (28.4), (28.5) in the restrictions (28.2), (28.3). This direct calculation gives

$$\eta_i[\tau_j] = x_{12}^2(\vartheta, \tau_j)n^2(\tau_j)b_i^2[\tau_j],$$

$$\xi_i[\tau_j] = x_{12}^2(\vartheta, \tau_j)k^2(\tau_j)a_i^2[\tau_j]. \tag{28.25}$$

Under the assumed conditions the following inequalities

$$\eta_2[\tau_j] \geq \eta_1[\tau_j] \geq 0 \tag{28.26}$$

$$\xi_1[\tau_j] \geq \xi_2[\tau_j] \geq 0, \qquad j = 1, \ldots, k \tag{28.27}$$

hold. The equalities in (28.26), (28.27) can hold only for the values τ_j for which $x_{12}(\vartheta, \tau_j) = 0$ or $k(\tau_j) = 0$ or $n(\tau_j) = 0$, where $k(\tau)$ and $n(\tau)$ are the functions in (28.1), (28.8).

We will describe construction of the functions $\varphi_j(\tau_*, m)$ only for the part

$$| \, m \, | \leq 1, \qquad 0 \leq m_1 \leq 1, \qquad 0 \leq m_2 \leq 1 \tag{28.28}$$

of the whole domain $|m| \leq 1$. In the other three quarters of the circle $|m| \leq 1$ the functions $\varphi_j(\tau_*, m)$ are constructed symmetrically with respect to the axis $m_1 = 0$ and $m_2 = 0$.

Consider the quantity

$$\nu(\tau_*, \tau_j, m) = \left[\frac{\partial}{\partial m_2}\Big(\Delta\psi_j(\tau_*, m)\Big)\right]_{|m|=1}. \tag{28.29}$$

Let us begin with the case when for the fixed m, $0 < m_1 < 1$, $0 < m_2$ the value $\nu(\tau_*, \tau_j, m)$ monotonically increases in j.

Then it follows from the geometric considerations that the functions $\varphi_j(\tau_*, m)$ have the following structure.

Construct the sequence of the points

$$m^{*T}[\tau_j] = \{\ m_1^*[\tau_j],\ m_2^*[\tau_j]\ \} \tag{28.30}$$

which satisfy the following equation

$$\nu(\tau_*, \tau_j, m) = 0 \tag{28.31}$$

under the conditions

$$|m| = 1, \qquad m_1 \geq 0, \qquad m_2 \geq 0\,. \tag{28.32}$$

Under our suppositions we have that for a fixed j, $j = 1, \ldots, k$ one of the following three cases holds.

1^0. The equation (28.31) under the conditions (28.32) has two solutions:

$$m' = \begin{bmatrix} m_1' \\ m_2' \end{bmatrix}, \quad m_1' = 1, \quad m_2' = 0,$$

$$m'' = \begin{bmatrix} m_1'' \\ m_2'' \end{bmatrix}, \quad m_1'' < 1, \quad m_2'' \leq 1. \tag{28.33}$$

In this case we take

$$m^*[\tau_j] = m''. \tag{28.34}$$

2^0. The equation (28.31) under the conditions (28.32) has one solution:

$$m' = \begin{bmatrix} m'_1 \\ m'_2 \end{bmatrix}, \quad m'_1 = 1, \quad m'_2 = 0, \tag{28.35}$$

and

$$\nu(\tau_*, \tau_j, m) > 0 \tag{28.36}$$

if $m_1 < 1$. Then we take

$$m^*[\tau_j] = \begin{bmatrix} m^*_1 \\ m^*_2 \end{bmatrix}, \quad m^*_1[\tau_j] = 0, \quad m^*_2[\tau_j] = 1. \tag{28.37}$$

3^0. The equation (28.31) under the conditions (28.32) has one solution

$$m' = \begin{bmatrix} m'_1 \\ m'_2 \end{bmatrix}, \quad m'_1 = 1, \quad m'_2 = 0 \tag{28.38}$$

and

$$\nu(\tau_*, \tau_j, m) < 0$$

if $0 \leq m_1 < 1$. Then we take

$$m^*[\tau_j] = m'. \tag{28.39}$$

Under our assumption that $\nu(\tau_*, \tau_j, m)$ is monotone the obtained sequence of the points $m^*[\tau_j]$, $j = k, k-1, \ldots, 1$ is such that the sequence

$$m^*_1[\tau_j], \quad j = 1, \ldots, k \tag{28.40}$$

does not increase in j.

Employing the points $m^*[\tau_j]$ we construct the sequence of the domains $\mathrm{I}(\tau_j)$, $\mathrm{II}(\tau_j)$ and $\mathrm{III}(\tau_j)$ (see Figure 28.1).

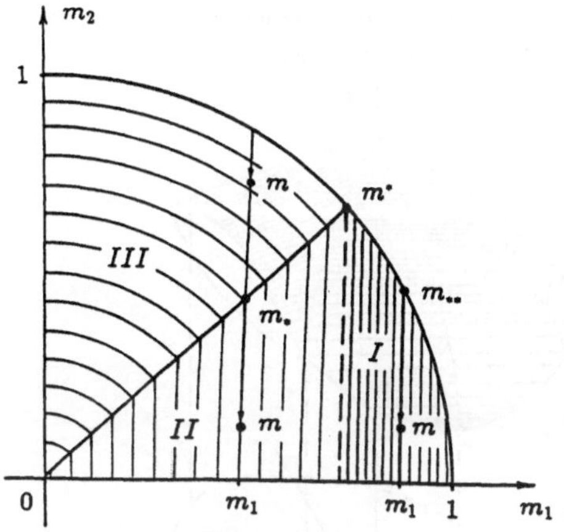

Figure 28.1

In the cases 2^0 and 3^0 some of these domains degenerate into intervals or points.

The recurrent sequence for the functions $\varphi_j(\tau_*, m)$, $j = k, \ldots, 1$ in the domain (28.28) is determined by the following equalities:

$$\varphi_k(\tau_*, m) = \begin{cases} \Delta\psi_k(\tau_*, m_{**}), & m_{**1} = m_1,\ m_{**2} = \sqrt{1 - m_{**1}^2} \\ & \text{in } \mathrm{I}(\tau_k); \\ \Delta\psi_k(\tau_*, m_*), & m_{*1} = m_1,\ m_{*2} = \dfrac{m_2^*[\tau_k] m_{*1}}{m_1^*[\tau_k]} \\ & \text{in } \mathrm{II}(\tau_k); \\ \Delta\psi_k(\tau_*, m), & \text{in } \mathrm{III}(\tau_k); \end{cases} \tag{28.41}$$

$$\varphi_j(\tau_*, m) = \begin{cases} \Delta\psi_j(\tau_*, m_{**}) + \varphi_{j+1}(\tau_*, m_{**}), & m_{**1} = m_1,\ m_{**2} = \sqrt{1 - m_{**1}^2} \\ & \text{in } \mathrm{I}(\tau_j); \\ \Delta\psi_j(\tau_*, m_*) + \varphi_{j+1}(\tau_*, m_*), & m_{*1} = m_1, m_{*2} = \dfrac{m_2^*[\tau_k] m_{*1}}{m_1^*[\tau_k]} \\ & \text{in } \mathrm{II}(\tau_k); \\ \Delta\psi_j(\tau_*, m) + \varphi_{j+1}(\tau_*, m) & \\ & \text{in } \mathrm{III}(\tau_k); \end{cases} \tag{28.42}$$

As is said above, in the other three quarters of the circle $| \ m \ | \leq 1$ the functions $\varphi_j(\tau_*, m)$ are constructed symmetrically with respect to the axes $m_1 = 0$ and $m_2 = 0$.

The properties of the roots of the equation (28.31) and the expressions for $\varphi_j(\tau_*, m)$ (28.41), (28.42) follow under our assumptions from the explicit expressions for $\Delta\psi(\tau_*, m)$ and $\nu(\tau_*, \tau_J, m)$. We omit here a detailed justification of this assertion.

Thus, in space $\{m_1, m_2, \varphi\}$ the surface $\varphi = \varphi_j(\tau_*, m)$ in the general case has the following form (see Figure 28.2).

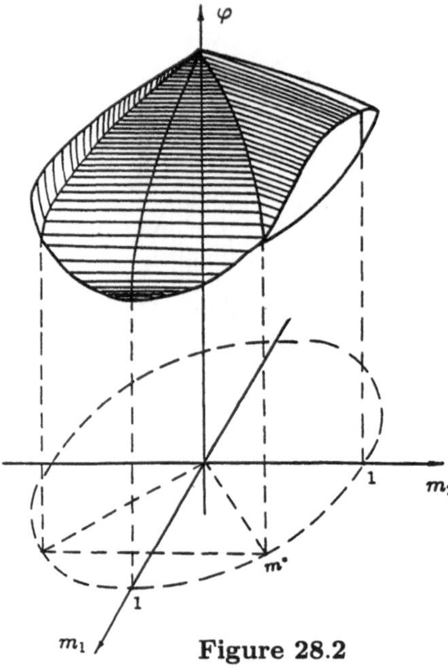

Figure 28.2

The recurrent sequence (28.41), (28.42) determines the algorithm that calculates the function $\kappa(\tau_*, m_*, \Delta)$ according to the equality

$$\kappa(\tau_*, m_*, \Delta) = \varphi_1(\tau_*, m), \quad m_*^T = \{m_1, 0, m_2, 0\} \tag{28.43}$$

for every required value m_*, $| m_* | \leq 1$.

Let us suppose now that the condition of the monotonicity of the value $\nu(\tau_*, \tau_j, m)$ (28.29) assumed above is not fulfilled. Then the previous construction is changed in the following way. The equation (28.31) which determines the points $m^*[\tau_j]$ is substituted by the equation constructed below.

For $j = k$ the vector $m^*[\tau_k]$ is calculated on the base of the equation (28.31) and in accordance with (28.32)–(28.39).

Let for $j > j_*$ the points $m^*[\tau_j]$ be already obtained. For the given values m, $| m | = 1$, $m_1 \leq m_1^*[\tau_{j_*+1}]$ and $j_* \in [1, \ k]$ we find the largest number j^* among the numbers $j > j_*$ such that for all $j \in [j_* + 1, \ j^*)$ the inequality

$$m_1^*[\tau_j] \geq m_1 \tag{28.44}$$

holds. But for $j = j^*$ this inequality is not valid.

Then we consider the point $m^*[\tau_{j_*}, m]$ determined by the equation

$$\nu_*(\tau_*, \tau_{j_*}, m) = \left[\frac{\partial}{\partial m_2}\Big(\sum_{j=j_*}^{j^*-1} \Delta\psi_j(\tau_*, m)\Big)\right]_{|m|=1} = 0. \tag{28.45}$$

If $m_1 > m_1^*[\tau_{j_*+1}]$ then we take

$$j^* = j_* + 1\,. \tag{28.46}$$

Then the equation (28.45) coincides with the equation (28.31).

The points $m^*[\tau_j]$ in the general case are determined by the equation (28.45) the same way as the corresponding points $m^*[\tau_j]$ are determined by the above equation (28.31).

Namely, let us consider for a fixed number j_* the point $m^T = \{m_1, m_2\}$, $\mid m \mid = 1$, in the domain (28.28) as a variable. If the value $\nu_*(\tau_*, \tau_{j_*}, m) > 0$ for all m, $m_1 < 1$ then we choose $m^{*T}[\tau_{j_*}] = \{0, 1\}$. If the value $\nu_*(\tau_*, \tau_{j_*}, m) \leq 0$ for all m, then we take $m^{*T}[\tau_{j_*}] = \{1, 0\}$. If neither of the two cases holds, we do the following. Let $m_1 \in [0,\ 1)$ decrease. For the point $m^*[\tau_{j_*}]$ we take the point m that satisfies the condition

$$m^*[\tau_{j_*}, m] = m. \tag{28.47}$$

In particular, if the value $\nu(\tau_*, \tau_j, m)$ monotonically increases in j then the value $m^*[\tau_{j_*}]$ determined by (28.47) coincides with the root of the equation (28.31).

When all the points $m^*[\tau_j]$, $j = 1, \ldots, k$ are constructed, we calculate the values of the functions $\varphi_1(\tau_*, m)$ for this or that fixed point m, $\mid m \mid \leq 1$ in the domain (28.28) using the same recurrent formulas (28.41), (28.42) from $j = k$ to $j = 1$. It is clear that in the general case this calculation is more complicated than in the case when the quantity $\nu(\tau_*, \tau_j, m)$ is monotone in j.

The algorithm block that calculates the quantities $\varphi_1(\tau_*, m)$ permits us to maximize the quantity $\widetilde{e}_*(\tau_*, w_*, m, \Delta)$ (28.23) that determines the program extremum

$$e_*(\tau_*, w_*, \Delta) = \max_{m_*}\{\tilde{e}_*(\tau_*, w_*, m, \Delta), \quad m_*^T = \{m_1, 0, m_2, 0\}. \tag{28.48}$$

Then we can calculate the vectors $s^{[u]a}(\tau_*, x_*, \varepsilon_u)$ and $s^{[v]a}(\tau_*, x_*, \varepsilon_v)$ which determine the approximating optimal control actions $u^a[t]$ and the approximating optimal disturbances $v^a[t]$. These vectors have a form

$$s^{[u]a}(\tau_*, x_*, \varepsilon_u) = R^{[u]}(\varepsilon_u, \tau_*)\frac{X^T(\vartheta, \tau_*)m_{*u}^0}{\mid X^T(\vartheta, \tau_*)m_{*u}^0 \mid} \tag{28.49}$$

if $\mid m_{*u}^0 \mid > 0$ or

$$s^{[u]a}(\tau_*, x_*, \varepsilon_u) = 0 \tag{28.50}$$

if $\mid m_{*u}^0 \mid = 0$,

$$s^{[v]a}(\tau_*, x_*, \varepsilon_v) = R^{[v]}(\varepsilon_v, \tau_*)\frac{X^T(\vartheta, \tau_*)m_{*v}^0}{\mid X^T(\vartheta, \tau_*)m_{*v}^0 \mid} \tag{28.51}$$

if $\mid m_{*v}^0 \mid > 0$ or

$$s^{[v]a}(\tau_*, x_*, \varepsilon_v) = 0 \tag{28.52}$$

if $\mid m_{*v}^0 \mid = 0$, where the vectors m_{*u}^0 and m_{*v}^0 are, respectively, the solutions of the following maximizing problems:

$$\begin{aligned}
&\langle\, m_{*u}^0,\ X(\vartheta, \tau_*)x_*\, \rangle + \varphi_1(\tau_*, m_{*u}^0) - \\
&-R^{[u]}(\varepsilon_u, \tau_*)[(x_{11}^2(\vartheta, \tau_*) + x_{12}^2(\vartheta, \tau_*))]^{1/2} \mid m_{*u}^0 \mid = \\
&= \max_{m_*}\langle\, m_*,\ X(\vartheta, \tau_*)x_*\, \rangle + \varphi_1(\tau_*, m_*) - \\
&- R^{[u]}(\varepsilon_u, \tau_*)[(x_{11}^2(\vartheta, \tau_*) + x_{12}^2(\vartheta, \tau_*))]^{1/2} \mid m_* \mid, \\
&\langle\, m_{*v}^0,\ X(\vartheta, \tau_*)x_*\, \rangle + \varphi_1(\tau_*, m_{*v}^0) + \\
&+R^{[v]}(\varepsilon_v, \tau_*)[(x_{11}^2(\vartheta, \tau_*) + x_{12}^2(\vartheta, \tau_*))]^{1/2} \mid m_{*v}^0 \mid = \\
&= \max_{m_*}\langle\, m_*,\ X(\vartheta, \tau_*)x_*\, \rangle + \varphi_1(\tau_*, m_*) +
\end{aligned} \tag{28.53}$$

$$+ R^{[v]}(\varepsilon_v, \tau_*)[(x_{11}^2(\vartheta, \tau_*) + x_{12}^2(\vartheta, \tau_*))]^{1/2} \mid m_* \mid . \qquad (28.54)$$

The optimal actions $u^a[t]$ and $v^a[t]$ are calculated using conditions similar to (8.11) and (8.16). These conditions in the considered case are

$$\langle\, s^{[u]a}(t, x, \varepsilon_u),\ B(t)u^a(t, x, \varepsilon_u)\, \rangle =$$

$$\min_{u \in P} \langle\, s^{[u]a}(t, x, \varepsilon_u),\ B(t)u\, \rangle, \qquad (28.55)$$

$$\langle\, s^{[v]a}(t, x, \varepsilon_v),\ C(t)v^a(t, x, \varepsilon_v)\, \rangle =$$

$$\max_{v \in Q} \langle\, s^{[v]a}(t, x, \varepsilon_v),\ C(t)v\, \rangle. \qquad (28.56)$$

From (28.49)–(28.52) and (28.55), (28.56) we obtain

$$u_1^a(t, x, \varepsilon_u) = -\mathrm{sign}[x_{12}(\vartheta, t)k(t)] \frac{m_{*u1}^0 a_1^2[t]}{[(m_{*u1}^0 a_1[t])^2 + (m_{*u2}^0 a_2[t])^2]^{1/2}}$$

$$u_2^a(t, x, \varepsilon_u) = -\mathrm{sign}[x_{12}(\vartheta, t)k(t)] \frac{m_{*u2}^0 a_2^2[t]}{[(m_{*u1}^0 a_1[t])^2 + (m_{*u2}^0 a_2[t])^2]^{1/2}}$$

if $\mid m_{*u}^0 \mid \neq 0$ and

$$u_1^a(t, x, \varepsilon_u) = 0, \quad u_2^a(t, x, \varepsilon_u) = 0 \qquad (28.57)$$

if $\mid m_{*u}^0 \mid = 0$

$$v_1^a(t, x, \varepsilon_v) = \mathrm{sign}[x_{12}(\vartheta, t)n(t)] \frac{m_{*v1}^0 b_1^2[t]}{[(m_{*v1}^0 b_1[t])^2 + (m_{*v2}^0 b_2[t])^2]^{1/2}},$$

$$v_2^a(t, x, \varepsilon_v) = \mathrm{sign}[x_{12}(\vartheta, t)n(t)] \frac{m_{*v2}^0 b_2^2[t]}{[(m_{*v1}^0 b_1[t])^2 + (m_{*v2}^0 b_2[t])^2]^{1/2}}$$

if $\mid m_{*v}^0 \mid \neq 0$ and

$$v_1^a(t, x, \varepsilon_v) = 0, \quad v_2^a(t, x, \varepsilon_v) = 0 \qquad (28.58)$$

if $\mid m_{*v}^0 \mid = 0$.

We also note, that for the stabilization of the considered process of control it can be expedient to spoil somewhat the values of the ensured results $\rho_u^0(t, w_*^{[u]})$

and $\rho_v^0(t, w_*^{[v]})$. Namely, when the motion of $w^{[u]}$-model is formed we can subject the control actions u to more strict conditions:

$$\frac{u_1^2}{a_1^2} + \frac{u_2^2}{a_2^2} \le (1 - \varepsilon_u^*), \qquad \varepsilon_u^* > 0 . \tag{28.59}$$

Conversely, when we form the motion of $w^{[v]}$-model we can subject the actions v to more strict conditions:

$$\frac{v_1^2}{b_1^2} + \frac{v_2^2}{b_2^2} \le (1 - \varepsilon_v^*), \qquad \varepsilon_v^* > 0 . \tag{28.60}$$

These restrictions provide a more stable following of the motions of $w^{[u]}$-model and $w^{[v]}$-model by the motion $x[t]$ of x-object.

The results of the simulation of the algorithm described in this section are represented in Section 1 in Figures 1.3–1.5.

29. Example with a quadratic quality index

In this section we consider one particular case of the quality index $\gamma_{(4)}$ (21.4). For this particular case we will describe the procedure of constructing the value $e_{*(4)}(\cdot)$ and the approximating optimal strategies $u^a_{*(4)}(\cdot) = u^a_{*(4)}(t, x, \varepsilon)$ and $v^a_{*(4)}(\cdot) = v^a_{*(4)}(t, x, \varepsilon)$ which approximate the saddle point in the corresponding differential game with $\gamma_{(4)}$. Examples of computer simulation of the control process will be presented, too.

We consider the control system described by the linear differential equation

$$\dot{x} = A(t)x + B(t)u + C(t)v, \qquad t_0 \le t < \vartheta \tag{29.1}$$

where $x \in \mathbf{R}^n$, $u \in \mathbf{R}^r$, $v \in \mathbf{R}^s$; $A(t)$, $B(t)$, and $C(t)$ are piecewise-continuous matrix-valued functions continuous from the right.

Let the quality index have a form

$$\gamma = | Dx[\vartheta] | + \int_{t_*}^{\vartheta} [\langle u[\tau], \Phi(\tau)u[\tau] \rangle - \langle v[\tau], \Psi(\tau)v[\tau] \rangle] d\tau \tag{29.2}$$

where $\Phi(t)$ and $\Psi(t)$ are symmetric continuous matrix-valued functions. In (29.2) the quadratic forms are positively definite, i.e.,

$$\langle\, u,\ \Phi(t)u\,\rangle \geq \alpha_u \mid u \mid^2, \qquad \alpha_u > 0\ -\ \text{const}, \tag{29.3}$$

$$\langle\, v,\ \Psi(t)v\,\rangle \geq \alpha_v \mid v \mid^2, \qquad \alpha_v > 0\ -\ \text{const}. \tag{29.4}$$

As above, the matrix D determines the seminorm $\mu(x) =\mid Dx \mid$.

Below we can restrict ourselves to the case in which $D = E$ is the unit matrix. Indeed, if this is not valid for the original problem, one can use the transformation

$$y = DX(\vartheta, t)x \tag{29.5}$$

and rename y as a new vector x. The dimension of the new phase vector x is equal the number ν of lines of the matrix D to be linear independent (see Section 10). The new vector x satisfies the equation

$$\dot{x} = DX(\vartheta, t)B(t)u + DX(\vartheta, t)C(t)v \tag{29.6}$$

which after renaming the matrices $B(t)$ and $C(t)$ takes the form (29.1), where $A(t)$ is the zero matrix. Due to the equality $X(\vartheta, \vartheta) = E$ and to the fact that the lines of D are linear independent we obtain a problem that is the same as the initial one but with the quality index $\gamma_{(4)}(\cdot)$ in the form

$$\gamma =\mid x[\vartheta] \mid + \int_{t_*}^{\vartheta} [\langle\, u[\tau],\ \Phi(\tau)u[\tau]\,\rangle - \langle\, v[\tau],\ \Psi(\tau)v[\tau]\,\rangle] d\tau. \tag{29.7}$$

We assume a priori that here we have no restrictions for control actions u and disturbances v in (29.1). But it is known that under the conditions (29.3) and (29.4) the optimal values $u^0(t, x, \varepsilon)$ and optimal disturbances $v^0(t, x, \varepsilon)$ which solve the considered problem for γ (29.2) are bounded a posteriori with respect to the Euclidean norm by some sufficiently large number M. The proof of this assertion is published in the book [77]. We can impose in our case the restrictions $\mid u \mid\leq M$, $\mid v \mid\leq M$. Thus, we can suppose a priori that the control actions u and the disturbances v satisfy the conditions

$$u \in P, \qquad P = \{u : \mid u \mid\leq M\}, \tag{29.8}$$

$$v \in Q, \quad Q = \{v : | v | \leq M\}. \tag{29.9}$$

These formal restrictions do not distort the initial problem.

Thus, the system (29.1), (29.8), (29.9) is a particular case of the system (10.1) under the restrictions (3.5), (3.6). It is obvious that for the system (29.1), (29.8), (29.9) with the functional (29.7) the saddle point condition in the small game (21.9) is valid.

The quality index γ (29.7) is a particular case of the quality index $\gamma_{(4)}$ (21.4). So, according to Theorem 22.1 (and to the result of [77], as was mentioned above) the differential game for the object (29.1), (29.8), (29.9) with the quality index γ (29.7) has the value $\rho^0(t,x)$ and a saddle point $\{u^0(\cdot), v^0(\cdot)\}$ which is a pair of pure optimal strategies $u^0(\cdot) = u^0(t,x,\varepsilon)$ and $v^0(\cdot) = v^0(t,x,\varepsilon)$.

We can propose, for instance, the following natural interpretation of the considered model example.

Let in some territory two points A^*, B^* be fixed. It is required that a certain load be transferred from the point A^* to the point B^* (see Figure 29.1).

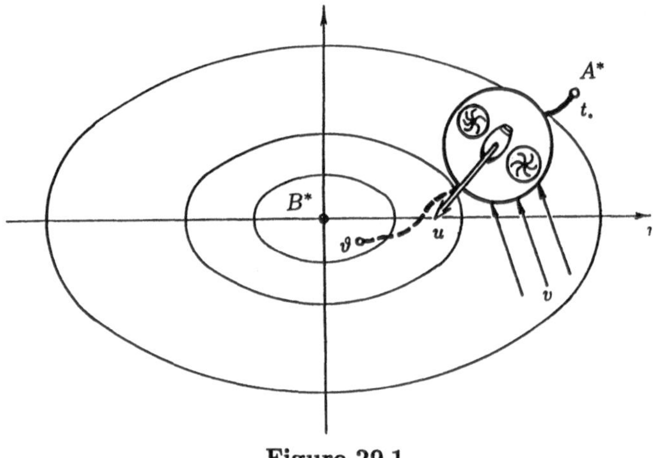

Figure 29.1

Let $r^T[t] = \{r_1[t], r_2[t]\}$ be the radius-vector of the load at the current time moment t. For example, the load can be fastened to some mobile platform. Then $r[t]$ can be the radius-vector of the center of mass of the platform with the load. The motion begins at the time moment t_* from the state $r[t_*] = r_A$, $\dot{r}^T[t_*] = \{0,0\}$. The moment of termination of the process ϑ is given. The desired terminal state is $r^T[\vartheta] = \{0,0\}$, $\dot{r}^T[\vartheta] = \{0,0\}$. For the transportation the owner of the load is ready to pay the amount of money $\mathcal{G}$ provided the conditions given above are fulfilled. However, if the conditions at the time moment $t = \vartheta$ are not fulfilled, then from the fee $\mathcal{G}$ the fine $\mathcal{F}$ is subtracted. This fine $\mathcal{F}$ is determined by the deviation of the values $r[\vartheta]$ and $\dot{r}[\vartheta]$ from the required values. We assume that the fine is calculated according to the formula

$$\mathcal{F} = \Big[\; e_1 \mid r[\vartheta] \mid^2 + e_2 \mid \dot{r}[\vartheta] \mid^2 \Big]^{1/2} \tag{29.10}$$

with the given coefficients $e_1 > 0$ and $e_2 \geq 0$.

Having the jet-propelled platform we intend to undertake this job. Our platform can move under the action of the thrust u produced by the jet engine. The platform is also subjected to the force v created by the wind. Assuming that the total mass of the platform and the load is $m[t]$ we write the equation of the motion in the form of Meshcherskii

$$m[t]\ddot{r} = \beta r + \alpha\dot{r} + \dot{m}[t]Ku + Nv\,. \tag{29.11}$$

About the wind we assume that at every time moment the vector v can take an arbitrary value. We assume also that at the current time moment we can assign to the thrust u an arbitrary value. The only additional restrictions are that the realizations $u[t]$, $v[t]$, $t_* \leq t < \vartheta$, should satisfy the conditions (29.8), (29.9) and be piecewise-continuous.

Suppose that the balance of money is determined by the following circumstances. If the engine produces the thrust $u[t]$, $t_* \leq t < \vartheta$, then the cost $\mathcal{E}$ of the energy spent is calculated by the formula

$$\mathcal{E} = \int_{t_*}^{\vartheta} \Big(\; \varphi_1[t]u_1^2[t] + \varphi_2[t]u_2^2[t] \; \Big)dt, \quad \varphi_i[t] \geq \alpha_u > 0\,. \tag{29.12}$$

We do not justify this formula with any physical reasons and take it as given. We can note, for example, that the value $u_i^2[t]$ may be interpreted as the power spent on the thrust $u_i[t]$. Then the function $\varphi_i[t] > 0$ can be considered as the cost of the unit of power. Thus, we assume here that the cost of power may depend on time $t \in [t_*, \; \vartheta]$.

Assume also that our platform is equipped with a wind-propelled generator. If the object is subjected to the actions $v[t]$, $t_* \leq t < \vartheta$, then this generator produces the energy whose cost $\mathcal{H}$ is calculated according to the formula

$$\mathcal{H} = \int_{t_*}^{\vartheta} \Big(\; \psi_1[t]v_1^2[t] + \psi_2[t]v_2^2[t] \; \Big)dt, \quad \psi_i[t] \geq \alpha_v > 0\,. \tag{29.13}$$

So, our profit $\mathcal{I}$ is determined by the equality

$$\mathcal{I} = \mathcal{G} - \mathcal{F} - \mathcal{E} + \mathcal{H}. \tag{29.14}$$

We are interested in finding a method U of control that should minimize the value

$$\gamma = \mathcal{F} + \mathcal{E} - \mathcal{H}. \tag{29.15}$$

This value is connected with our profit by the formula

$$\mathcal{I} = \mathcal{G} - \gamma. \tag{29.16}$$

Thus, the value γ is the quality index in this example.

If in (29.10) we have $e_1 > 0$ and $e_2 > 0$, then it is desirable to deliver the platform in a neighbourhood of the point B^* with a sufficiently small final speed. In this case the matrix D in (29.2) is the 4-dimensional unit matrix E. If $e_1 > 0$, $e_2 = 0$, then it is desirable only to bring the platform in a neighbourhood of the point B^*. Here the terminal speed does not matter. In this case the matrix D in (29.2) is of dimension 2×4.

Thus, we come to the problem on the minimax of the quality index

$$\gamma = \Big(e_1 \mid r[\vartheta] \mid^2 + e_2 \mid \dot{r}[\vartheta] \mid^2 \Big)^{1/2} + \int_{t_*}^{\vartheta} \Big[\varphi_1[t] u_1^2[t] +$$

$$+ \varphi_2[t] u_2^2[t] \Big] dt - \int_{t_*}^{\vartheta} \Big[\psi_1[t] v_1^2[t] + \psi_2[t] v_2^2[t] \Big] dt \tag{29.17}$$

for the process of control for the system (29.11). Thus, we can consider this problem as a particular case of the model problem dealt with in this section.

Let us continue our considerations concerned with the general control problem for the system (29.1)–(29.9).

Using the constructions from Sections 22 and 23 we now describe the computation of the stochastic program extremum $e_*^{[c]}(\tau_*, w_*, \Delta)$ which provides the approximation of the value of the game $\rho^0(t_*, x_*)$ for γ (29.2). Thus we construct also the approximating optimal strategies $u^a(\cdot)$ and $v^a(\cdot)$ basing on the value $e_*^{[c]}(\cdot)$.

The functional γ (29.2) and the corresponding program extremum $e_*(\cdot)$ were introduced in the book [77]. In that book the construction of the value $e_*(\cdot)$ is presented for a concrete functional (29.2). In [77] the computation of the value $e_*(\cdot)$ is based on some artificial construction. This construction essentially employs some special expression for the conditional expectation $m(\tau,\omega)$ considered there, which turns out to be the martingale

$$m(\tau,\omega) = m(\xi_1,\ldots,\xi_j) = m_* + \sum_{i=1}^{j} a(\xi_1,\ldots,\xi_i), \qquad \tau_j \leq \tau < \tau_{j+1},$$

$$j = 1,\ldots,k, \qquad M\{a(\xi_1,\ldots,\xi_j) \mid \xi_1,\ldots,\xi_{j-1}\} = a(\xi_1,\ldots,\xi_{j-1}) \,. \tag{29.18}$$

In this section we show that the calculation of the desired value $e_*^{[c]}(\cdot)$ can be obtained by the uniform method given in the present book for the functionals of the considered type $\gamma_{(4)}$.

Consider the $\widehat{w}$-model described by the system of differential equations:

$$\dot{w} = A(\tau)w + B(\tau)u + C(\tau)v, \tag{29.19}$$

$$\dot{\widehat{w}}_{n+1} = \langle\, u,\ \Phi(\tau)u \,\rangle - \langle\, v,\ \Psi(\tau)v \,\rangle \tag{29.20}$$

where $w \in \mathbf{R}^n$, $\widehat{w}_{n+1} \in \mathbf{R}^1$, $\widehat{w} = \{w, \widehat{w}_{n+1}\}$. Let $\{\tau_*, \widehat{w}_*\} = \{\tau_*, w_*, \widehat{w}_{*n+1}\}$ be the initial position for the model (29.19), (29.20). The probability basis for this model is the same as in Section 12.

The quantity $e_*(\cdot) = e_*^{[c]}(\tau_*, w_*, \Delta)$, $\widehat{w}_{*n+1} = c$ has a form

$$e_*^{[c]}(\tau_*, w_*, \Delta) = \sup_{\| l(\cdot) \|^* \leq 1} \Big[\langle\, m_*,\ X(\vartheta,\tau_*)w_* \,\rangle +$$

$$+ \sum_{j=1}^{k} (\tau_{j+1} - \tau_j) M\Big\{\min_u \max_v \Big[\langle\, m(\tau_j,\omega),\ X(\vartheta,\tau_j)(B(\tau_j)u + C(\tau_j)v) \,\rangle +$$

$$+ \langle\, u,\ \Phi(\tau_j)u \,\rangle - \langle\, v,\ \Psi(\tau_j)v \,\rangle\Big]\Big\}\Big] + c =$$

$$= \sup_{\| l(\cdot) \|^* \leq 1} \Big[\langle\, m_*,\ X(\vartheta,\tau_*)w_* \,\rangle +$$

$$+ \sum_{j=1}^{k} (\tau_{j+1} - \tau_j) M\Big\{\langle\, m(\tau_j,\omega),\ N_*(\tau_j)m(\tau_j,\omega) \,\rangle\Big\}\Big] + c\,. \tag{29.21}$$

Here $X(t,\tau)$ is the fundamental matrix for the equation $dw/dt = A(t)w$; $l(\cdot) = \{l_i(\omega),\ \omega \in \Omega,\ i = 1,\dots,n\}$ is the vector random variable with the norm

$$\| l(\cdot) \|^* = \operatorname{vrai}\max_{\omega \in \Omega} | l(\omega) | . \tag{29.22}$$

In (29.21) we have

$$N_*(\tau) = \frac{1}{4} X(\vartheta,\tau)\Big[C(\tau)\Psi^{-1}(\tau)C^T(\tau) - B(\tau)\Phi^{-1}(\tau)B^T(\tau)\Big]X^T(\vartheta,\tau), \tag{29.23}$$

$$m_* = M\{l(\omega)\}; \quad m(\tau_j,\omega) = M\{l(\xi_1,\dots,\xi_k) \mid \xi_1,\dots,\xi_j\}, \quad j = 1,\dots,k. \tag{29.24}$$

In (29.23) $\Phi^{-1}(\tau)$ and $\Psi^{-1}(\tau)$ are the inverse matrices for $\Phi(\tau)$ and $\Psi(\tau)$ in (29.3) and (29.4), respectively.

According to (23.15), (23.20) the value $e_*(\cdot)$ (29.21) is defined by the following equality:

$$e_*^{[c]}(\tau_*, w_*, \Delta) = \max_{|m|\le 1}\Big[\langle\, m,\ X(\vartheta,\tau_*)w_*\,\rangle + \varphi_1(\tau_*, m)\Big] + c \tag{29.25}$$

where $m \in \mathbf{R}^n$.

In (29.25) the quantity $\varphi_1(\tau_*, m)$ is calculated in accordance with the following procedure.

Let us denote for the chosen partition $\Delta = \Delta\{\tau_j\},\ j = 1,\dots,k$

$$F(\tau_j) = \sum_{i=j}^{k} N_*(\tau_i)(\tau_{i+1} - \tau_i) \tag{29.26}$$

where $N_*(\tau)$ is a matrix-function (29.23).

We take

$$\varphi_{k+1}(\tau_*, m) = 0, \quad \Delta\psi_k(\tau_*, m) = (\tau_{k+1} - \tau_k)\langle\, m,\ N_*(\tau_k)m\,\rangle = \langle\, m,\ F(\tau_k)m\,\rangle,$$

$$\psi_k(\tau_*, m) = \Delta\psi_k(\tau_*, m), \qquad \varphi_k(\tau_*, m) = \overline{\psi}_k(\tau_*, m), \qquad |\, m \,| \leq 1\,. \tag{29.27}$$

The function $\psi_k(\tau_*, m)$ in (29.27) is a quadratic form. Let $Q_j,\ j = 1, \ldots, k$ be the orthogonal matrix of the transformation of the variables

$$m = Q_j \tilde{m}, \qquad |\, m \,| = |\, Q_j \tilde{m} \,| = |\, \tilde{m} \,| \tag{29.28}$$

which reduces the quadratic forms $\langle\, m,\ F(\tau_j) m \,\rangle$ to the canonical form:

$$\langle\, m,\ F(\tau_j) m \,\rangle = \langle\, Q_j \tilde{m},\ F(\tau_j) Q_j \tilde{m} \,\rangle = \lambda_1^{(j)} \tilde{m}_1^2 + \ldots + \lambda_n^{(j)} \tilde{m}_n^2\,. \tag{29.29}$$

Here the real numbers $\lambda_i^{(j)},\ i = 1, \ldots, n$ are the eigen-values of the matrix $F(\tau_j)$ (29.26), $j = 1, \ldots, k$.

We denote

$$\lambda_j^* = \max \Big[\big(\max_{l=j,\ldots,k} \ \max_{i=1,\ldots,n} \ \lambda_i^{(l)} \big),\ 0 \Big], \qquad j = 1, \ldots, k\,. \tag{29.30}$$

The equality

$$\psi_k(\tau_*, m) = \psi_k(\tau_*, Q_k \tilde{m}) = \lambda_1^{(k)} \tilde{m}_1^2 + \ldots + \lambda_n^{(k)} \tilde{m}_n^2 \tag{29.31}$$

holds.

Denote

$$\lambda^* = \max_{i=1,\ldots,n} \lambda_i^{(k)}\,. \tag{29.32}$$

We consider two cases. In the first case $\lambda^* \leq 0$. Then the function $\psi_k(\tau_*, m)$ is of constant sign negative quadratic form, i.e., $\psi_k(\tau_*, m)$ is a concave function. Therefore, the equality

$$\varphi_k(\tau_*, m) = \overline{\psi}_k(\tau_*, m) = \psi_k(\tau_*, m), \qquad |\, m \,| \leq 1 \tag{29.33}$$

is valid.

Consider now the second case $\lambda^* > 0$. In this case we construct $\varphi_k(\tau_*, m)$ in the following way. Suppose

$$\lambda^* = \lambda_n^{(k)}. \tag{29.34}$$

Then according to (29.27)–(29.29), (29.32) we take

$$\varphi_k(\tau_*, m) = \varphi_k(\tau_*, Q_k\tilde{m}) = \lambda_1^{(k)}\tilde{m}_1^2 + \ldots + \lambda_{n-1}^{(k)}\tilde{m}_{n-1}^2 + \lambda_n^{(k)}\tilde{m}_n^2 +$$

$$+\lambda^*(1 - \tilde{m}_1^2 - \ldots - \tilde{m}_{n-1}^2) - \lambda^*\tilde{m}_n^2 = \psi_k(\tau_*, Q_k\tilde{m}) - \lambda^* \mid \tilde{m} \mid^2 + \lambda^* =$$
$$= \psi_k(\tau_*, m) - \lambda^* \mid m \mid^2 + \lambda^*. \tag{29.35}$$

Here we used (29.34) and the equality

$$\mid \tilde{m} \mid^2 = \tilde{m}_1^2 + \cdots \tilde{m}_{n-1}^2 + \tilde{m}_n^2. \tag{29.36}$$

Let us show now that the function $\varphi_k(\tau_*, m)$ (29.35) is an upper convex hull of the function $\psi_k(\tau_*, m)$. We have to check that the function $\varphi_k(\tau_*, m)$ (29.35) satisfies conditions 1–3 (see (13.11)–(13.15)). According to (29.35) we have $\varphi_k(\tau_*, m) = \lambda^* - \lambda^* \mid m \mid^2 + \psi_k(\tau_*, m) = \lambda^*(1 - \mid m \mid^2) + \psi_k(\tau_*, m) \geq \psi_k(\tau_*, m)$ because $\lambda^* > 0$ and $\mid m \mid \leq 1$. Therefore, condition 1 is valid.

Futher, taking into account (29.27), (29.29), (29.32) and (29.35) we obtain

$$\varphi_k(\tau_*, m) = -\lambda^*(\tilde{m}_1^2 + \cdots + \tilde{m}_n^2) + \lambda_1^{(k)}\tilde{m}_1^2 + \cdots + \lambda_n^{(k)}\tilde{m}_n^2 + \lambda^* =$$

$$= (\lambda_1^{(k)} - \lambda^*)\tilde{m}_1^2 + \cdots + (\lambda_n^{(k)} - \lambda^*)\tilde{m}_n^2 + \lambda^* \tag{29.37}$$

and $\lambda_i^{(k)} \leq \lambda^*$. Therefore, the function $\varphi_k(\tau_*, m)$ is concave. And thus, this function satisfies condition 2. Let us check that condition 3 is also fulfilled. For the arbitrary vector m_*, $\mid m_* \mid \leq 1$ the following representation

$$m_* = \lambda m' + (1 - \lambda)m'', \quad \mid m' \mid = 1, \quad \mid m'' \mid = 1, \quad 0 \leq \lambda \leq 1 \tag{29.38}$$

is valid, where in the case $| m_* |< 1$ we take

$$m' = Q_k\widetilde{m}', \qquad \widetilde{m}' = \{ \widetilde{m}_1, \dots, \widetilde{m}_{n-1}, \widetilde{m}'_n \},$$

$$\widetilde{m}'_n = \left(1 - \widetilde{m}_1^2 - \dots - \widetilde{m}_{n-1}^2 \right)^{1/2} \tag{29.39}$$

$$m'' = Q_k\widetilde{m}'', \qquad \widetilde{m}'' = \{ \widetilde{m}_1, \dots, \widetilde{m}_{n-1}, \widetilde{m}''_n \},$$

$$\widetilde{m}''_n = -\left(1 - \widetilde{m}_1^2 - \dots - \widetilde{m}_{n-1}^2 \right)^{1/2} \tag{29.40}$$

$$\widetilde{m} = \{ \widetilde{m}_1, \dots, \widetilde{m}_{n-1}, \widetilde{m}_n \} = Q_k^{-1} m_* \tag{29.41}$$

$$\lambda = | \widetilde{m}_n - \widetilde{m}''_n || \widetilde{m}'_n - \widetilde{m}''_n |^{-1}, \quad 0 < \lambda < 1 . \tag{29.42}$$

In the case $| m_* |= 1$ we can assume $m_* = m'$, $\lambda = 1$.

According to (29.35), (29.38)–(29.42) the following equalities

$$\varphi_k(\tau_*, m_*) = \varphi_k(\tau_*, m') - \varphi_k(\tau_*, m'') =$$

$$= \psi_k(\tau_*, m') = \psi_k(\tau_*, m'') \tag{29.43}$$

hold.

For any arbitrary concave function $\zeta(\tau_*, m)$ majorizing the function $\psi_k(\tau_*, m)$ we have according to (29.35), (29.39), (29.40), (29.43) the following relations:

$$\zeta(\tau_*, m_*) = \zeta(\tau_*, \lambda m' + (1 - \lambda)m'') \geq$$

$$\geq \lambda\zeta(\tau_*, m') + (1 - \lambda)\zeta(\tau_*, m'') \geq$$
$$\geq \lambda\psi_k(\tau_*, m') + (1 - \lambda)\psi_k(\tau_*, m'') =$$
$$= \psi_k(\tau_*, m') = \varphi_k(\tau_*, m_*) . \tag{29.44}$$

Therefore the condition 3 (13.13) is valid.

Using (29.33) and (29.35) we obtain that the function $\varphi_k(\tau_*, m)$ is determined by the equality

$$\varphi_k(\tau_*, m) = \langle m, F(\tau_k)m \rangle - \lambda_k^* | m |^2 + \lambda_k^* \tag{29.45}$$

in both cases $\lambda^* \leq 0$ and $\lambda^* > 0$. We would like to note that in the case $\lambda^* \leq 0$ the equality (29.45) for the function $\varphi_k(\tau_*, m)$ coincides with the equality (29.33) because in this case $\lambda_k^* = 0$ according to (29.30), (29.32).

Consider the induction in j from $j = k$ to $j = 1$. Suppose that the function $\varphi_{j+1}(\tau_*, m)$ is already constructed and has the form

$$\varphi_{j+1}(\tau_*, m) = \langle\, m,\ F(\tau_{j+1})m \,\rangle - \lambda_{j+1}^* \mid m \mid^2 + \lambda_{j+1}^* \,. \tag{29.46}$$

Then

$$\begin{aligned} \Delta\psi_j(\tau_*, m) &= (\tau_{j+1} - \tau_j)\langle\, m,\ N_*(\tau_j)m \,\rangle, \\ \psi_j(\tau_*, m) &= \Delta\psi_j(\tau_*, m) + \varphi_{j+1}(\tau_*, m), \\ \varphi_j(\tau_*, m) &= \overline{\psi}_j(\tau_*, m), \qquad \mid m \mid \leq 1\,. \end{aligned} \tag{29.47}$$

Thus, according to (29.26), (29.46), (29.47) we calculate the function

$$\psi_j(\tau_*, m) = \langle\, m,\ F(\tau_j)m \,\rangle - \lambda_{j+1}^* \mid m \mid^2 + \lambda_{j+1}^* \,. \tag{29.48}$$

The function $\psi_j(\tau_*, m)$ (29.48) is a quadratic form, too. Therefore, according to (29.28), (29.29) and (29.48) we have

$$\begin{aligned} \psi_j(\tau_*, m) = \psi_j(Q_j\widetilde{m}) = \lambda_1^{(j)}\widetilde{m}_1^2 + \ldots + \lambda_n^{(j)}\widetilde{m}_n^2 - \\ - \lambda_{j+1}^* \mid \widetilde{m} \mid^2 + \lambda_{j+1}^* \,. \end{aligned} \tag{29.49}$$

Denote

$$\lambda^* = \max_{i=1,\ldots,n} \lambda_i^{(j)}\,. \tag{29.50}$$

We consider again two cases. In the first case $\lambda^* \leq \lambda_{j+1}^*$. Then the function $\psi_j(\tau_*, m)$ (29.48), (29.49) is concave. That is, the equalities

$$\varphi_j(\tau_*, m) = \overline{\psi}_j(\tau_*, m) = \psi_j(\tau_*, m), \qquad \mid m \mid \leq 1 \tag{29.51}$$

hold. And in this case, according to (29.30), (29.48), (29.50) and, consequently, according to $\lambda_j^* = \lambda_{j+1}^*$ we have

$$\varphi_j(\tau_*, m) = \langle\, m,\ F(\tau_j)m\, \rangle - \lambda_j^* \mid m \mid^2 + \lambda_j^*. \tag{29.52}$$

In the second case $\lambda^* > \lambda_{j+1}^*$. Then $\varphi_j(\tau_*, m)$ is constructed as follows. According to (29.28)–(29.30) and (29.47), (29.50) we take

$$\varphi_j(\tau_*, m) = \langle\, m,\ F(\tau_j)m\, \rangle - \lambda_j^* \mid m \mid^2 + \lambda_j^*\,. \tag{29.53}$$

Repeating with understandable changes the considerations given above for the case $j = k$, we see that the function $\varphi_j(\tau_*, m)$ (29.53) is an upper convex hull for the function $\psi_j(\tau_*, m)$ (29.48), (28.49) in the region $\mid m \mid \leq 1$. Then it follows from (29.45), (29.46), (29.52), (29.53) by induction in j that the function $\varphi_j(\tau_*, m)$ is determined by the formula

$$\varphi_j(\tau_*, m) = \langle\, m,\ F(\tau_j)m\, \rangle - \lambda_j^* \mid m \mid^2 + \lambda_j^* \tag{29.54}$$

for $j = 1, \ldots, k$. Let us recall that according to (29.30) $\lambda_j^* \geq 0$ for all $j = 1, \ldots, k$.

According to (29.25), (29.54) we obtain that the following equality

$$e_*^{[c]}(\tau_*, w_*, \Delta) = \max_{|m|\leq 1}\Big[\langle\, m,\ X(\vartheta, \tau_*)w_*\, \rangle + \langle\, m,\ F(\tau_*)m\, \rangle -$$

$$- \lambda_1^* \mid m \mid^2\Big] + \lambda_1^* + c \quad (\lambda_1^* \geq 0) \tag{29.55}$$

is valid.

In (29.55) the maximum really exists because the maximized expression is a continuous concave function of m.

Thus, in the considered case the approximating optimal control action $u^a(t_i, x[t_i], \varepsilon)$ and the approximating optimal disturbance $v^a(t_i, x[t_i], \varepsilon)$ are determined by the equalities

$$u^a(t_i, x[t_i], \varepsilon) = -\frac{1}{2}\Phi^{-1}(t_i)B^T(t_i)X^T(\vartheta, t_i)\, m_u^0(t_i, x[t_i], \varepsilon) \tag{29.56}$$

$$v^a(t_i, x[t_i], \varepsilon) = \frac{1}{2}\Psi^{-1}(t_i)C^T(t_i)X^T(\vartheta, t_i)\, m_v^0(t_i, x[t_i], \varepsilon) \tag{29.57}$$

where $m_u^0 = m_u^0(t_i, x[t_i], \varepsilon)$ and $m_v^0 = m_v^0(t_i, x[t_i], \varepsilon)$ are the maximizing vectors:

$$\begin{aligned} &\langle\, m_u^0,\ X(\vartheta, t_i)x[t_i]\,\rangle + \langle\, m_u^0,\ F(t_i)m_u^0\,\rangle - \lambda_i^* \mid m_u^0 \mid^2 - \\ &\quad -R(\varepsilon, t_i)[1+ \mid X^T(\vartheta, t_i)m_u^0 \mid^2]^{1/2} = \\ &= \max_{|m|\le 1}\Big[\langle\, m,\ X(\vartheta, t_i)x[t_i]\,\rangle + \langle\, m,\ F(t_i)m\,\rangle - \lambda_i^* \mid m \mid^2 - \\ &\quad - R(\varepsilon, t_i)[1+ \mid X^T(\vartheta, t_i)m \mid^2]^{1/2}\Big] \end{aligned} \tag{29.58}$$

and

$$\begin{aligned} &\langle\, m_v^0,\ X(\vartheta, t_i)x[t_i]\,\rangle + \langle\, m_v^0,\ F(t_i)m_v^0\,\rangle - \lambda_i^* \mid m_u^0 \mid^2 + \\ &\quad +R(\varepsilon, t_i)[1+ \mid X^T(\vartheta, t_i)m_v^0 \mid^2]^{1/2} = \\ &= \max_{|m|\le 1}\Big[\langle\, m,\ X(\vartheta, t_i)x[t_i]\,\rangle + \langle\, m,\ F(t_i)m\,\rangle - \lambda_i^* \mid m \mid^2 + \\ &\quad + R(\varepsilon, t_i)[1+ \mid X^T(\vartheta, t_i)m \mid^2]^{1/2}\Big] \end{aligned} \tag{29.59}$$

where $R(\varepsilon, t_i)$ is a quantity in (22.13). The values (29.56), (29.57) and relations (29.58), (29.59) are obtained from the conditions (8.11), (8.16), (22.16)–(22.22) with the substitution $\rho_{(4)}^0(\cdot)$ by $e_*^{[c]}(\cdot)$ and $\widehat{s}^{[u]0}$ by $\widehat{s}^{[u]a}$ and with the help of the interchanging of operations minimum in $\widehat{s}^{[u]a}$ and maximum in m similar to the calculations (18.1)–(18.15) in Section 18.

Let us note the following. In the considered case we can construct the optimal actions using the value of the game $\rho_{(4)}^0(\tau_*, w_*)$ itself instead of the program extremum $e_{*(4)}^{[c]}(\tau_*, w_*, \Delta)$ which is only an approximation to this value. According to (11.15), (23.22) the following equalities are valid: $\rho_{(4)}^0(\tau_*, w_*) = \lim \rho_{*(4)}^0(\tau_*, w_*)$, $\rho_{*(4)}^0(\tau_*, w_*) = \lim e_{*(4)}(\tau_*, w_*, \Delta_\delta)$. To obtain this limit value it suffices to substitute the integral sums in the formulas by the corresponding integrals. Thus, for example, instead of the matrix $F(\tau_i)$ we have to use the matrix $F(\tau_i) = \int_{\tau_i}^{\vartheta} N_*(\tau)d\tau$. And so on.

In conclusion of this section we consider three simple examples of computer simulation of model control processes.

Let us consider the results of simulation of the control process that corresponds to the above-given model of delivering a platform. Assume that the equation (29.11) has the form

$$\ddot{r} = \left[\begin{bmatrix} \beta_1 & 0 \\ 0 & \beta_2 \end{bmatrix} r - \begin{bmatrix} \alpha & 0 \\ 0 & \alpha \end{bmatrix} \dot{r} + \begin{bmatrix} c_1 & 0 \\ 0 & c_2 \end{bmatrix} v\right] \exp(\lambda(t - t_0))/m_0 -$$

$$- \lambda \begin{bmatrix} b_1 & 0 \\ 0 & b_2 \end{bmatrix} u, \qquad t_0 \leq t < \vartheta\,. \tag{29.60}$$

This means that the variable mass is

$$m[t] = m_0 \exp(-\lambda(t - t_0)). \tag{29.61}$$

Let the quality index γ have the form

$$\gamma = (r_1^2[\vartheta] + r_2^2[\vartheta])^{1/2} + \int_{t_*}^{\vartheta} [\, \langle\, u(\tau), \begin{bmatrix} \varphi_1 & 0 \\ 0 & \varphi_2 \end{bmatrix} u(\tau) \,\rangle -$$

$$- \langle\, v(\tau), \begin{bmatrix} \psi_1 & 0 \\ 0 & \psi_2 \end{bmatrix} v(\tau) \,\rangle \,]d\tau\,. \tag{29.62}$$

The process was simulated for the following data:

The parameters of the system were

$$\lambda = 0.20 \quad m_0 = 4.00 \quad \beta_1 = 0.20 \quad \beta_2 = 0.80 \quad \alpha = 1.00$$

$$b_1 = b_2 = 40.00 \quad c_1 = 9.00 \quad c_2 = 26.00$$

$$\varphi_1 = 2.00 \quad \varphi_2 = 0.40 \quad \psi_1 = 2.00 \quad \psi_2 = 1.00\,.$$

For the initial conditions we took

$$t_* = t_0 = 0.00, \qquad r_{1*} = r_{2*} = 1.00, \qquad \dot{r}_{1*} = \dot{r}_{2*} = 0.00\,.$$

The final moment was $\vartheta = 4.00$. The a priori calculated approximating value $\rho^a(t_*, r_*, \dot{r}_*)$ was

$$\rho^a(t_*, r_*, \dot{r}_*) = 0.25.$$

In the case of the approximating optimal control actions and the approximating optimal disturbances the simulation of the process gave

$$\gamma_{u^a, v^a} = 0.32.$$

The corresponding motion is shown in Figure 29.2.

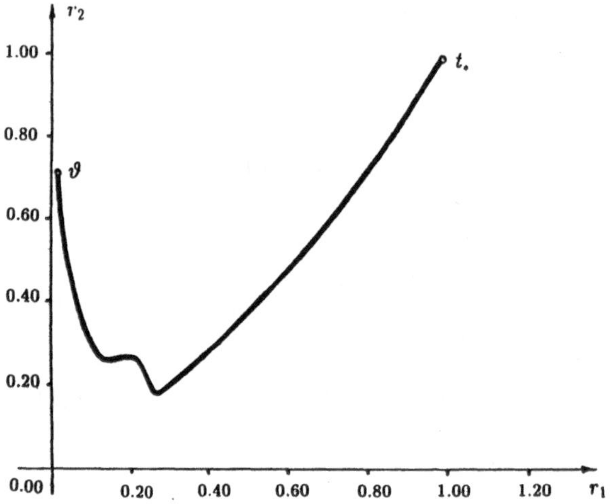

Figure 29.2

The graphs of approximating optimal control actions are shown in Figures 29.3 and 29.4, and the graphs of approximating optimal disturbances in Figures 29.5 and 29.6.

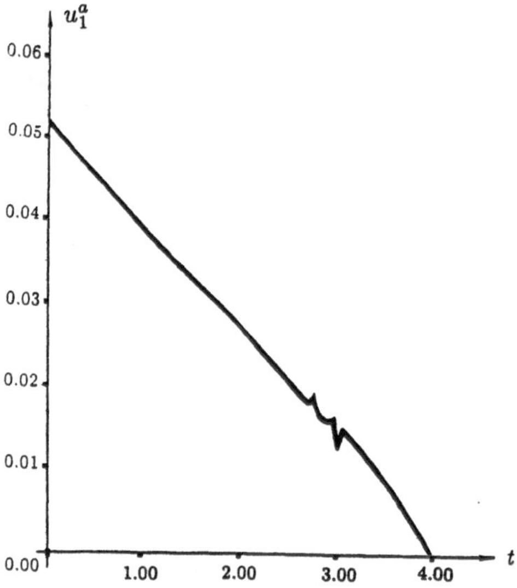

Figure 29.3

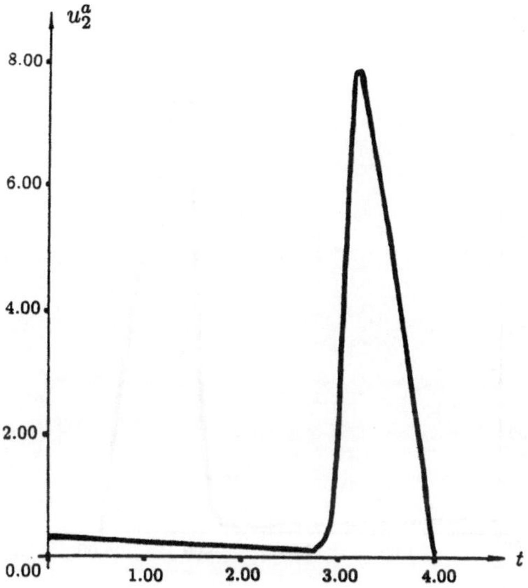

Figure 29.4

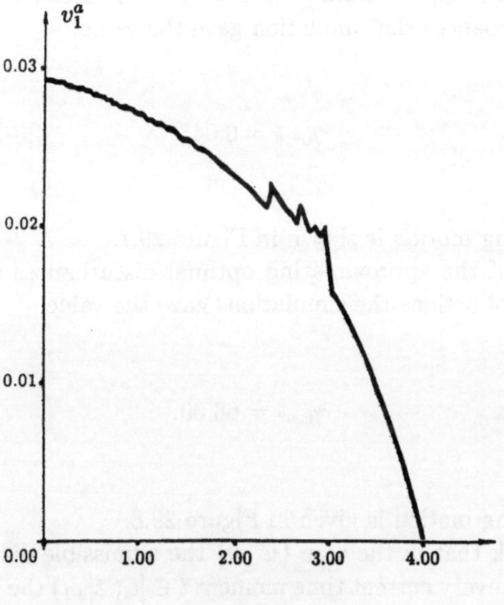

Figure 29.5

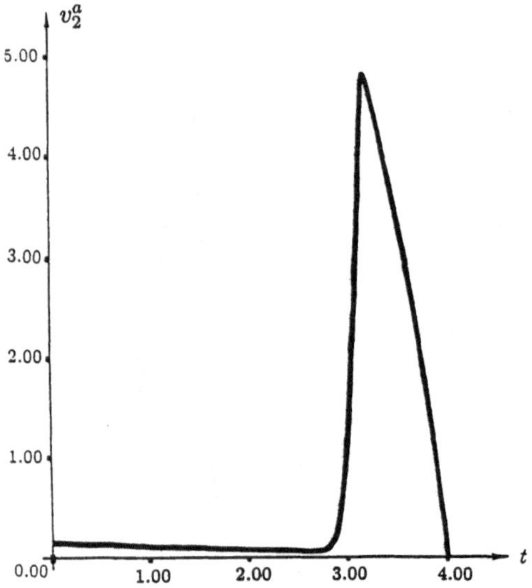

Figure 29.6

In the case of the approximating optimal control actions and some chosen admissible disturbances the simulation gave the value

$$\gamma_{u^a,v} = 0.045.$$

The corresponding motion is shown in Figure 29.7.

In the case of the approximating optimal disturbances and some chosen admissible control actions the simulations gave the value

$$\gamma_{u,v^a} = 66.00.$$

The corresponding motion is given in Figure 29.8.

Let us remark that in the case $\{u^a, v\}$ the admissible disturbances v were chosen so that at every current time moment $t \in [t_i,\ t_{i+1})$ the vector $v[t] = v[t_i]$ was collinear with the current radius-vector $r[t_i]$ and directed from the point B^*. The modulus of the vector $v[t_i]$ was equal to the mean value of the modulus of the approximating optimal disturbances $v^a[t],\ t \in [t_*, \vartheta)$.

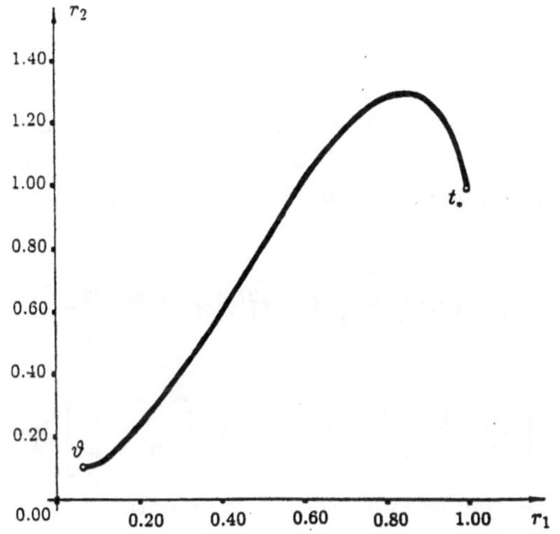

Figure 29.7

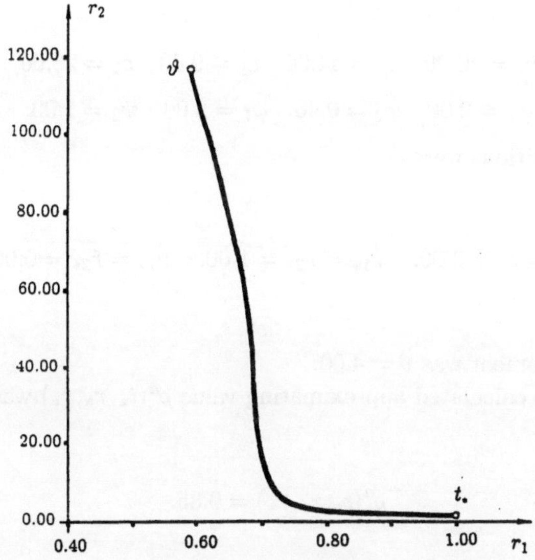

Figure 29.8

In the case $\{u, v^a\}$ the admissible control actions u at every time moment $t \in [t_i,\ t_{i+1})$ were chosen so that the vector $u[t] = u[t_i]$ was collinear with the current radius-vector $r[t_i]$ and directed to the point B^*. The modulus of

$u[t_i]$ was equal to the mean value of the modulus of the approximating optimal control actions $u^a[t]$, $t \in [t_*, \vartheta)$.

Let us consider the second example of simulation of a control process that corresponds to the model of delivering a platform. However, now we assume that in the quality index (29.17) both numbers e_1 and e_2 are positive.

Assume that the equation (29.11) has the form (29.60); the mass $m[t]$ varies according to (29.61). The quality index γ has the form

$$\gamma = (r_1^2[\vartheta] + r_2^2[\vartheta] + \dot{r}_1^2[\vartheta] + \dot{r}_2^2[\vartheta])^{1/2} +$$

$$+ \int_{t_*}^{\vartheta} [\langle\, u(\tau), \begin{bmatrix} \varphi_1 & 0 \\ 0 & \varphi_2 \end{bmatrix} u(\tau) \,\rangle - \langle\, v(\tau), \begin{bmatrix} \psi_1 & 0 \\ 0 & \psi_2 \end{bmatrix} v(\tau) \,\rangle] d\tau \qquad (29.63)$$

The process was simulated for the following data:

The parameters of the system were

$$\lambda = 0.20 \quad m_0 = 4.00 \quad \beta_1 = 0.20 \quad \beta_2 = 0.80 \quad \alpha = 1.00$$

$$b_1 = 40.00 \quad b_2 = 44.00 \quad c_1 = 9.00 \quad c_2 = 26.00$$

$$\varphi_1 = 2.00 \quad \varphi_2 = 0.40 \quad \psi_1 = 2.00 \quad \psi_2 = 1.00.$$

The initial conditions were

$$t_* = t_0 = 0.00, \quad r_{1*} = r_{2*} = 1.00, \quad \dot{r}_{1*} = \dot{r}_{2*} = 0.00.$$

The final moment was $\vartheta = 4.00$.

The a priori calculated approximating value $\rho^a(t_*, r_*, \dot{r}_*)$ was

$$\rho^a(t_*, r_*, \dot{r}_*) = 0.35.$$

In the case of the approximating optimal control actions and the approximating optimal disturbances the simulation of the process gave

$$\gamma_{u^a, v^a} = 0.39.$$

The corresponding motion $\{r_1[t], r_2[t]\}$, $t_* \le t \le \vartheta$ is shown in Figure 29.9. The curve that represents the corresponding velocities $\{\dot r_1[t], \dot r_2[t]\}$, $t_* \le t \le \vartheta$ of the platform is depicted in Figure 29.10.

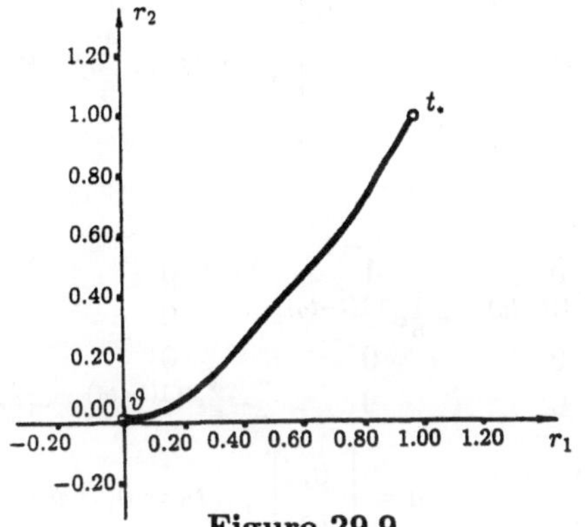

Figure 29.9

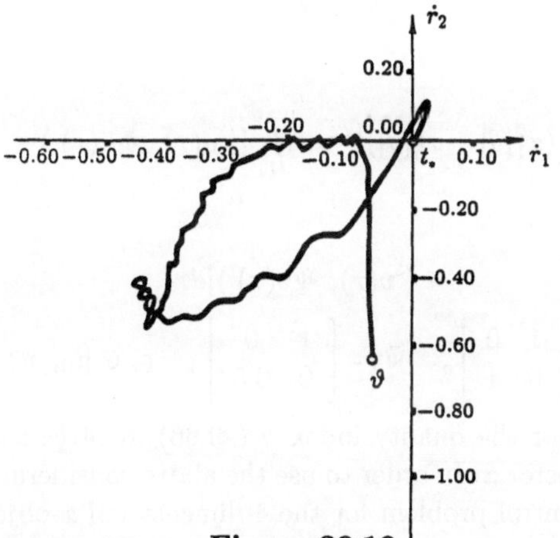

Figure 29.10

Let us consider now the third example of simulation of a control process with the functional $\gamma_{(4)}(\cdot)$ (29.2). This example has an abstract character. Consider the system

$$\dot x = A(t)x + B(t)u + C(t)v, \quad t_0 \le t < \vartheta \tag{29.64}$$

where

$$x = \begin{bmatrix} x_1 \\ x_2 \\ x_3 \\ x_4 \end{bmatrix}, \quad B = \begin{bmatrix} 0 & 0 \\ -1 & 0 \\ 0 & 0 \\ 0 & -0.5 \end{bmatrix}, \quad C(t) = \frac{1}{5}e^{0.1(t-t_0)} \begin{bmatrix} 0 & 0 \\ 6 & 0 \\ 0 & 0 \\ 0 & 3 \end{bmatrix},$$

$$A(t) = \begin{bmatrix} 0 & 1 & 0 & 0 \\ \frac{2}{5}e^{0.1(t-t_0)} & -\frac{1}{5}e^{0.1(t-t_0)} & 0 & 0 \\ 0 & 0 & 0 & 1 \\ 0 & 0 & -\frac{2}{5}e^{0.1(t-t_0)} & -\frac{2}{5}e^{0.1(t-t_0)} \end{bmatrix}$$

$$u = \begin{bmatrix} u_1 \\ u_2 \end{bmatrix}, \quad v = \begin{bmatrix} v_1 \\ v_2 \end{bmatrix}, \quad t_0 = 0, \quad \vartheta = 2. \tag{29.65}$$

The quality index is

$$\gamma = (x_1^2[\vartheta] + x_3^2[\vartheta])^{1/2} + \int_{t_*}^{\vartheta} [\langle\, u(\tau),\ \Phi u(\tau)\, \rangle -$$

$$- \langle\, v(\tau),\ \Psi v(\tau)\, \rangle] d\tau \tag{29.66}$$

$$\Phi = \begin{bmatrix} 1 & 0 \\ 0 & 1 \end{bmatrix}, \quad \Psi = \begin{bmatrix} 1 & 0 \\ 0 & 0.5 \end{bmatrix}, \quad t_* \in [t_0,\ \vartheta). \tag{29.67}$$

The expression for the quality index γ (29.66) involves the semi-norm of the 4-dimensional vector x. In order to use the above considerations we have to reduce the initial control problem for the 4-dimensional x-object to a control problem for an x-object with a new 2-dimensional phase vector x (see (29.5)–(29.7)). The considered control process was computer-simulated for the initial data

$$t_* = 0, \qquad x_* = \{2,\ -1,\ 1,\ 1\} \tag{29.68}$$

and for the parameters $\varepsilon = 0.07$ and $\delta = (\vartheta - t_*)/100 = 0.02$.

The continuous curve in Figure 29.11 represents the motion of the object (29.64) in plane $\{x_1, x_3\}$ with $\{u^a(\cdot) = u^a(t, x, \varepsilon),\ v^a(\cdot) = v^a(t, x, \varepsilon)\}$. (The obtained value of the quality index γ of (29.66) is close to the value of the game $\rho^a(t_*, x_*) = 2.41$); the dashed line represents the motion of the object with $u^a(\cdot)$, $v(\cdot) \equiv 0 \neq v^a(\cdot)$ ($\gamma = 1.81 < \rho^a(t_*, x_*)$); the dotted line corresponds to the motion with $u(\cdot) \equiv 0 \neq u^a(\cdot)$ ($\gamma = 2.95 > \rho^a(t_*, x_*)$). The continuous line in Figure 29.12 represents the motion with $u^a(\cdot)$, $v(\cdot) = \{v_1^a(\cdot) + 2\cos(3t),\ v_2^a(\cdot)\} \neq v^a(\cdot)$ ($\gamma = -1.41 < \rho^a(t_*, x_*)$); the dashed line depicts the motion with $u(\cdot) = \{u_1^a(\cdot) + 2\cos(3t),\ u_2^a(\cdot)\} \neq u^a(\cdot),\ v^a(\cdot)$ ($\gamma = 6.22 > \rho^a(t_*, x_*) = 2.41$).

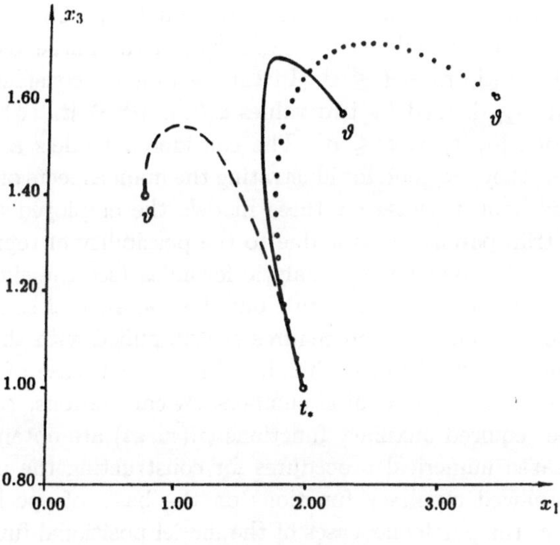

Figure 29.11

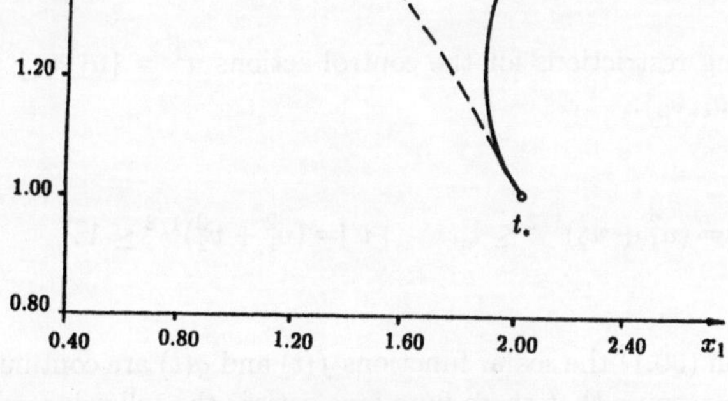

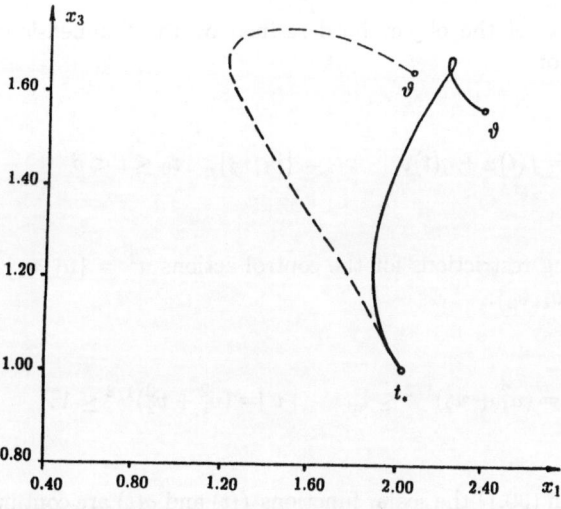

Figure 29.12

30. Examples of constructing optimal strategies for positional functionals

In this section on concrete model examples we illustrate the construction of the approximating optimal actions $u^a[t]$ and $v^a[t]$ in the cases of the positional functionals $\gamma_{(1)}$ and $\gamma_{(2)}$. One example of this kind for the functional $\gamma_{(1)}$ was considered above in Section 28. The functional $\gamma_{(1)}$ (28.9) in Section 28 was determined by the terminational value $x[\vartheta]$ of the phase vector x of the controlled motion $x[t]$, $t_* \leq t \leq \vartheta$. In this section we consider the quality indices $\gamma_{(1)}$ and $\gamma_{(2)}$ defined by two values $x[t^{[1]}]$, $t^{[1]} \in [t_*, \vartheta)$, and $x[\vartheta]$ of the phase vector $x[t]$, $t_* \leq t \leq \vartheta$. The considered models are sufficiently simple. However, they are good for illustrating the main aspects of the methods proposed in this book because for these models the employed constructions are sufficiently transparent. This is due to the possibility of representing the basic elements of the solutions by analytic formulas (see the similar cases in Sections 28 and 29). This does not rule out the necessity of some computer-realized procedures. But these procedures are simplified with the help of the mentioned formulas. On the other hand, in the general case of more or less serious problems the solution requires cumbersome calculations. Namely, in the general case the required auxiliary functions $\varphi_j(\tau_*, m)$ are obtained with the help of complicated numerical procedures for constructing the upper convex hulls of the employed auxiliary functions on the basis of special computer programs. But in the particular cases of the model positional functionals $\gamma_{(1)}$ and $\gamma_{(2)}$ considered in this section, we obtain auxiliary constructions that have sufficiently clear analytical and geometrical sense.

Let us begin with the model problem for the case of the positional functional $\gamma_{(1)}$ (see (10.3)).

Let the motion of the object be described by the 2-dimensional vector differential equation

$$\ddot{r} = f(t)u + g(t)v, \quad r^T = \{r_1, r_2\}, \quad t_0 \leq t < \vartheta \tag{30.1}$$

under the following restrictions for the control actions $u^T = \{u_1, u_2\}$ and disturbances $v^T = \{v_1, v_2\}$:

$$| u |= (u_1^2 + u_2^2)^{1/2} \leq 1, \qquad | v |= (v_1^2 + v_2^2)^{1/2} \leq 1\,. \tag{30.2}$$

In the equation (30.1) the scalar functions $f(t)$ and $g(t)$ are continuous and non-negative. We assume that these functions satisfy the following condition. The equation

$$f(t) = g(t) \tag{30.3}$$

in the time interval $t_0 \leq t \leq \vartheta$ has one root $t = \tilde{t}$ such that

$$f(t) > g(t) \quad \text{for } t < \tilde{t}$$

$$f(t) < g(t) \quad \text{for } \tilde{t} < t\,. \tag{30.4}$$

We choose the quality index

$$\gamma(r[t_*[\cdot]\vartheta]) = \gamma_{(1)}(r[t_*[\cdot]\vartheta]) = | \, r[t^{[1]}] \, | + | \, r[\vartheta] \, | \tag{30.5}$$

where the symbol $|\, r|$, as everywhere in this book, denotes the Euclidean norm of vector $r^T = \{r_1, r_2\}$. The time moment $t^{[1]} \in [t_0, \ \vartheta)$ is given.

Denote $x_1 = r_1, \ x_2 = \dot{r}_1, \ x_3 = r_2, \ x_4 = \dot{r}_2$. The equation (30.1) is reduced to the normal system of four equations:

$$\begin{aligned} \dot{x}_1 &= x_2 \\ \dot{x}_2 &= f(t)u_1 + g(t)v_1 \\ \dot{x}_3 &= x_4 \\ \dot{x}_4 &= f(t)u_2 + g(t)v_2\,. \end{aligned} \tag{30.6}$$

The quality index γ (30.5) takes a form

$$\gamma(x[t_*[\cdot]\vartheta]) = \left(x_1^2[t^{[1]}] + x_3^2[t^{[1]}]\right)^{1/2} +$$

$$+ \left(x_1^2[t^{[2]}] + x_3^2[t^{[2]}]\right)^{1/2}, \quad t^{[2]} = \vartheta\,. \tag{30.7}$$

We restrict ourselves to the case when the inequalities

$$t_0 < t^{[1]} < \tilde{t} < \vartheta \tag{30.8}$$

are valid.

As in other model problems in this book we will construct approximating actions $u^a[t]$ and $v^a[t]$ on the basis of the program extremum $e_{(1)}(\tau_*, w_*, \Delta)$ (see Sections 12 and 13). In the considered case the approximating functional $\gamma_{*(1)}$ (11.9) coincides with the initial functional $\gamma_{(1)}$ (30.7). Therefore the program extrema $e_{*(1)}(\tau_*, w_*, \Delta)$ (13.5) for $\gamma_{*(1)}$ and $e_{(1)}(\tau_*, w_*, \Delta)$ for $\gamma_{(1)}$ coincide, too. So according to (12.14), (13.1)–(13.10) we have

$$e_{(1)}(\tau_*, w_*, \Delta) = \max_{m^{(i)}, m_j^{(i)}} \Big[(\sum_{i=g}^{2} m^{(i)T}) \widetilde{w}_* +$$

$$+ M\{\sum_{j=1}^{k} (\tau_{j+1} - \tau_j) \max_v \min_u (\sum_{i=d(j)}^{2} m_j^{(i)T}) f(\tau_j, u, v)\} \Big] =$$

$$= \max_{m^*, m_*} [m^{*T} \widetilde{w}_* + k_{(1)}(\tau_*, m_*, \Delta)] \tag{30.9}$$

where $m^{(i)}$ and $m_j^{(i)}$, $i = 1, 2,\ j = 1, \dots, k$ are 4-dimensional vectors.

The fundamental matrix $X(t, \tau)$ for the homogeneous system that corresponds to (30.6) has a form

$$X(t, \tau) = \begin{pmatrix} 1 & t-\tau & 0 & 0 \\ 0 & 1 & 0 & 0 \\ 0 & 0 & 1 & t-\tau \\ 0 & 0 & 0 & 1 \end{pmatrix}. \tag{30.10}$$

The matrices $D^{[i]}$, $i = 1, 2$ which determine the semi-norm

$$\mu^{[i]}(x) = | D^{[i]} x | = (x_1^2 + x_3^2)^{1/2} \tag{30.11}$$

are

$$D^{[i]} = D = \begin{pmatrix} 1 & 0 & 0 & 0 \\ 0 & 0 & 1 & 0 \end{pmatrix}, \quad i = 1, 2 \tag{30.12}$$

and so we have

$$\begin{bmatrix} x_1 \\ x_3 \end{bmatrix} = Dx = \begin{pmatrix} 1 & 0 & 0 & 0 \\ 0 & 0 & 1 & 0 \end{pmatrix} \begin{bmatrix} x_1 \\ x_2 \\ x_3 \\ x_4 \end{bmatrix}. \tag{30.13}$$

The quantities that determine the program extremum $e_{(1)}(\tau_*, w_*, \Delta)$ (30.9) are calculated according to the formulas (13.1)–(13.6). However, let us remark that, below, the values m and l do not have any probabilistic sense. They are considered not as expectations, but as some deterministic arguments (see the remark concerning (13.5), (13.6) in Section 13, p. 86). Using (30.10), (30.12) we obtain

$$\widetilde{w}_* = X(\vartheta, \tau_*) w_* = \begin{bmatrix} w_{*1} + w_{*2}(\vartheta - \tau_*) \\ w_{*2} \\ w_{*3} + w_{*4}(\vartheta - \tau_*) \\ w_{*4} \end{bmatrix} \tag{30.14}$$

$$\widetilde{f}(\tau, u, v) = X(\vartheta, \tau) f(\tau, u, v) =$$

$$= X(\vartheta, \tau) \begin{bmatrix} 0 \\ f(\tau)u_1 + g(\tau)v_1 \\ 0 \\ f(\tau)u_2 + g(\tau)v_2 \end{bmatrix} = \begin{bmatrix} (\vartheta - \tau)(f(\tau)u_1 + g(\tau)v_1) \\ f(\tau)u_1 + g(\tau)v_1 \\ (\vartheta - \tau)(f(\tau)u_2 + g(\tau)v_2) \\ f(\tau)u_2 + g(\tau)v_2 \end{bmatrix} \tag{30.15}$$

$$m^{[i]} = X^T(t^{[i]}, \vartheta) D^T l^{[i]} =$$

$$= \begin{pmatrix} 1 & t^{[i]} - \vartheta & 0 & 0 \\ 0 & 1 & 0 & 0 \\ 0 & 0 & 1 & t^{[i]} - \vartheta \\ 0 & 0 & 0 & 1 \end{pmatrix}^T \begin{pmatrix} 1 & 0 & 0 & 0 \\ 0 & 0 & 1 & 0 \end{pmatrix}^T \begin{bmatrix} l_1^{[i]} \\ l_2^{[i]} \end{bmatrix} =$$

$$= \begin{bmatrix} l_1^{[i]} \\ (t^{[i]} - \vartheta) l_1^{[i]} \\ l_2^{[i]} \\ (t^{[i]} - \vartheta) l_2^{[i]} \end{bmatrix}, \qquad | l^{[i]} | \leq 1, \quad i = 1, 2. \tag{30.16}$$

As usual, we assume $\tau_1 = \tau_*$.
For the quantity $\Delta\psi_j(\tau_*, m)$ (13.25) we obtain

$$\Delta\psi_j(\tau_*, m) = (\tau_{j+1} - \tau_j) \max_v \min_u m^{[2]T} \widetilde{f}(\tau_j, u, v) =$$

$$= (\tau_{j+1} - \tau_j)(g(\tau_j) - f(\tau_j))(t^{[2]} - \tau_j) \mid l^{[2]} \mid \tag{30.17}$$

for

$$\tau_j \geq t^{[1]}, \qquad \mid l^{[2]} \mid \leq 1 \tag{30.18}$$

and

$$\begin{aligned}\Delta\psi_j(\tau_*, m) = (\tau_{j+1} - \tau_j) \max_v \min_u (m^{[1]} + m^{[2]})^T \widetilde{f}(\tau_j, u, v) = \\ = (\tau_{j+1} - \tau_j)(g(\tau_j) - f(\tau_j)) * \\ * \mid (t^{[2]} - \tau_j) l^{[2]} + (t^{[1]} - \tau_j) l^{[1]} \mid\end{aligned} \tag{30.19}$$

for

$$\tau_j < t^{[1]}, \qquad \mid l^{[1]} \mid \leq 1, \qquad \mid l^{[2]} \mid \leq 1 . \tag{30.20}$$

From (30.17)–(30.20), according to the procedure described in Section 13, we obtain the following expressions for the value $\varphi_j(\tau_*, m)$:

$$\varphi_{k+1}(\tau_*, m) = 0,$$

$$\begin{aligned}\varphi_j(\tau_*, m) = \overline{\psi}_j(\tau_*, m), \\ \psi_j(\tau_*, m) = \Delta\psi_j(\tau_*, m) + \varphi_{j+1}(\tau_*, m).\end{aligned} \tag{30.21}$$

We have to consider three cases. In the first case $\widetilde{t} \leq \tau_*$, where $\widetilde{t}$ is a given time moment in (30.3), (30.4) and (30.8). Then according to (30.16)–(30.18) and (30.21) we have

$$\varphi_j(\tau_*, m) = \sum_{s=j}^{k} (\tau_{s+1} - \tau_s)(g(\tau_s) - f(\tau_s))(t^{[2]} - \tau_s),$$

$$t^{[2]} = \vartheta, \qquad 1 \leq j \leq k$$

$$m = m^{[2]} = \begin{bmatrix} l_1^{[2]} \\ 0 \\ l_2^{[2]} \\ 0 \end{bmatrix} . \tag{30.22}$$

Let us consider now the second case $t^{[1]} \leq \tau_* < \widetilde{t}$. Then using again (30.16)–(30.18), (30.21) we obtain

$$\varphi_j(\tau_*, m) = \sum_{s=j}^{\widetilde{j}-1} (\tau_{s+1} - \tau_s)(g(\tau_s) - f(\tau_s))(t^{[2]} - \tau_s) \mid l^{[2]} \mid +$$

$$+\sum_{s=\tilde{j}}^{k}(\tau_{s+1}-\tau_s)(g(\tau_s)-f(\tau_s))(t^{[2]}-\tau_s),$$

$$1 \leq j < \tilde{j},$$

$$m = m^{[2]} = \begin{bmatrix} l_1^{[2]} \\ 0 \\ l_2^{[2]} \\ 0 \end{bmatrix}, \quad | \, l^{[2]} \, | \leq 1. \tag{30.23}$$

In the third case $t_0 \leq \tau_* < t^{[1]}$. Then

$$\varphi_j(\tau_*, m) = \sum_{s=j}^{j_1-1}(\tau_{s+1}-\tau_s)(g(\tau_s)-f(\tau_s))(t^{[2]}-\tau_s)*$$

$$* \, | \, l^{[2]} + (t^{[1]}-\tau_s)l^{[1]} \, | \, +$$

$$+\sum_{s=j_1}^{\tilde{j}-1}(\tau_{s+1}-\tau_s)(g(\tau_s)-f(\tau_s))(t^{[2]}-\tau_s) \, | \, l^{[2]} \, | \, +$$

$$+\sum_{s=\tilde{j}}^{k}(\tau_{s+1}-\tau_s)(g(\tau_s)-f(\tau_s))(t^{[2]}-\tau_s),$$

$$1 \leq j < j_1,$$

$$m = m^{[2]} + m^{[1]} = \begin{bmatrix} l_1^{[2]} + l_1^{[1]} \\ 0 + (t^{[1]}-\vartheta)l_1^{[1]} \\ l_2^{[2]} + l_2^{[1]} \\ 0 + (t^{[1]}-\vartheta)l_2^{[1]} \end{bmatrix},$$

$$| \, l^{[2]} \, | \leq 1, \quad | \, l^{[1]} \, | \leq 1. \tag{30.24}$$

Let us take into consideration the following circumstance. We denote by j_1 the number that satisfies the condition $\tau_{j_1} = t^{[1]}$. When we calculate the function $\varphi_j(\tau_*, m)$ for $j = j_1 - 1$ in the set $G_j(\tau_*) = \{m\}$, where m is given by (30.24), we have to maximize the corresponding quantity with respect to the component $m_* \in G_2(\tau_*)$, i.e., to the component $m_* = m^{[2]} \in G_j(\tau_*)$, i.e., to the vector

$$m_* = \begin{bmatrix} l_1^{[2]} \\ 0 \\ l_2^{[2]} \\ 0 \end{bmatrix}, \quad | \, l^{[2]} \, | \leq 1 \tag{30.25}$$

for the fixed vector m (see Section 13, (13.28)–(13.32)). However, in the considered case this operation of maximization is degenerated. Really, both components $m^{[1]}$ and $m^{[2]} = m_*$ of the vector $m = m^{[2]} + m^{[1]}$ are determined for a fixed vector $m \in G_j(\tau_*)$ uniquely. Indeed, the equality

$$m^{[2]} + m^{[1]} = \begin{bmatrix} l_1^{[2]} + l_1^{[1]} \\ (t^{[1]} - \vartheta) l_1^{[1]} \\ l_2^{[2]} + l_2^{[1]} \\ (t^{[1]} - \vartheta) l_2^{[2]} \end{bmatrix} = \begin{bmatrix} m_1 \\ m_2 \\ m_3 \\ m_4 \end{bmatrix} \tag{30.26}$$

for a given $m_i, i = 1, \ldots, 4$ is solved uniquely with respect to $l_r^{[s]}$, $s = 1, 2$, $r = 1, 2$.

Thus, for the quantity $e_{(1)}(\tau_*, w_*, \Delta)$ we obtain

$$e_{(1)}(\tau_*, w_*, \Delta) = \max_{m_*} \left[m_*^T \widetilde{w}_* + \varphi_1(\tau_*, m_*) \right] \tag{30.27}$$

for

$$t^{[1]} < \tau_*, \qquad m_* = m^{[2]} = \begin{bmatrix} l_1^{[2]} \\ 0 \\ l_2^{[2]} \\ 0 \end{bmatrix}, \qquad | \, l^{[2]} \, | \leq 1 \tag{30.28}$$

where the quantity $\varphi_1(\tau_*, m)$ is determined by the equality (30.22) for $\widetilde{t} \leq \tau_*$ and by the equality (30.23) for $t^{[1]} < \tau_* < \widetilde{t}$.

In the case $\tau_* = t^{[1]}$ we have

$$e_{(1)}(\tau_*, w_*, \Delta) = \max_{m^*, m_*} \left(m^{*T} \widetilde{w}_* + \varphi_1(\tau_*, m_*) \right) \tag{30.29}$$

where the quantity $\varphi_1(\tau_*, m)$ is determined by the equality (30.23) and

$$m_* = m^{[2]} = \begin{bmatrix} l_1^{[2]} \\ 0 \\ l_2^{[2]} \\ 0 \end{bmatrix}, \qquad m^* = m^{[2]} + m^{[1]} = \begin{bmatrix} l_1^{[2]} + l_1^{[1]} \\ (t^{[1]} - \vartheta) l_1^{[1]} \\ l_2^{[2]} + l_2^{[1]} \\ (t^{[1]} - \vartheta) l_2^{[1]} \end{bmatrix},$$

$$| l^{[2]} | \leq 1, \quad | l^{[1]} | \leq 1. \tag{30.30}$$

In the case $\tau_* < t^{[1]}$ we have

$$e_{(1)}(\tau_*, w_*, \Delta) = \max_{m^*} \Big[\, m^{*T} \widetilde{w}_* + \varphi_1(\tau_*, m^*) \, \Big] \tag{30.31}$$

where the quantity $\varphi_1(\tau_*, m)$ is determined by the equalities (30.21)–(30.26) and

$$m^* = \begin{bmatrix} l_1^{[2]} + l_1^{[1]} \\ (t^{[1]} - \vartheta) l_1^{[1]} \\ l_2^{[2]} + l_2^{[1]} \\ (t^{[1]} - \vartheta) l_2^{[1]} \end{bmatrix}. \tag{30.32}$$

We have calculated the program extremum $e_{(1)}(\cdot)$ and now will calculate the approximating optimal actions $u^a[t]$ and $v^a[t]$. In accordance with the proposed general theory and the positional properties of functional $\gamma_{(1)}(\cdot)$ for the informational image $y[\cdot]$ one can take the position $\{t, x[t]\}$ realized at the current time moment t. Therefore, to construct the actions $u^a[t] = u^a[t_i]$, $v^a[t] = v^a[t_i]$, $t_i \leq t < t_{i+1}$ one has to employ the values $x[t_i]$ and the corresponding accompanying values $w^{[u]0}(t_i, x[t_i])$, $w^{[v]0}(t_i, x[t_i])$. Here we assume that the accompanying points $w^{[u]0}(\cdot)$ and $w^{[v]0}(\cdot)$ are subjected to the conditions

$$| w^{[u]0}(t_i, x[t_i], \varepsilon_u) - x[t_i] | \leq R^{[u]}(\varepsilon_u, t_i) = \varepsilon[t_i], \tag{30.33}$$

$$| w^{[v]0}(t_i, x[t_i], \varepsilon_v) - x[t_i] | \leq R^{[v]}(\varepsilon_v, t_i) = \varepsilon[t_i], \tag{30.34}$$

$$\varepsilon[t] = \varepsilon \exp \lambda_\varepsilon (t - t_0), \quad \varepsilon > 0, \quad \lambda_\varepsilon > 0. \tag{30.35}$$

First of all, we have to calculate the approximating functions

$$\rho_u^a(t_*, x_*) = e_{(1)}(t_*, w_u^0(t_*, x_*), \Delta), \tag{30.36}$$

$$\rho_v^a(t_*, x_*) = e_{(1)}(t_*, w_v^0(t_*, x_*), \Delta). \tag{30.37}$$

These functions approximate the value of the game $\rho^0(t_*, x_*)$.

Let us denote

$$s_* = x_* - w^{[u]}. \tag{30.38}$$

For $t_* \in [\widetilde{t}, \vartheta)$ we have

$$\rho_u^a(t_*, x_*) = \max_{|l^{[2]}|\le 1} \min_{|s_*|\le \varepsilon[t_*]} \Big\{ l_1^{[2]} *$$

$$* [x_{*1} - s_{*1} + (\vartheta - t_*)(x_{*2} - s_{*2})] + l_2^{[2]} [x_{*3} - s_{*3} + (\vartheta - t_*)(x_{*4} - s_{*4})] \Big\} +$$

$$+ \sum_{j=1}^{k} (\tau_{j+1} - \tau_j)(\vartheta - \tau_j)(g(\tau_j) - f(\tau_j)) =$$

$$= \max_{|l^{[2]}|\le 1} \Big\{ l_1^{[2]} [x_{*1} + (\vartheta - t_*)x_{*2}] +$$

$$+ l_2^{[2]} [x_{*3} + (\vartheta - t_*)x_{*4}] - \varepsilon[t_*] \mid l^{[2]} \mid [1 + (\vartheta - t_*)^2]^{1/2} +$$

$$+ \sum_{j=1}^{k} (\tau_{j+1} - \tau_j)(\vartheta - \tau_j)(g(\tau_j) - f(\tau_j)) \Big\}. \tag{30.39}$$

Below we omit the explicit description of the minimization in s_*. We give only the final expressions for the function $\rho_u^a(\cdot)$ and $s^{[u]a}(\cdot)$ obtained as the result of this operation. For $t_* \in (t^{[1]}, \widetilde{t})$ we obtain

$$\rho_u^a(t_*, x_*) = \max_{|l^{[2]}|\le 1} \Big\{ l_1^{[2]} [x_{*1} + (\vartheta - t_*)x_{*2}] +$$

$$+ l_2^{[2]} [x_{*3} + (\vartheta - t_*)x_{*4}] - \varepsilon[t_*] \mid l^{[2]} \mid [1 + (\vartheta - t_*)^2]^{1/2} +$$

$$+ \mid l^{[2]} \mid \sum_{j=1}^{\widetilde{j}-1} (\tau_{j+1} - \tau_j)(\vartheta - \tau_j)(g(\tau_j) - f(\tau_j)) +$$

$$+ \sum_{j=\widetilde{j}}^{k} (\tau_{j+1} - \tau_j)(\vartheta - \tau_j)(g(\tau_j) - f(\tau_j)) \Big\}. \tag{30.40}$$

For $t_* = t^{[1]}$ we have

$$\rho_u^a(t_*, x_*) = \max_{\substack{|l^{[1]}|\le 1, \\ |l^{[2]}|\le 1}} \Big\{ (l_1^{[2]} + l_1^{[1]})[x_{*1} + (\vartheta - t_*)x_{*2}] +$$

$$+ l_1^{[1]}(t^{[1]} - \vartheta)x_{*2} + (l_2^{[2]} + l_2^{[1]})[x_{*3} + (\vartheta - t_*)x_{*4}] +$$

$$+l_2^{[1]}(t^{[1]} - \vartheta)x_{*4} - \varepsilon[t_*][\mid l^{[1]} + l^{[2]} \mid^2 +$$
$$+ \mid l^{[1]}(t^{[1]} - t_*) + l^{[2]}(\vartheta - t_*) \mid^2]^{1/2} +$$
$$+ \mid l^{[2]} \mid \sum_{j=1}^{\widetilde{j}-1}(\tau_{j+1} - \tau_j)(\vartheta - \tau_j)(g(\tau_j) - f(\tau_j)) +$$
$$+ \sum_{j=\widetilde{j}}^{k}(\tau_{j+1} - \tau_j)(\vartheta - \tau_j)(g(\tau_j) - f(\tau_j)) \Big\}. \tag{30.41}$$

For $t_0 \leq t_* < t^{[1]}$ we come to

$$\rho_u^a(t_*, x_*) = \max_{|l^{[1]}|\leq 1,\ |l^{[2]}|\leq 1} \Big\{ (l_1^{[2]} + l_1^{[1]})[x_{*1} + (\vartheta - t_*)x_{*2}] +$$
$$+l_1^{[1]}(t^{[1]} - \vartheta)x_{*2} + (l_2^{[2]} + l_2^{[1]})[x_{*3} + (\vartheta - t_*)x_{*4}] +$$
$$+l_2^{[1]}(t^{[1]} - \vartheta)x_{*4} - \varepsilon[t_*][\mid l^{[1]} + l^{[2]} \mid^2 +$$
$$+ \mid l^{[1]}(t^{[1]} - t_*) + l^{[2]}(\vartheta - t_*) \mid^2]^{1/2} +$$
$$+ \sum_{j=1}^{j_1-1}(\tau_{j+1} - \tau_j)(g(\tau_j) - f(\tau_j)) \mid l^{[2]}(\vartheta - \tau_j) + l^{[1]}(t^{[1]} - \tau_j) \mid +$$
$$+ \mid l^{[2]} \mid \sum_{j=j_1}^{\widetilde{j}-1}(\tau_{j+1} - \tau_j)(\vartheta - \tau_j)(g(\tau_j) - f(\tau_j)) +$$
$$+ \sum_{j=\widetilde{j}}^{k}(\tau_{j+1} - \tau_j)(\vartheta - \tau_j)(g(\tau_j) - f(\tau_j)) \Big\}. \tag{30.42}$$

Similar relations can be obtained for the value $\rho_v^a(t_*, x_*)$ (30.37) that approximates the value of the game $\rho^0(t_*, x_*)$. This function $\rho_v^a(t_*, x_*)$ is used to form the actions $v^a[t]$. The only difference is that in the relations (30.39)–(30.42) the terms containing $\varepsilon[t_*]$ should have the plus sign.

The maximizing values $l_u^{[1]0}$, $l_u^{[2]0}$ given in (30.39)–(30.42) and the corresponding maximizing values $l_v^{[1]0}$, $l_v^{[2]0}$, defined by the similar relations for $\rho_v^a(\cdot)$, determine the approximating optimal actions $u^a[t]$ and $v^a[t]$. Namely, the actions $u^a[t]$ and $v^a[t]$ are determined in the following way.

According to (18.6)–(18.15) these actions $u^a[t]$ and $v^a[t]$ are calculated with the help of the following conditions. We have

$$s^{[u]a}(t_i, x[t_i], \varepsilon_u) = \varepsilon[t_i]\frac{X^T(\vartheta, t_i)m_u^0[t_i]}{\mid X^T(\vartheta, t_i)m_u^0[t_i] \mid} \tag{30.43}$$

if $\mid m_u^0[t_i] \mid > 0$ or

$$s^{[u]a}(t_i, x[t_i], \varepsilon_u) = 0 \tag{30.44}$$

if $\mid m_u^0[t_i] \mid = 0$. Here $m_u^0[t_i]$ is the maximizing vector in (30.39)–(30.42). So, according to (30.28), (30.30), (30.32) we obtain

$$m_u^0[t_i] = \begin{bmatrix} l_{u2}^{[2]0}[t_i] \\ 0 \\ l_{u1}^{[2]0}[t_i] \\ 0 \end{bmatrix}, \quad t^{[1]} < t_i, \tag{30.45}$$

$$m_u^0[t_i] = \begin{bmatrix} l_{u1}^{[2]0}[t_i] + l_{u1}^{[1]0}[t_i] \\ (t^{[1]} - \vartheta) l_{u1}^{[1]0}[t_i] \\ l_{u2}^{[2]0}[t_i] + l_{u2}^{[1]0}[t_i] \\ (t^{[1]} - \vartheta) l_{u2}^{[1]0}[t_i] \end{bmatrix}, \quad t_0 \le t_i \le t^{[1]}. \tag{30.46}$$

Similar to (8.11) we come to the conditions for $u^a[t] = u^a[t_i],\ t_i \le t < t_{i+1}$

$$\langle\, s^{[u]a}(t_i, x[t_i], \varepsilon_u), \begin{bmatrix} 0 \\ f[t_i]u_1^a[t_i] \\ 0 \\ f[t_i]u_2^a[t_i] \end{bmatrix} \rangle =$$

$$= \min_{|u|\le 1} \langle\, s^{[u]a}(t_i, x[t_i], \varepsilon_u), \begin{bmatrix} 0 \\ f[t_i]u_1 \\ 0 \\ f[t_i]u_2 \end{bmatrix} \rangle. \tag{30.47}$$

Thus we obtain

$$u^a[t_i] = -l_u^{[2]0}[t_i] \tag{30.48}$$

for $t^{[1]} < t_i$. And for $t_0 \le t_i \le t^{[1]}$ we have

$$u^a[t_i] = -\frac{l_u^{[1]0}[t_i](t^{[1]} - t_i) + l_u^{[2]0}[t_i](\vartheta - t_i)}{\mid l_u^{[1]0}[t_i](t^{[1]} - t_i) + l_u^{[2]0}[t_i](\vartheta - t_i) \mid} \tag{30.49}$$

if $\mid l_u^{[1]0}[t_i](t^{[1]} - t_i) + l_u^{[2]0}[t_i](\vartheta - t_i) \mid > 0$ or

$$u^a[t_i] = 0 \tag{30.50}$$

if $| \, l_u^{[1]0}[t_i](t^{[1]} - t_i) + l_u^{[2]0}[t_i](\vartheta - t_i) \, | = 0$.

Let $m_v^0[t_i]$ be the vectors given by equalities similar to (30.45), (30.46). To obtain these equalities one has to substitute in (30.45), (30.46) the lower index u by v. We emphasize also that the values $l_v^{[i]0}[t_*]$, $i = 1, 2$ are the solutions of maximum problems similar to the problems (30.39)–(30.42), but now in the corresponding expressions the terms with $\varepsilon[t_*]$ have the plus sign. The vectors

$$s^{[v]a}(t_i, x[t_i], \varepsilon_v) = w_v^0(t_i, x[t_i], \varepsilon_v) - x[t_i] \tag{30.51}$$

are determined by equalities similar to (30.43), (30.44) where only the symbol u is substituted by v. The conditions for approximating optimal disturbances $v^a[t] = v^a[t_i], \;\; t_i \le t < t_{i+1}$ are

$$\langle \; s^{[v]a}(t_i, x[t_i], \varepsilon_v) \begin{bmatrix} 0 \\ g[t_i]v_1^a[t_i] \\ 0 \\ g[t_i]v_2^a[t_i] \end{bmatrix} \rangle =$$

$$= \max_{|v| \le 1} \langle \; s^{[v]a}(t_i, x[t_i], \varepsilon_v), \begin{bmatrix} 0 \\ g[t_i]v_1 \\ 0 \\ g[t_i]v_2 \end{bmatrix} \rangle. \tag{30.52}$$

Then the approximating optimal disturbances $v^a[t_i]$ are determined by the following equalities:

$$v^a[t_i] = l_v^{[2]0}[t_i] \tag{30.53}$$

for $t^{[1]} < t_i$. And for $t_0 \le t_i \le t^{[1]}$ we have

$$v^a[t_i] = \frac{l_v^{[1]0}[t_i](t^{[1]} - t_i) + l_v^{[2]0}[t_i](\vartheta - t_i)}{| \, l_v^{[1]0}[t_i](t^{[1]} - t_i) + l_v^{[2]0}[t_i](\vartheta - t_i) \, |} \tag{30.54}$$

if $| \, l_v^{[1]0}[t_i](t^{[1]} - t_i) + l_v^{[2]0}[t_i](\vartheta - t_i) \, | > 0$ or

$$v^a[t_i] = 0 \tag{30.55}$$

if $| l_v^{[1]0}[t_i](t^{[1]} - t_i) + l_v^{[2]0}[t_i](\vartheta - t_i) |= 0$.

Let us note the following circumstance. The approximating optimal actions $u^a[t]$ (30.48)–(30.50) and $v^a[t]$ (30.53)–(30.55) are constructed on the basis of the program extremum $e_{(1)}(\tau_*, w_*, \Delta)$. They are formed in the process of control by the approximating optimal control law U_Δ^a and by the optimal law V_Δ^a for the disturbances. These laws act in real time t in the discrete scheme with the partition $\Delta\{t_i\}$ that coincides with the partition $\Delta\{\tau_j\}$ for the imaginary time τ from the auxiliary constructions. Let us emphasize, however, that for every current time moment t_i the moments τ_j are numerated anew so that $\tau_1 = t_i$. This scheme can be realized in practice, and for a sufficiently small partition step $\delta = \max_j(\tau_{j+1} - \tau_j)$ of the partition $\Delta = \Delta_\delta$ it provides a good approximation for the value of the game $\rho_{(1)}^0(t_*, x_*)$.

However, in the considered case we can construct the optimal actions $u^0[t]$ and $v^0[t]$ using the value of the game $\rho_{(1)}^0(\tau_*, w_*)$ itself instead of the program extremum $e_{(1)}(\tau_*, w_*, \Delta)$ that only approximates this value $\rho_{(1)}^0(t_*, x_*)$. In fact, according to Section 17 (see (17.12), (17.13)) the following equality

$$\rho_{(1)}^0(\tau_*, w_*) = \lim_{\delta \to 0} e_{(1)}(\tau_*, w_*, \Delta_\delta) \tag{30.56}$$

is valid.

This limit value $\rho_{(1)}^0(\tau_*, w_*)$ (30.56) can be obtained naturally in the following way. It suffices just to substitute the integral sums in the formulas for $e_{(1)}(\tau_*, w_*, \Delta_\delta)$ by the corresponding integrals. Thus we obtain, for example, for $\tau_* < t^{[1]}$ the following expression for the value of the game

$$\rho_{(1)}^0(\tau_*, w_*) = \max_{m^*} \Big(m^{*T} \widetilde{w}_*[\tau_*] + \int_{\tau_*}^{t^{[1]}} | l^{[2]}(\vartheta - \tau) +$$

$$+ l^{[1]}(t^{[1]} - \tau) | (g(\tau) - f(\tau))d\tau + | l^{[2]} | \int_{t^{[1]}}^{\widetilde{t}} (\vartheta - \tau)(g(\tau) -$$

$$- f(\tau))d\tau + \int_{\widetilde{t}}^{\vartheta} (\vartheta - \tau)(g(\tau) - f(\tau))d\tau \Big), \tag{30.57}$$

where

$$m^* = m^{[1]} + m^{[2]}, \quad m^{[1]} = \begin{bmatrix} l_1^{[1]} \\ (t^{[1]} - \vartheta) l_1^{[1]} \\ l_2^{[1]} \\ (t^{[1]} - \vartheta) l_2^{[1]} \end{bmatrix}, \quad m^{[2]} = \begin{bmatrix} l_1^{[2]} \\ 0 \\ l_2^{[2]} \\ 0 \end{bmatrix},$$

$$|\, l^{[1]} \,| \le 1, \qquad |\, l^{[2]} \,| \le 1. \tag{30.58}$$

And so on. Then the optimal control actions $u^0[t]$ and optimal disturbances $v^0[t]$ are determined by the following equalities:

$$u^0[t] = u^0(t_i, x[t_i], \varepsilon), \qquad v^0[t] = v^0(t_i, x[t_i], \varepsilon),$$

$$t_i \le t < t_{i+1} \tag{30.59}$$

where the optimal strategies $u^0(\cdot) = u^0(t, x, \varepsilon)$ and $v^0(\cdot) = v^0(t, x, \varepsilon)$ are determined by the conditions of the *extremal shift* (see (8.11), (8.16)) to the accompanying points $w^{[u]0}(\cdot)$ and $w^{[v]0}(\cdot)$ which are calculated with the help of the function $\rho^0_{(1)}(t, x)$. If the corresponding integrals can be effectively calculated in analytic form, then using $\rho^0(\tau_*, w_*)$ instead of $e_{(1)}(\tau_*, w_*, \Delta)$ we simplify the calculation of the optimal actions $u^0[t]$ and $v^0[t]$. However, the direct realization of the precise strategies $u^0(\cdot)$ and $v^0(\cdot)$, as a rule, implies a discrete time scheme.

Now we consider a concrete model problem with the quality index $\gamma_{(2)}$ (see (10.4)).

Assume again that the motion of the controlled object is described by the differential equation (30.1) under the restrictions (30.2). And let all the previous assumptions on the considered controlled system be valid with only the exception that we consider now another form of the functional γ. Namely, instead of the quality index $\gamma = \gamma_{(1)}$ (30.5) we take now the quality index

$$\gamma(r[t_*[\cdot]\vartheta]) = \gamma_{(2)}(r[t_*[\cdot]\vartheta]) = \max\Big(\,|\, r[t^{[1]}] \,|, \,|\, r[t^{[2]}] \,|\, \Big) \tag{30.60}$$

where $t^{[2]} = \vartheta, \ t^{[1]} \in [t_0, \ \vartheta)$.

In order to construct approximating optimal actions $u^a[t]$ and $v^a[t]$ we will follow the procedures described in Sections 14 and 18.

Let us again construct the functions $\varphi_j(\tau_*, m, \nu)$, $j = 1, \ldots, k$ according to Section 14. The quantities $\tilde{f}(\cdot)$, $m^{[i]}$, $l^{[i]}$ and $\Delta\psi_j(\tau_*, m, \nu)$ used in the construction of the functions $\varphi_j(\tau_*, m, \nu)$ are determined here by the same expressions as above in the case of the functional $\gamma_{(1)}$ (30.5), (30.7). The only difference is concerned with the definition of the sets of vectors $m = m^{[2]}$ and vectors $m = m^{[2]} + m^{[1]}$. Namely, now we consider the sets of 4-dimensional vectors $m = m^{[2]}$ that correspond to the values ν

$$m = m^{[2]} = \begin{bmatrix} l_1^{[2]} \\ 0 \\ l_2^{[2]} \\ 0 \end{bmatrix}, \qquad | l^{[2]} | \leq \nu, \qquad \nu \in [0,\ 1] \tag{30.61}$$

and the sets of 4-dimensional vectors $m = m^{[2]} + m^{[1]}$

$$m = m^{[2]} + m^{[1]} = \begin{bmatrix} l_1^{[2]} + l_1^{[1]} \\ (t^{[1]} - \vartheta) l_1^{[1]} \\ l_2^{[2]} + l_2^{[1]} \\ (t^{[1]} - \vartheta) l_2^{[1]} \end{bmatrix}, \qquad | l^{[1]} | + | l^{[2]} | \leq 1,$$

$$| l^{[2]} | \leq \nu, \qquad \nu \in [0,\ 1]\,. \tag{30.62}$$

Acting as described in Section 14 and calculating the functions $\varphi_j(\tau_*, m, \nu)$, which determine the program extremum

$$e_{(2)}(\tau_*, w_*, \Delta) = \max_{m^*, m_*, \nu} \left(m^{*T} \tilde{w}_* + \varphi_1(\tau_*, m_*, \nu)\right) \tag{30.63}$$

and repeating with understandable changes the above calculations for the case of $\gamma_{(1)}$ (30.7), we obtain the following expression for the program extremum $e_{(2)}(\tau_*, w_*, \Delta)$. Let us consider the possible cases.

Consider at first two cases: $t^{[1]} < \tau_* < \tilde{t}$ and $\tilde{t} \leq \tau_* < \vartheta$, where $t^{[1]}$ and $\tilde{t}$ are the time moments given in (30.4), (30.8). In these cases the quantity $e_{(2)}(\tau_*, w_*, \Delta)$ (30.63) is determined by the equalities (30.22), (30.23), (30.27), (30.28). In addition, we have $\nu = \nu^0 = 1$ and $\varphi_j(\tau_*, m, \nu) = \varphi_j(\tau_*, m, \nu^0) = \varphi_j(\tau_*, m)$.

For $\tau_* = t^{[1]}$ we have

$$e_{(2)}(\tau_*, w_*, \Delta) = \max_{m^*, m_*, \nu} \left[m^{*T} \widetilde{w}_* + \varphi_1(\tau_*, m_*, \nu) \right] \tag{30.64}$$

where $\varphi_1(\tau_*, m, \nu)$ is determined by the equality

$$\varphi_1(\tau_*, m, \nu) = \nu \sum_{s=\widetilde{j}}^{k} (\tau_{s+1} - \tau_s)(g(\tau_s) - f(\tau_s))(t^{[2]} - \tau_s) +$$

$$+ \sum_{s=1}^{\widetilde{j}-1} (\tau_{s+1} - \tau_s)(g(\tau_s) - f(\tau_s))(t^{[2]} - \tau_s) \mid l^{[2]} \mid \tag{30.65}$$

$$m_* = \begin{bmatrix} l_1^{[2]} \\ 0 \\ l_2^{[2]} \\ 0 \end{bmatrix}, \quad m^* = \begin{bmatrix} l_1^{[2]} + l_2^{[1]} \\ (t^{[1]} - \vartheta) l_1^{[1]} \\ l_2^{[2]} + l_2^{[1]} \\ (t^{[1]} - \vartheta) l_2^{[1]} \end{bmatrix}$$

$$\mid l^{[1]} \mid \leq 1 - \nu, \quad \mid l^{[2]} \mid \leq \nu, \quad \nu \in [0, 1]. \tag{30.66}$$

For $\tau_* < t^{[1]}$ we have

$$e_{(2)}(\tau_*, w_*, \Delta) = \max_{m^*} \left[m^{*T} \widetilde{w}_* + \varphi_1(\tau_*, m^*, 1) \right]. \tag{30.67}$$

In this case the function $\varphi_1(\tau_*, m, 1)$ is determined by the equality

$$\varphi_1(\tau_*, m, 1) = \max_{\nu} \Big[\nu \sum_{s=\widetilde{j}}^{k} (\tau_{s+1} - \tau_s)(g(\tau_s) - f(\tau_s))(t^{[2]} - \tau_s) +$$

$$+ \sum_{s=j_1}^{\widetilde{j}-1} (\tau_{s+1} - \tau_s)(g(\tau_s) - f(\tau_s))(t^{[2]} - \tau_s) \mid l^{[2]} \mid +$$

$$+ \sum_{s=1}^{j_1-1} (\tau_{s+1} - \tau_s)(g(\tau_s) - f(\tau_s)) \mid (t^{[2]} - \tau_s) l^{[2]} +$$

$$+ (t^{[1]} - \tau_s) l^{[1]} \,|\Big], \tag{30.68}$$

$$|\, l^{[2]} \,| \le \nu, \qquad |\, l^{[1]} \,| \le 1 - \nu\,. \tag{30.69}$$

Here the vectors m^*, m_* and the value ν are determined by the conditions

$$m^* = m_* = m^{[2]} + m^{[1]} = \begin{bmatrix} l_1^{[2]} + l_1^{[1]} \\ (t^{[1]} - \vartheta) l_1^{[1]} \\ l_2^{[2]} + l_2^{[1]} \\ (t^{[1]} - \vartheta) l_2^{[1]} \end{bmatrix}, \tag{30.70}$$

$$|\, l^{[2]} \,| \le \nu, \qquad \nu \in [0,\ 1], \qquad |\, l^{[1]} \,| \le 1 - \nu\,. \tag{30.71}$$

The approximating optimal actions $u^a[t]$ and $v^a[t]$ according to Section 8 (see (8.11), (8.16)) are determined again by the formulas (30.48)–(30.50) and (30.53)–(30.55), respectively. But now in these formulas the vectors $l_u^{[i]0}$ and $l_v^{[i]0}$, $i = 1, 2$ are the solutions of the maximum problems similar to the problems (30.36)–(30.42). We have only the following difference. Instead of $e_{(1)}(\cdot)$ we use $e_{(2)}(\cdot)$ and the restrictions $|\, l_u^{[i]} \,| \le 1$ and $|\, l_v^{[i]} \,| \le 1, i = 1, 2$ are substituted by the restrictions $|\, l_u^{[2]} \,| \le \nu$, $|\, l_v^{[2]} \,| \le \nu$, $|\, l_u^{[1]} \,| + |\, l_u^{[2]} \,| \le 1$, $|\, l_v^{[1]} \,| + |\, l_v^{[2]} \,| \le 1$.

Let us note the following circumstance.

As well as in the case of the quality index $\gamma_{(1)}$ (30.5), here in the considered case of $\gamma_{(2)}$ (30.60) the optimal control actions $u^0[t]$ and optimal disturbances $v^0[t]$ can be obtained directly from the precise optimal extremal strategies $u_e^0(\cdot) = u_e^0(t, x, \varepsilon)$ and $v_e^0(t, x, \varepsilon)$ according to the corresponding formulas (see Section 8). These strategies $u_e^0(\cdot)$ and $v_e^0(\cdot)$ are determined by the conditions (8.11), (8.16) of the extremal shift to the accompanying points $w^{[u]0}(\cdot)$ and $w^{[v]0}(\cdot)$. These points are found on the basis of the function $\rho_{(2)}^0(\tau_*, w_*)$ according to the conditions (8.9), (8.10) and (8.14), (8.15). The function $\rho_{(2)}^0(\tau_*, w_*)$ satisfies the following equality:

$$\rho_{(2)}^0(\tau_*, w_*) = \lim_{\delta \to 0} e_{(2)}(\tau_*, w_*, \Delta_\delta)\,. \tag{30.72}$$

This limit is calculated again in the following way. In expressions similar to (30.27), (30.28) and in expressions (30.64), (30.67), for the quantity $e_{(2)}(\tau_*, w_*, \Delta)$ the integral sums are changed for the integrals. We obtain, for example, for $\tau_* < t^{[1]}$ the following expression for the value of the game:

$$\rho^0_{(2)}(\tau_*, w_*) = \max_{m^*,\nu} \Big[m^{*T}\widetilde{w}_*[\tau_*] +$$

$$+ \int_{\tau_*}^{t^{[1]}} | l^{[2]}(\vartheta - \tau) + l^{[1]}(t^{[1]} - \tau) | (\vartheta - \tau)(g(\tau) - f(\tau))d\tau +$$

$$+ | l^{[2]} | \int_{t^{[1]}}^{\widetilde{t}} (\vartheta - \tau)(g(\tau) - f(\tau))d\tau +$$

$$+ \nu \int_{\widetilde{t}}^{\vartheta} (\vartheta - \tau)(g(\tau) - f(\tau))d\tau \Big], \tag{30.73}$$

$$m^* = m^{[1]} + m^{[2]}, \qquad m^{[1]} = \begin{bmatrix} l_1^{[1]} \\ (t^{[1]} - \vartheta)l_1^{[1]} \\ l_2^{[1]} \\ (t^{[1]} - \vartheta)l_2^{[1]} \end{bmatrix},$$

$$m^{[2]} = \begin{bmatrix} l_1^{2]} \\ 0 \\ l_2^{[2]} \\ 0 \end{bmatrix}, \qquad | l^{[1]} | + | l^{[2]} | \leq 1,$$

$$| l^{[2]} | \leq \nu, \qquad \nu \in [0,\ 1]. \tag{30.74}$$

And similarly in other cases. The further constructions that should be described here for the case of $\gamma_{(2)}$ repeat with understandable changes the corresponding constructions for $\gamma_{(1)}$. These constructions are omitted here.

The processes of control considered in this section were simulated for the following parameters:

$$t_0 = 0, \qquad t^{[1]} = 0.5, \qquad \widetilde{t} = 0.6, \qquad t^{[2]} = \vartheta = 1 \tag{30.75}$$

and the functions $g(t)$ and $f(t)$ in (30.1) of the form

$$f(t) = 2.5 - 2.5t, \qquad g(t) = 1\,. \tag{30.76}$$

For the quality index $\gamma_{(1)}$ (30.5) and the initial data

$$r_{*1} = -0.5, \quad \dot{r}_{*1} = -0.1, \quad r_{*2} = 0.1, \quad \dot{r}_{*2} = 0.6 \tag{30.77}$$

the following results were obtained. The corresponding value of the game calculated before the simulation was $\rho^0_{(1)}(t_*, r_*) = 1.12$. The continuous curve in Figure 30.1 represents in the plane $\{r_1, r_2\}$ the motion of the object (30.1) generated by the approximating optimal strategies $u^a_{(1)}(\cdot) = u^a_{(1)}(t, x, \varepsilon)$ and $v^a_{(1)}(\cdot) = v^a_{(1)}(t, x, \varepsilon)$. In this case we obtained $\gamma_{(1)} = 1.14 \approx \rho^0_{(1)}(t_*, r_*) = 1.12$. The curve in Figure 30.2 represents the motion generated by the strategy $u^a_{(1)}(\cdot)$ and some strategy $v_{(1)}(\cdot) \neq v^a_{(1)}(\cdot)$. Here we obtain $\gamma_{(1)} = 0.65 < \rho^0_{(1)}(t_*, r_*)$. Figure 30.3 shows the motion of the object (30.1) for some strategy $u_{(1)}(\cdot) \neq u^a_{(1)}(\cdot)$ and the strategy $v^a_{(1)}(\cdot)$. Here $\gamma_{(1)} = 1.96 > \rho^0_{(1)}(t_*, r_*)$.

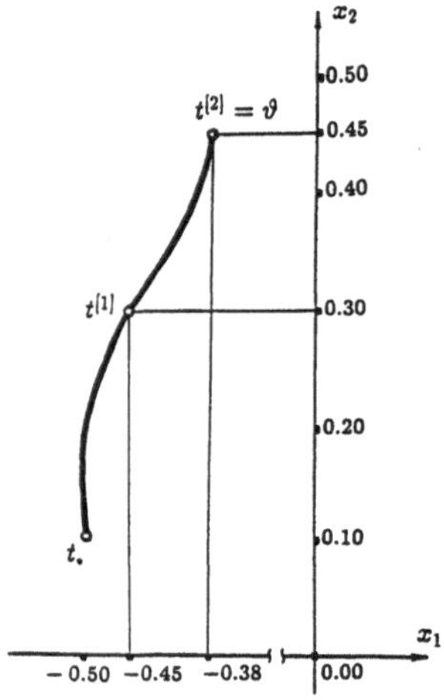

Figure 30.1

In the case of the quality index $\gamma_{(2)}$ (30.60) for the given initial data (30.75)–(30.77) the following results were obtained. The a priori value of the game was $\rho^0_{(2)}(t_*, r_*) = 0.58$. The curve in Figure 30.4 represents the motion of the object (30.1) generated by the approximating optimal strategies $u^a_{(2)}(\cdot) = u^a_{(2)}(t, x, \varepsilon)$, $v^a_{(2)}(\cdot) = v^a_{(2)}(t, x, \varepsilon)$ $[\gamma_{(2)} = 0.59 \approx \rho^0_{(2)}(t_*, r_*)]$. The curve in Figure 30.5 corresponds to the strategy $u^a_{(2)}(\cdot)$ and some strategy $v_{(2)}(\cdot) \neq v^a_{(2)}(\cdot)$. We obtained $\gamma_{(2)} = 0.48 < \rho^0_{(2)}(t_*, r_*)$. Figure 30.6 shows the motion of the object (30.1) for some strategy $u_{(2)}(\cdot) \neq u^a_{(2)}(\cdot)$ and the strategy $v^a_{(2)}(\cdot)$. In this case $\gamma_{(2)} = 1.07 > \rho^0_{(2)}(t_*, r_*)$.

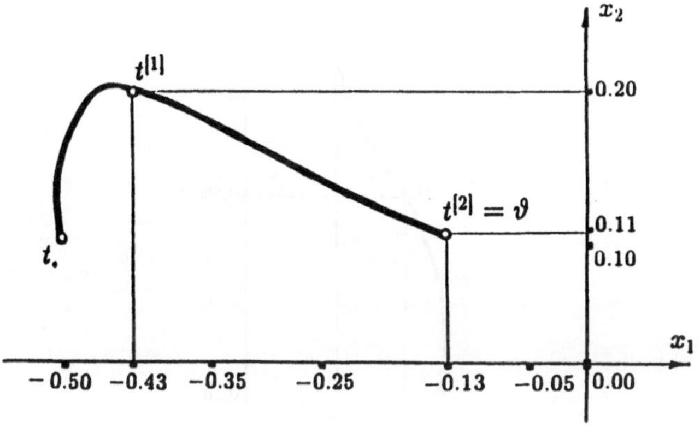

Figure 30.2

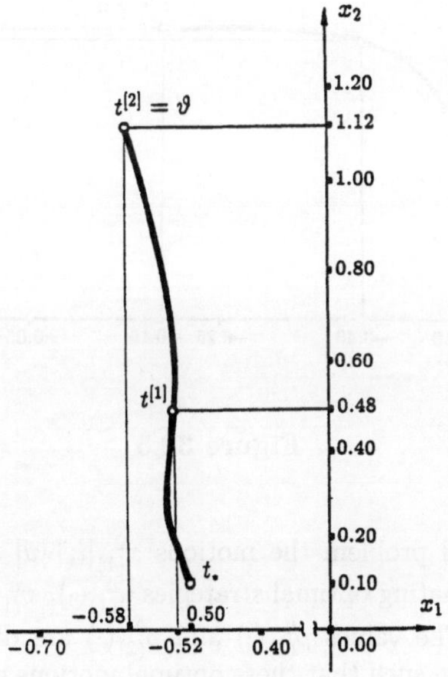

Figure 30.3

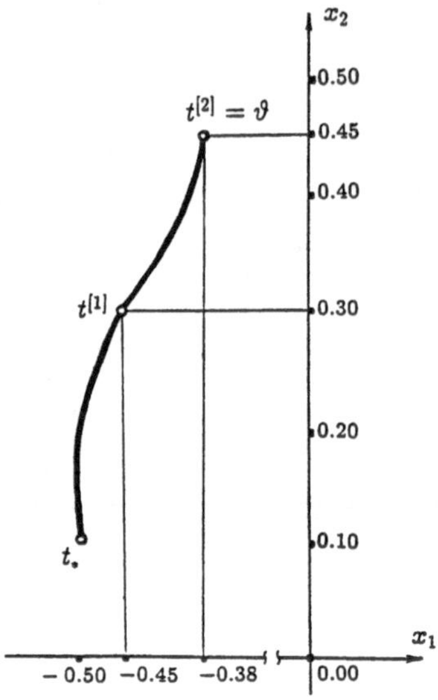

Figure 30.4

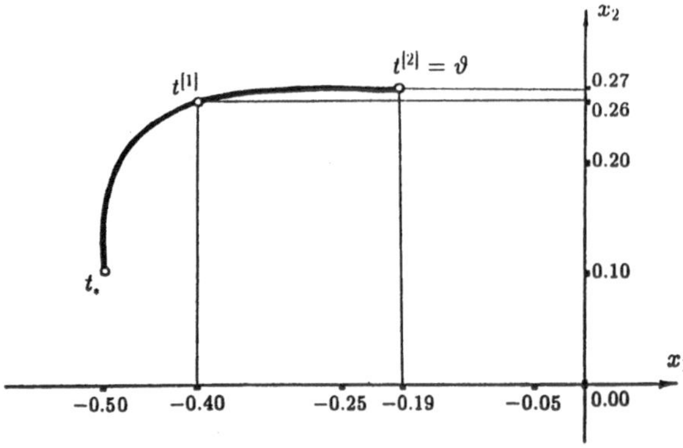

Figure 30.5

In the considered problem the motions $x^0_{(1)}[t_*[\cdot]\vartheta]$ and $x^0_{(2)}[t_*[\cdot]\vartheta]$ generated by the approximating optimal strategies $u^a_{(1)}(\cdot)$, $v^a_{(1)}(\cdot)$ and $u^a_{(2)}(\cdot)$, $v^a_{(2)}(\cdot)$ actually coincide. The values $\rho^0_{(1)}(\cdot)$ and $\rho^0_{(2)}(\cdot)$ are different, though. But examples can be given such that these optimal motions are also essentially different. Such an example is presented in Figure 30.7. Here the problems have the same parameters and differ only in the choice of functionals $\gamma_{(1)}(\cdot)$ and $\gamma_{(2)}(\cdot)$, but we obtain essentially different motions $x^0_{(1)}(\cdot)$, $x^0_{(2)}(\cdot)$. However, in this example the values $\rho^0_{(1)}(\cdot)$ and $\rho^0_{(2)}(\cdot)$ do not differ much. The functionals $\gamma_{(1)}(\cdot),\ \gamma_{(2)}(\cdot)$ have the form

$$\gamma_{(1)}(x[t_*[\cdot]\vartheta]) = \mid x[t^{[1]}] \mid + \mid x[\vartheta] \mid, \tag{30.78}$$

$$\gamma_{(2)}(x[t_*[\cdot]\vartheta]) = \max(\mid x[t^{[1]}] \mid,\ \mid x[\vartheta] \mid). \tag{30.79}$$

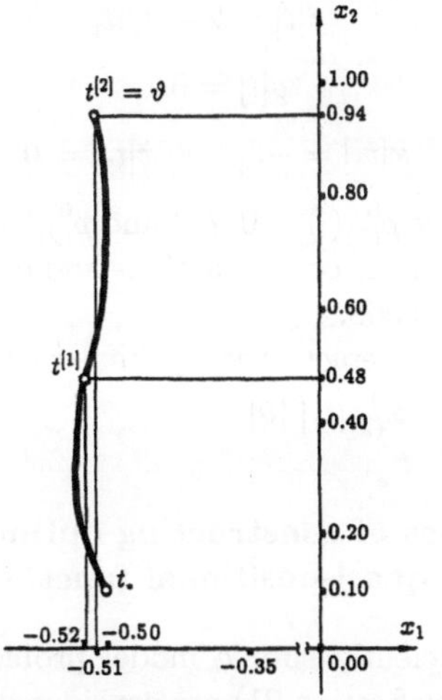

Figure 30.6

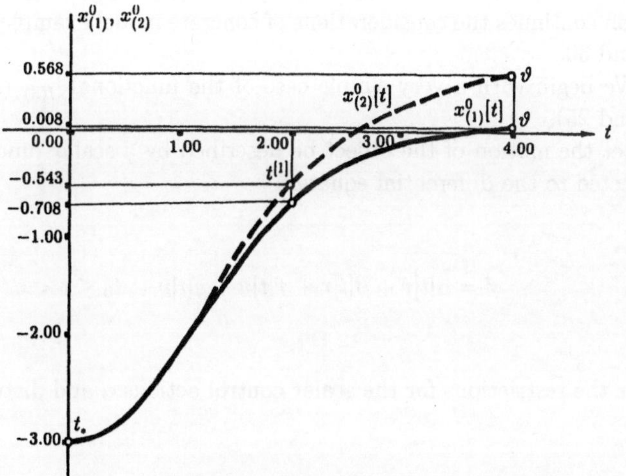

Figure 30.7

Here $x[t]$ is a scalar variable, $t_* = 0,\ t^{[1]} = 2,\ \vartheta = 4$.
The considered process is described by the equation

$$\ddot{x} = f[t]u + g[t]v \tag{30.80}$$

where $f(t)$ and $g(t)$ and the initial conditions are determined by the following equalities:

$$f[t] = 2 - 0.5t, \tag{30.81}$$

$$g[t] = 0.1, \tag{30.82}$$

$$x[t_*] = -3, \qquad \dot{x}[t_*] = 0. \tag{30.83}$$

The values of the game are $\rho^0_{(1)}(\cdot) = 0,711$ and $\rho^0_{(2)}(\cdot) = 0.552$. The simulations for the approximating optimal control actions and disturbances give the values $\gamma_{(1)}(\cdot) = 0.716,\ \gamma_{(2)}(\cdot) = 0.568$.

The continuous curve depicts the optimal motion $x^0_{(1)}[t_*[\cdot]\vartheta]$; dash line shows the optimal motion $x^0_{(2)}[t_*[\cdot]\vartheta]$.

31. Examples of constructing optimal strategies for quasi-positional functionals

In this section two sufficiently simple model problems of control for quasi-positional functionals (see Section 21) are considered. They illustrate the procedures described in Section 25. Thus, we obtain a concrete description of approximating optimal strategies in the considered cases. The results of simulation of the processes of control on the basis of these strategies are given. This section continues the considerations of concrete model examples of Sections 28, 29, and 30.

We begin with a very simple case of the functional $\gamma_{(3)}$ (see Sections 21, 24, and 25).

Let the motion of the object be described by a scalar function $r[t]$ that is subjected to the differential equation

$$\ddot{r} = \alpha[t]\dot{r} + \beta[t]r + f[t]u + g[t]v, \quad t_0 \le t < \vartheta \tag{31.1}$$

under the restrictions for the scalar control actions u and disturbances v

$$|\, u \,| \le 1, \qquad |\, v \,| \le 1\,. \tag{31.2}$$

We take the quality index

$$\gamma = \gamma_{(3)}(r[t_*^0[\cdot]\vartheta]) = \max\Big(\mid r[t^{[1]}]\mid, \mid r[t^{[2]}]\mid\Big) + \mid r[t^{[3]}]\mid . \tag{31.3}$$

As in Section 30 we assume that the continuous functions $f[t]$ and $g[t]$ satisfy the following conditions:

$$f[t] > g[t] \geq 0, \quad t_*^0 \leq t < \tilde{t} < \vartheta,$$

$$\begin{aligned} g[t] > f[t] \geq 0, \quad \tilde{t} < t \leq \vartheta, \\ f[\tilde{t}] = g[\tilde{t}]. \end{aligned} \tag{31.4}$$

Let the moments $t^{[i]}$, $i = 1,2,3$ in (31.3) satisfy the conditions

$$t_0 \leq t_*^0 = t^{[1]} < t^{[2]} = \tilde{t} < t^{[3]} = \vartheta . \tag{31.5}$$

In the concrete cases considered in this section it is convenient to enumerate the moments $t^{[i]}$ in a somewhat different way than in Sections 24 and 25. For instance, the moments $t^{[i]}$, $i = 1,2,3$ (31.3), (31.5) would be enumerated in Section 25 as $t^{[3]} = t^{[1]}_{(1)}$, $t^{[1]} = t^{[1]}_{(2)}$, $t^{[2]} = t^{[2]}_{(2)}$.

Let $x_1 = r$, $x_2 = \dot{r}$. The equation (31.1) is equivalent to the following normal system of two equations:

$$\begin{cases} \dot{x}_1 = x_2 \\ \dot{x}_2 = \beta[t]x_1 + \alpha[t]x_2 + f[t]u + g[t]v . \end{cases} \tag{31.6}$$

Let us recall the remark in Section 25 about the introduction of the deterministic vectors $m^{[i]}$ and $l^{[i]}$. These deterministic vectors determine the value $e_{*(3)}(\cdot)$ in accordance with (25.5)–(25.13). Let us introduce these vectors here.

Let $X(t,\tau)$ be the fundamental matrix of the homogeneous part of system (31.6):

$$X(t,\tau) = \begin{bmatrix} x_{11}(t,\tau) & x_{12}(t,\tau) \\ x_{21}(t,\tau) & x_{22}(t,\tau) \end{bmatrix} . \tag{31.7}$$

The matrices $D^{[i]}$, $i = 1,2,3$ that we have to use according to (13.3), (13.4) are here the following:

$$D^{[i]} = D = [1,\ 0]\,. \tag{31.8}$$

According to the formulas (13.1)–(13.4), (25.5)–(25.13) and the above remark about the vectors $m^{[i]}$ and $l^{[i]}$ we take

$$\widetilde{w}_* = \begin{bmatrix} x_{11}(\vartheta,\tau_*)w_{*1} + x_{12}(\vartheta,\tau_*)w_{*2} \\ x_{21}(\vartheta,\tau_*)w_{*1} + x_{22}(\vartheta,\tau_*)w_{*2} \end{bmatrix},$$

$$w_* = w[\tau_*], \qquad \widetilde{w}_* = \widetilde{w}[\tau_*], \tag{31.9}$$

$$\widetilde{f}(\tau,u,v) = \begin{bmatrix} x_{12}(\vartheta,\tau)(f[\tau]u + g[\tau]v) \\ x_{22}(\vartheta,\tau)(f[\tau]u + g[\tau]v) \end{bmatrix}, \tag{31.10}$$

$$m^{[i]} = \begin{bmatrix} x_{11}(t^{[i]},\vartheta) & x_{21}(t^{[i]},\vartheta) \\ x_{12}(t^{[i]},\vartheta) & x_{22}(t^{[i]},\vartheta) \end{bmatrix} \begin{bmatrix} 1 \\ 0 \end{bmatrix} l^{[i]} =$$

$$= \begin{bmatrix} x_{11}(t^{[i]},\vartheta) \\ x_{12}(t^{[i]},\vartheta) \end{bmatrix} l^{[i]}. \tag{31.11}$$

We have to calculate the quantity (see Section 25, (25.8), (25.14))

$$e_{(3)}(\tau_*, w[t_*^0[\cdot]\tau_*], \Delta) =$$

$$\max_{l^{[i]},m_*,\nu} \Big[\, m_*^T\widetilde{w}_* + \sum_{i=1}^{g(\tau_*)} l^{[i]}D^{[i]}w[t^{[i]}] + \kappa(\tau_*, m_*, \nu, \Delta) \,\Big] =$$

$$= \max_{l^{[i]},m_*,\nu} \Big[\, m_*^T\widetilde{w}_* + \sum_{i=1}^{g(\tau_*)} l^{[i]}D^{[i]}w[t^{[i]}] + \varphi_1(\tau_*, m_*, \nu) \,\Big] \tag{31.12}$$

where

$$g(\tau_*) = \max_{t^{[i]}} i, \qquad t^{[i]} \le \tau_*, \qquad i = 1,2\,. \tag{31.13}$$

At first we calculate the quantities $\varphi_j(\tau_*, m, \nu)$. Here as in Section 30, the functions $\varphi_j(\tau_*, m, \nu)$ can be calculated in the limit form that involves the integrals. Let $\tau_* \in [\widetilde{t},\ \vartheta]$, $t^{[3]} = \vartheta$. Then

$$m_*^T = m^{[3]T} = \{\, l^{[3]}x_{11}(t^{[3]},\vartheta),\ l^{[3]}x_{12}(t^{[3]},\vartheta) \,\},$$

$$| l^{[3]} | \leq 1, \quad t^{[2]} = \widetilde{t} \leq \tau_* \leq \vartheta. \tag{31.14}$$

And after corresponding calculations we obtain

$$\psi_j(\tau_*, m_*) = \int_{\tau_j}^{\tau_{j+1}} \min_u \max_v \left[l^{[3]}(f[\tau]u + g[\tau]v) \right] x_{12}(\vartheta, \tau) d\tau +$$

$$+\varphi_{j+1}(\tau_*, m_*, \nu) =$$

$$= \int_{\tau_j}^{\tau_{j+1}} (g[\tau] - f[\tau]) \mid l^{[3]} x_{12}(\vartheta, \tau) \mid d\tau + \varphi_{j+1}(\tau_*, m_*, \nu) \tag{31.15}$$

and

$$\varphi_j(\tau_*, m_*, \nu) = \int_{\tau_j}^{\vartheta} (g[\tau] - f[\tau]) \mid x_{12}(\vartheta, \tau) \mid d\tau \,. \tag{31.16}$$

Let $\tau_* \in [t^{[1]},\ t^{[2]})$, $t^{[1]} = t_*^0$. Then

$$m_* = m^{[2]} + m^{[3]},$$

$$m^{[i]T} = \{ l^{[i]} x_{11}(t^{[i]}, \vartheta),\ l^{[i]} x_{12}(t^{[i]}, \vartheta) \}, \quad i = 2, 3,$$

$$| l^{[3]} | \leq 1, \quad | l^{[2]} | \leq \nu, \quad \nu \in [0,\, 1]. \tag{31.17}$$

So we obtain

$$\psi_j(\tau_*, m_*, \nu) = \int_{\tau_j}^{\tau_{j+1}} \min_u \max_v \left[(l^{[3]} x_{12}(\vartheta, \tau) + l^{[2]} x_{12}(t^{[2]}, \tau)) * \right.$$

$$\left. * \, (f[\tau]u + g[\tau]v) \right] d\tau + \varphi_{j+1}(\tau_*, m_*, \nu) \tag{31.18}$$

and

$$\varphi_j(\tau_*, m_*, \nu) = \int_{\tau_j}^{t^{[2]}} | \, l^{[3]} x_{12}(\vartheta, \tau) + l^{[2]} x_{12}(t^{[2]}, \tau) \, | \, (g[\tau] - f[\tau]) d\tau +$$

$$+ \int_{t^{[2]}}^{\vartheta} (g[\tau] - f[\tau]) \mid x_{12}(\vartheta, \tau) \mid d\tau \,. \tag{31.19}$$

Now let us calculate the quantity

$$e_{(3)}(\tau_*, w[t_*^0[\cdot]\tau_*]) = \lim_{\delta\to 0} e_{(3)}(\tau_*, w[t_*^0[\cdot]\tau_*], \Delta_\delta) \qquad (31.20)$$

where

$$\delta = \max_j(\tau_{j+1} - \tau_j) . \qquad (31.21)$$

Let $\tau_* \in [t^{[2]},\ \vartheta]$. Then according to (31.12)–(31.16) we obtain

$$e_{(3)}(\tau_*, w[t^{[1]}], w[t^{[2]}], w_*) =$$

$$= \max_{\substack{|l^{[3]}|\le 1,\\ |l^{[1]}|+|l^{[2]}|\le 1}} \Big[l^{[1]}w_1[t^{[1]}] + l^{[2]}w_1[t^{[2]}] +$$

$$+ l^{[3]}\,(x_{11}(\vartheta,\tau_*)w_{*1} + x_{12}(\vartheta,\tau_*)w_{*2}) +$$

$$+ \int_{\tau_*}^{\vartheta} (g[\tau] - f[\tau])\ |\ x_{12}(\vartheta,\tau)\ |\ d\tau\ \Big] . \qquad (31.22)$$

Let $\tau_* \in [t^{[1]},\ t^{[2]})$. In this case according to (31.17)–(31.21) we have

$$e_{(3)}(\tau_*, w[t^{[1]}], w_*) =$$

$$= \max_{\substack{|l^{[3]}|\le 1,\\ |l^{[1]}|+|l^{[2]}|\le 1}} \Big[l^{[1]}w_1[t^{[1]}] + l^{[2]}(x_{11}(t^{[2]},\tau_*)w_{*1} +$$

$$+ x_{12}(t^{[2]},\tau_*)w_{*2}) + l^{[3]}(x_{11}(\vartheta,\tau_*)w_{*1} + x_{12}(\vartheta,\tau_*)w_{*2}) +$$

$$+ \int_{\tau_*}^{t^{[2]}} (g[\tau] - f[\tau])\ |\ l^{[2]}x_{12}(t^{[2]},\tau) + l^{[3]}x_{12}(\vartheta,\tau)\ |\ d\tau +$$

$$+ \int_{t^{[2]}}^{\vartheta} (g[\tau] - f[\tau])\ |\ x_{12}(\vartheta,\tau)\ |\ d\tau\ \Big] . \qquad (31.23)$$

Let us remark the following. In the considered case the value $e_{(3)}(\cdot)$ for the given τ_* is, generally speaking, determined by the history of the motion $w[\tau]$, $t_*^0 \le \tau$ (Sections 24 and 25). Therefore, we should have denoted this value $e_{(3)}(\cdot)$ in (31.22), (31.23) in the form $e_{(3)}(\tau_*, w[t_0^*[\cdot]\tau_*])$. However, for $\tau_* \in [t^{[2]},\ \vartheta)$ the value $e_{(3)}(\cdot)$ actually depends only on $w[t^{[1]}]$, $w[t^{[2]}]$ and w_*.

Therefore, we denote this value for $\tau_* \in [t^{[2]},\ \vartheta)$ as $e_{(3)}(\tau_*, w[t^{[1]}], w[t^{[2]}], w_*)$, where $w_* = w[\tau_*]$. For $\tau_* \in [t^{[1]},\ t^{[2]})$ the considered value $e_{(3)}(\cdot)$ actually depends on the realized vectors $w[t^{[1]}]$ and $w_* = w[\tau_*]$. Therefore, in this case we use the notation $e_{(3)}(\tau_*, w[t^{[1]}], w_*)$.

We have calculated the program extremum $e_{(3)}(\cdot)$ and now will calculate the approximating optimal actions $u^a[t]$ and $v^a[t]$. In accordance with the proposed general theory and the remark made above in this section we have the following. For the sufficient informational image $y[\cdot]$ in the considered case one can take the history of the motion $x[t_*^0[\cdot]t]$ that has been realized up to the current time moment t. The actions $u^a[t_i]$ and $v^a[t_i]$ are constructed on the basis of the accompanying histories $w^{[u]0}(t_i, x[t_*^0[\cdot]t_i], \varepsilon)$, $w^{[v]0}(t_i, x[t_*^0[\cdot]t_i], \varepsilon)$, $t_*^0 \le \tau \le t_i$ (see Sections 24 and 25). However, due to the above remark, in our concrete case of the functional $\gamma_{(3)}$ it suffices to use only a small part of the whole available information on the histories $x[t_*^0[\cdot]t_i]$ and $w^{[u]0}(t_i, x[t_*^0[\cdot]t_i], \varepsilon)$, $w^{[v]0}(t_i, x[t_*^0[\cdot]t_i], \varepsilon)$. For $t_i \in (t^{[2]},\ \vartheta)$ it suffices to use only the past values $x[t^{[1]}]$, $x[t^{[2]}]$, the current value $x[t_i]$, the accompanying values $w^{[u]0}(t^{[1]}, x[t^{[1]}], \varepsilon)$, $w^{[u]0}(t^{[2]}, x[t^{[2]}], \varepsilon)$, $w^{[v]0}(t^{[1]}, x[t^{[1]}], \varepsilon)$, $w^{[v]0}(t^{[2]}, x[t^{[2]}], \varepsilon)$ as well as the accompanying values $w^{[u]0}(t_i, x[t_i], \varepsilon)$, $w^{[v]0}(t_i, x[t_i], \varepsilon)$; for $t_i \in (t^{[1]},\ t^{[2]}]$ it is enough to employ the values $x[t_i]$, $x[t^{[1]}]$ and the corresponding accompanying values $w^{[u]0}(t_i, x[t_i], \varepsilon)$, $w^{[u]0}(t^{[1]},\ x[t^{[1]}], \varepsilon)$, $w^{[v]0}(t_i, x[t_i], \varepsilon)$, $w^{[v]0}(t^{[1]}, x[t^{[1]}], \varepsilon)$. Here the accompanying points $w^{[u]0}(t_i, \cdot)$, $w^{[v]0}(t_i, \cdot)$ and $w^{[u]0}(t^{[j]}, \cdot)$, $w^{[v]0}(t^{[j]}, \cdot)$, $j = 1, 2$, are subjected to the conditions

$$| w^{[u]0}(t^{[j]}, x[t^{[j]}], \varepsilon) - x[t^{[j]}] | \le R^{[u]}(\varepsilon_u, t^{[j]}) = \varepsilon[t^{[j]}], \tag{31.24}$$

$$| w^{[u]0}(t_i, x[t_i], \varepsilon) - x[t_i] | \le R^{[u]}(\varepsilon_u, t_i) = \varepsilon[t_i], \tag{31.25}$$

$$| w^{[v]0}(t^{[j]}, x[t^{[j]}], \varepsilon) - x[t^{[j]}] | \le R^{[v]}(\varepsilon_v, t^{[j]}) = \varepsilon[t^{[j]}], \tag{31.26}$$

$$| w^{[v]0}(t_i, x[t_i], \varepsilon) - x[t_i] | \le R^{[v]}(\varepsilon_v, t_i) = \varepsilon[t_i] \tag{31.27}$$

where we assume

$$\varepsilon[t] = \varepsilon \exp \lambda_\varepsilon (t - t_0), \quad \varepsilon > 0, \quad \lambda_\varepsilon > 0. \tag{31.28}$$

Let us now calculate the actions $u^a[t]$ and $v^a[t]$. We use the notation $x_* = x[t_*]$, $w_* = w[t_*]$, $x_*^0 = x[t_*^0]$, $w_*^0 = w[t_*^0]$.

Firstly, we have to calculate the approximating functions $\rho_u^a(t_*, x[t_*^0[\cdot]t_*], \varepsilon)$, $\rho_v^a(t_*, x[t_*^0[\cdot]\ t_*], \varepsilon)$.

The function $\rho_u^a(t_*, x[t_*^0[\cdot]t_*], \varepsilon) = \rho_u^a(t_*, w_u^0(t_*, x[t_*^0[\cdot]t_*], \varepsilon))$ approximates the value of the game $\rho^0(t_*, x[t_*^0[\cdot]t_*])$. This approximation is used to form the actions $u^a[t]$. The function $\rho_u^a(t_*, x[t_*^0[\cdot]t_*], \varepsilon)$ is determined by the following equalities.

Let $[t^{[i]}]$, $w_*^{[u]0}$ be shortened notations of the elements of the corresponding accompanying history $w^{[u]0}(t_*, x[t_*^0[\cdot]t_*], \varepsilon)$, $t^{[i]} \in [t_*^0,\ t_*]$, $w_*^{[u]0} = w^{[u]0}[t_*]$. Let us denote

$$s^{[i]} = x[t^{[i]}] - w^{[u]}[t^{[i]}], \quad i = 1, 2,$$

$$s_* = x_* - w_*^{[u]}. \tag{31.29}$$

For $t_* \in (t^{[2]},\ \vartheta)$ we have

$$\rho_u^a(t_*, x[t^{[1]}], x[t^{[2]}], x_*, \varepsilon) =$$

$$= e_{(3)}(t_*, w^{[u]0}[t^{[1]}], w^{[u]0}[t^{[2]}], w_*^{[u]0}) =$$

$$= \max_{\substack{|l^{[3]}|\leq 1,\\ |l^{[1]}|+|l^{[2]}|\leq 1}} \min_{\substack{|s^{[1]}|\leq\varepsilon[t^{[1]}],\\ |s^{[2]}|\leq\varepsilon[t^{[2]}],\\ |s_*|\leq\varepsilon[t_*]}}$$

$$\Big[l^{[1]}(x_1[t^{[1]}] - s_1^{[1]}) + l^{[2]}(x_1[t^{[2]}] - s_1^{[2]}) +$$

$$+ l^{[3]}\left[x_{11}(\vartheta, t_*)(x_{*1} - s_{*1}) + x_{12}(\vartheta, t_*)(x_{*2} - s_{*2})\right] + \varphi_1(t_*, m^{[3]})\ \Big] =$$

$$= \max_{\substack{|l^{[3]}|\leq 1,\\ |l^{[1]}|+|l^{[2]}|\leq 1}} \Big[\ l^{[1]}x_1[t^{[1]}] + l^{[2]}x_1[t^{[2]}] +$$

$$+ l^{[3]}[x_{11}(\vartheta, t_*)x_{*1} + x_{12}(\vartheta, t_*)x_{*2}] -$$

$$- \mid l^{[1]} \mid \varepsilon[t^{[1]}] - \mid l^{[2]} \mid \varepsilon[t^{[2]}] - (x_{11}^2(\vartheta, t_*) + x_{12}^2(\vartheta, t_*))^{1/2} \mid l^{[3]} \mid \varepsilon[t_*] +$$

$$+ \int_{t_*}^{\vartheta} (g[\tau] - f[\tau]) \mid x_{12}(\vartheta, \tau) \mid d\tau\ \Big]. \tag{31.30}$$

Below we omit the explicit description of the minimization in s. We give only the expression for the function ρ_u^a that is obtained as a result of this operation.

So, for $t_* = t^{[2]}$ we obtain

$$\rho_u^a(t_*, x[t^{[1]}], x_*, \varepsilon) =$$

$$= e_{(3)}(t_*, w^{[u]0}[t^{[1]}], w_*^{[u]0}) =$$

$$= \max_{\substack{|l^{[3]}|\leq 1,\\ |l^{[1]}|+|l^{[2]}|\leq 1}} \Big[\ l^{[1]}x_1[t^{[1]}] + l^{[2]}x_{*1} +$$

$$+l^{[3]}[x_{11}(\vartheta,t_*)x_{*1}+x_{12}(\vartheta,t_*)x_{*2}]-\mid l^{[1]}\mid \varepsilon[t^{[1]}]-$$
$$-[(l^{[2]}+l^{[3]}x_{11}(\vartheta,t_*))^2+(l^{[3]}x_{12}(\vartheta,t_*))^2]^{1/2}\varepsilon[t_*]+$$
$$+\int_{t_*}^{\vartheta}(g[\tau]-f[\tau])\mid x_{12}(\vartheta,\tau)\mid d\tau\Big]. \qquad (31.31)$$

For $t_* \in (t^{[1]},\ t^{[2]})$ we come to

$$\rho_u^a(t_*, x[t^{[1]}], x_*, \varepsilon) =$$

$$= e_{(3)}(t_*, w^{[u]0}[t^{[1]}], w_*^{[u]0}) =$$

$$= \max_{\substack{|l^{[3]}|\le 1,\\ |l^{[1]}|+|l^{[2]}|\le 1}} \Big[\ l^{[1]}x_1[t^{[1]}]+$$

$$+l^{[2]}(x_{11}(t^{[2]},t_*)x_{*1}+x_{12}(t^{[2]},t_*)x_{*2})+$$
$$+l^{[3]}[x_{11}(\vartheta,t_*)x_{*1}+x_{12}(\vartheta,t_*)x_{*2}]-\mid l^{[1]}\mid \varepsilon[t^{[1]}]-$$
$$-[(l^{[2]}x_{11}(t^{[2]},t_*)+l^{[3]}x_{11}(\vartheta,t_*))^2+$$
$$+(l^{[2]}x_{12}(t^{[2]},t_*)+l^{[3]}x_{12}(\vartheta,t_*))^2]^{1/2}\varepsilon[t_*]+$$
$$+\int_{t_*}^{t^{[2]}}(g[\tau]-f[\tau])\mid l^{[2]}x_{12}(t^{[2]},\tau)+l^{[3]}x_{12}(\vartheta,\tau)\mid d\tau+$$
$$+\int_{t^{[2]}}^{\vartheta}(g[\tau]-f[\tau])\mid x_{12}(\vartheta,\tau)\mid d\tau\Big]. \qquad (31.32)$$

And for $t_* = t^{[1]} = t_*^0$ we derive

$$\rho_u^a(t_*, x_*, \varepsilon) = e_{(3)}(t_*, w_*^{[u]0}) =$$

$$= \max_{\substack{|l^{[3]}|\le 1,\\ |l^{[1]}|+|l^{[2]}|\le 1}} \Big[\ l^{[1]}x_{*1}+$$

$$+l^{[2]}(x_{11}(t^{[2]},t_*)x_{*1}+x_{12}(t^{[2]},t_*)x_{*2})+$$
$$+l^{[3]}[x_{11}(\vartheta,t_*)x_{*1}+x_{12}(\vartheta,t_*)x_{*2}]-$$
$$-[(l^{[1]}+l^{[2]}x_{11}(t^{[2]},t_*)+l^{[3]}x_{11}(\vartheta,t_*))^2+$$
$$+(l^{[2]}x_{12}(t^{[2]},t_*)+l^{[3]}x_{12}(\vartheta,t_*))^2]^{1/2}\varepsilon[t_*]+$$

$$+\int_{t_*}^{t^{[2]}}(g[\tau]-f[\tau])\mid l^{[2]}x_{12}(t^{[2]},\tau)+l^{[3]}x_{12}(\vartheta,\tau)\mid d\tau+$$

$$+\int_{t^{[2]}}^{\vartheta}(g[\tau]-f[\tau])\mid x_{12}(\vartheta,\tau)\mid d\tau\Big]. \qquad (31.33)$$

Similar relations can be obtained for the value $\rho_v^a(t_*, x[t_*^0[\cdot]t_*], \varepsilon) = \rho^a(t_*, w_v^0(t_*, x[t_*^0[\cdot]t_*], \varepsilon))$ that approximates the value of the game $\rho^0(t_*, x[t_*^0[\cdot]t_*])$. This function $\rho_v^a(t_*, x[t_*^0[\cdot]t_*], \varepsilon)$ is used to form the actions $v^a[t]$. The only difference is that in the relations (31.30)–(31.33) the terms containing $\varepsilon[t^{[1]}]$, $\varepsilon[t^{[2]}]$ and $\varepsilon[t_*]$ should have the plus sign.

The maximizing values $l_u^{[1]0}$, $l_u^{[2]0}$ and $l_u^{[3]0}$ given in (31.30)–(31.33) for $\rho_u^a(\cdot)$ and the corresponding maximizing values $l_v^{[1]0}$, $l_v^{[2]0}$ and $l_v^{[3]0}$, defined by the similar relations for $\rho_v^a(\cdot)$, determine the approximating optimal actions $u^a[t]$ and $v^a[t]$. Namely, the actions $u^a[t]$ and $v^a[t]$ are determined in the following way.

According to (25.12), (25.15)–(25.20), (31.11) these actions $u^a[t]$ and $v^a[t]$ are calculated with the help of the following conditions. Let $m_u^0[t_i]$ be the vectors

$$m_u^0[t_i] = \begin{bmatrix} l_u^{[3]0}x_{11}(\vartheta, t_i) \\ l_u^{[3]0}x_{12}(\vartheta, t_i) \end{bmatrix}, \quad t_i \in (t^{[2]},\ \vartheta),$$

$$m_u^0[t_i] = \begin{bmatrix} l_u^{[2]0}x_{11}(t^{[2]}, t_i) + l_u^{[3]0}x_{11}(\vartheta, t_i) \\ l_u^{[2]0}x_{12}(t^{[2]}, t_i) + l_u^{[3]0}x_{12}(\vartheta, t_i) \end{bmatrix}, \quad t_i \in (t^{[1]},\ t^{[2]}],$$

$$m_u^0[t_i] = \begin{bmatrix} l_u^{[1]0} + l_u^{[2]0}x_{11}(t^{[2]}, t_i) + l_u^{[3]0}x_{11}(\vartheta, t_i) \\ l_u^{[2]0}x_{12}(t^{[2]}, t_i) + l_u^{[3]0}x_{12}(\vartheta, t_i) \end{bmatrix}, \quad t_i = t^{[1]} \qquad (31.34)$$

where $l_u^{[i]0}$ are the maximizing scalars in (31.30)–(31.33).

The scalar $u^a[t_i]$ satisfies the condition

$$\langle\ m_u^0[t_i], \begin{bmatrix} 0 \\ f[t_i]u^a[t_i] \end{bmatrix} \rangle = \min_{|u|\le 1} \langle\ m_u^0[t_i], \begin{bmatrix} 0 \\ f[t_i]u \end{bmatrix} \rangle. \qquad (31.35)$$

That is, the desired approximating optimal control actions $u^a[t] = u^a[t_i]$, $t_i \le t < t_{i+1}$ are determined by the equalities

$$u^a[t_i] = -\ \text{sign}\ (l_u^{[3]0}x_{12}(\vartheta, t_i)),$$

$$t_i \in (t^{[2]},\ \vartheta),$$

$$u^a[t_i] = -\ \text{sign}\ (l_u^{[2]0}x_{12}(t^{[2]}, t_i) + l_u^{[3]0}x_{12}(\vartheta, t_i)),$$

$$t_i \in [t^{[1]},\ t^{[2]}]. \tag{31.36}$$

Let $m_v^0[t_i]$ be the vectors given by equalities similar to (31.34). To obtain these equalities one has to substitute in (31.34) the lower index u by v. We emphasize also that the values $l_v^{[i]0}$, $i = 1, 2, 3$ are the solutions of the maximum problems similar to the problems (31.30)–(31.33), but now in the corresponding expressions the terms with $\varepsilon[t_i]$ $\varepsilon[t^{[1]}]$, $\varepsilon[t^{[2]}]$ have the plus sign.

Then the approximating optimal disturbances $v^a[t] = v^a[t_i]$, $t_i \le t < t_{i+1}$ are determined by the following condition:

$$\langle\ m_v^0[t_i],\ \begin{bmatrix} 0 \\ g[t_i]v^a[t_i] \end{bmatrix} \rangle = \max_{|v|\le 1} \langle\ m_v^0[t_i],\ \begin{bmatrix} 0 \\ g[t_i]v \end{bmatrix} \rangle \tag{31.37}$$

and have the form

$$\begin{aligned} v^a[t_i] = {} & \text{sign } (l_v^{[3]0} x_{12}(\vartheta, t_i)), \\ & t_i \in (t^{[2]},\ \vartheta), \\ v^a[t_i] = {} & \text{sign } (l_v^{[2]0} x_{12}(t^{[2]}, t_i) + l_v^{[3]0} x_{12}(\vartheta, t_i)), \\ & t_i \in [t^{[1]},\ t^{[2]}]. \end{aligned} \tag{31.38}$$

Thus, in the considered concrete case, to form the approximating optimal control actions $u^a[t_i]$ and approximating optimal disturbances $v^a[t_i]$ at the time moments $t_i \in (t^{[1]},\ \vartheta)$, one has to use also some information about the past of the motion $x[t]$ at $t < t_i$. Here this information is reduced only to the value $x[t^{[1]}]$ in the case $t_i \in (t^{[1]},\ t^{[2]}]$ and to the values $x[t^{[1]}]$, $x[t^{[2]}]$ in the case $t_i \in (t^{[2]},\ \vartheta)$. This scarce information on the past of the motion turns out to be sufficient due to the fact that the quality index $\gamma = \gamma_{(3)}$ (31.3) has a very simple form. It is clear that for control problems with more complicated functionals $\gamma_{(3)}$ (21.1) a more detailed information on the history of the motion $x[t_*^0[\cdot]t_i]$ might be essential.

Consider now the second example of a model control problem with the quality index $\gamma_{(3)}$. This example is somewhat more complicated than the previous one.

Let the object be described again by the vector differential equation (31.1). But now $r^T = \{r_1, r_2\}$, $u^T = \{u_1, u_2\}$, $v^T = \{v_1, v_2\}$ are 2-dimensional vectors. The scalar functions $f[t]$ and $g[t]$ satisfy conditions similar to the ones in the previous example. The vectors of the control actions u and the disturbances v are constrained by the conditions (31.2), where $\mid u \mid$ and $\mid v \mid$ are the Euclidean norms.

We take the quality index

$$\gamma = \gamma_{(3)}(r[t_*^0[\cdot]\vartheta]) = \mid r_1[t^{[1]}] \mid +$$

$$+\max\Big(\mid \dot{r}_1[t^{[2]}]\mid,\ \mid \dot{r}_2[t^{[3]}]\mid\Big)+\mid r_2[t^{[4]}]\mid \tag{31.39}$$

where the moments $t^{[i]}$, $i = 1, \ldots, 4$ satisfy the following conditions

$$t_0 = t_*^0 \leq t^{[1]} < t^{[2]} < t^{[3]} = \tilde{t} < t^{[4]} = \vartheta\,. \tag{31.40}$$

Here as above $\tilde{t}$ is the time moment where the difference $g[t] - f[t]$ changes its sign from minus to plus as t increases (see (31.4)).

Let $x_1 = r_1$, $x_2 = \dot{r}_1$, $x_3 = r_2$, $x_4 = \dot{r}_2$. The vector equation (31.1) is now equivalent to the following normal system of four equations:

$$\begin{aligned}
\dot{x}_1 &= x_2\\
\dot{x}_2 &= \beta[t]x_1 + \alpha[t]x_2 + f[t]u_1 + g[t]v_1\\
\dot{x}_3 &= x_4\\
\dot{x}_4 &= \beta[t]x_3 + \alpha[t]x_4 + f[t]u_2 + g[t]v_2\,.
\end{aligned} \tag{31.41}$$

The quality index $\gamma_{(3)}$ (31.39) takes the form

$$\gamma_{(3)}(x[t_*^0[\cdot]\vartheta]) = \mid x_1[t^{[1]}]\mid +$$

$$+\max\Big(\ x_2[t^{[2]}]\mid,\ \mid x_4[t^{[3]}]\mid\Big)+\mid x_3[t^{[4]}]\mid\,. \tag{31.42}$$

The fundamental matrix $X(t,\tau)$ has the following block structure:

$$X(t,\tau) = \begin{pmatrix} x_{11}(t,\tau) & x_{12}(t,\tau) & 0 & 0\\ x_{21}(t,\tau) & x_{22}(t,\tau) & 0 & 0\\ 0 & 0 & x_{11}(t,\tau) & x_{12}(t,\tau)\\ 0 & 0 & x_{21}(t,\tau) & x_{22}(t,\tau) \end{pmatrix}. \tag{31.43}$$

According to the formulas (13.1)–(13.4) and (25.5)–(25.13) we have

$$\widetilde{w}_* = X(\vartheta, \tau_*) w_* = \begin{bmatrix} x_{11}(\vartheta, \tau_*) w_{*1} + x_{12}(\vartheta, \tau_*) w_{*2} \\ x_{21}(\vartheta, \tau_*) w_{*1} + x_{22}(\vartheta, \tau_*) w_{*2} \\ x_{11}(\vartheta, \tau_*) w_{*3} + x_{12}(\vartheta, \tau_*) w_{*4} \\ x_{21}(\vartheta, \tau_*) w_{*3} + x_{22}(\vartheta, \tau_*) w_{*4} \end{bmatrix} \tag{31.44}$$

$$\widetilde{f}(\tau, u, v) = X(\vartheta, \tau) \begin{bmatrix} 0 \\ f[\tau]u_1 + g[\tau]v_1 \\ 0 \\ f[\tau]u_2 + g[\tau]v_2 \end{bmatrix} =$$

$$= \begin{bmatrix} x_{12}(\vartheta, \tau)(f[\tau]u_1 + g[\tau]v_1) \\ x_{22}(\vartheta, \tau)(f[\tau]u_1 + g[\tau]v_1) \\ x_{12}(\vartheta, \tau)(f[\tau]u_2 + g[\tau]v_2) \\ x_{22}(\vartheta, \tau)(f[\tau]u_2 + g[\tau]v_2) \end{bmatrix}. \tag{31.45}$$

The matrices $D^{[i]}$, $i = 1, \ldots, 4$ are the following ones:

$$D^{[1]} = (\,1\,0\,0\,0\,), \qquad D^{[2]} = (\,0\,1\,0\,0\,),$$

$$D^{[3]} = (\,0\,0\,0\,1\,), \qquad D^{[4]} = (\,0\,0\,1\,0\,). \tag{31.46}$$

Below, we use the 4-dimensional vectors $m^{[i]}$ and the scalars $l^{[i]}$, $i = 1, 2, 3, 4$, which correspond to the vectors $m^{(i)}$ (13.3) and the scalars $l^{(i)}$ in (13.4). In accordance with (25.12) and the remark at the beginning of this section about the enumeration of the moments $t^{[i]}$ the vectors $m^{[i]}$ have the form

$$m^{[i]} = X^T(t^{[i]}, \vartheta) D^{[i]T} l^{[i]}. \tag{31.47}$$

Here the scalars $l^{[i]}$, $i = 1, 2, 3, 4$ satisfy the conditions

$$|\, l^{[1]} \,| \leq 1, \quad |\, l^{[2]} \,| + |\, l^{[3]} \,| \leq 1, \quad |\, l^{[4]} \,| \leq 1\,. \tag{31.48}$$

The calculations of the program extremum $e_{(3)}(\tau_*, w[t^{[1]}], w[t^{[2]}], w[t^{[3]}], w_*)$ for $\tau_* \geq t^{[3]}$, $w_* = w[\tau_*]$, $e_{(3)}(\tau_*, w[t^{[1]}], w[t^{[2]}], w_*)$ for $\tau_* \in [t^{[2]},\ t^{[3]})$, $e_{(3)}(\tau_*, w[t^{[1]}], w_*)$

for $\tau_* \in [t^{[1]},\ t^{[2]})$ and $e_{(3)}(\tau_*, w_*)$ for $\tau_* \in [t_*^0,\ t^{[1]})$ actually repeat the calculations of the program extremum $e_{(3)}(\cdot)$ in the previous example. These calculations differ only in details connected with the increase of the dimension of the system of equations and with the increase of the number of the time moments $t^{[i]}$ in the quality index γ.

Similarly, the calculations of the quantities $\rho_u^a(t_*, x[t_*^0[\cdot]t_*], \varepsilon)$, $\rho_v^a(t_*, x[t_*^0[\cdot]t_*], \varepsilon)$, which approximate the value of the game $\rho^0(t_*, x[t_*^0[\cdot]t_*])$, actually repeat the corresponding calculations in the previous example. Therefore, we omit here the mentioned calculations and present only the final formulas for the values $\rho_u^a(t_*, x[t_*^0[\cdot]t_*], \varepsilon) = e_{(3)}(t_*, w^{[u]0}[t_*^0[\cdot]t_*], \Delta)$ and $\rho_v^a(t_*, x[t_*^0[\cdot]t_*], \varepsilon) = e_{(3)}(t_*, w^{[v]0}[t_*^0[\cdot]t_*], \Delta)$.

For $t^{[3]} < t_*,\ t^{[3]} = \widetilde{t}$ we have

$$\rho_u^a(t_*, x[t^{[1]}], x[t^{[2]}], x[t^{[3]}], x_*, \varepsilon) =$$

$$= e_{(3)}(t_*, w^{[u]0}[t^{[1]}], w^{[u]0}[t^{[2]}], w^{[u]0}[t^{[3]}], w_*^{[u]0}) =$$

$$= \max_{\substack{|l^{[1]}|\le 1,\\ |l^{[2]}|+|l^{[3]}|\le 1,\\ |l^{[4]}|\le 1}} \Big[l^{[1]}x_1[t^{[1]}] + l^{[2]}x_2[t^{[2]}] + l^{[3]}x_4[t^{[3]}] +$$

$$+ l^{[4]}\Big(x_{11}(\vartheta, t_*)x_{*3} + x_{12}(\vartheta, t_*)x_{*4} \Big) -$$

$$- \mid l^{[1]} \mid \varepsilon[t^{[1]}] - \mid l^{[2]} \mid \varepsilon[t^{[2]}] - \mid l^{[3]} \mid \varepsilon[t^{[3]}] -$$

$$- \mid l^{[4]} \mid \Big(x_{11}^2(\vartheta, t_*) + x_{12}^2(\vartheta, t_*) \Big)^{1/2} \varepsilon[t_*] +$$

$$+ \int_{t_*}^{\vartheta} (g[\tau] - f[\tau]) \mid x_{12}(\vartheta, \tau) \mid d\tau \Big]. \tag{31.49}$$

For $t_* = t^{[3]}$ we have

$$\rho_u^a(t_*, x[t^{[1]}], x[t^{[2]}], x_*, \varepsilon) =$$

$$= e_{(3)}(t_*, w^{[u]0}[t^{[1]}], w^{[u]0}[t^{[2]}], w_*^{[u]0}) =$$

$$= \max_{\substack{|l^{[1]}|\le 1,\\ |l^{[2]}|+|l^{[3]}|\le 1,\\ |l^{[4]}|\le 1}} \Big[l^{[1]}x_1[t^{[1]}] + l^{[2]}x_2[t^{[2]}] + l^{[3]}x_{*4} +$$

$$+ l^{[4]}\Big(x_{11}(\vartheta, t_*)x_{*3} + x_{12}(\vartheta, t_*)x_{*4} \Big) -$$

$$- \mid l^{[1]} \mid \varepsilon[t^{[1]}] - \mid l^{[2]} \mid \varepsilon[t^{[2]}] -$$
$$-\Big((l^{[4]} x_{11}(\vartheta, t_*))^2 + (l^{[3]} + l^{[4]} x_{12}(\vartheta, t_*))^2 \Big)^{1/2} \varepsilon[t_*] +$$
$$+ \int_{t_*}^{\vartheta} (g[\tau] - f[\tau]) \mid x_{12}(\vartheta, \tau) \mid d\tau \Big]. \tag{31.50}$$

For $t^{[2]} < t_* < t^{[3]}$ we have

$$\rho_u^a(t_*, x[t^{[1]}], x[t^{[2]}], x_*, \varepsilon) =$$

$$= e_{(3)}(t_*, w^{[u]0}[t^{[1]}], w^{[u]0}[t^{[2]}], w_*^{[u]0}) =$$
$$= \max_{\substack{|l^{[1]}| \le 1, \\ |l^{[2]}| + |l^{[3]}| \le 1, \\ |l^{[4]}| \le 1}} \Big[l^{[1]} x_1[t^{[1]}] + l^{[2]} x_2[t^{[2]}] +$$
$$+ l^{[3]}\Big(x_{21}(t^{[3]}, t_*) x_{*3} + x_{22}(t^{[3]}, t_*) x_{*4} \Big) +$$
$$+ l^{[4]}\Big(x_{11}(\vartheta, t_*) x_{*3} + x_{12}(\vartheta, t_*) x_{*4} \Big) -$$
$$- \mid l^{[1]} \mid \varepsilon[t^{[1]}] - \mid l^{[2]} \mid \varepsilon[t^{[2]}] -$$
$$-\Big((l^{[3]} x_{21}(t^{[3]}, t_*) + l^{[4]} x_{11}(\vartheta, t_*))^2 +$$
$$+ (l^{[3]} x_{22}(t^{[3]}, t_*) + l^{[4]} x_{12}(\vartheta, t_*))^2 \Big)^{1/2} \varepsilon[t_*] +$$
$$+ \int_{t_*}^{t^{[3]}} (g[\tau] - f[\tau]) \mid l^{[3]} x_{22}(t^{[3]}, \tau) + l^{[4]} x_{12}(\vartheta, \tau) \mid d\tau +$$
$$+ \int_{t^{[3]}}^{\vartheta} (g[\tau] - f[\tau]) \mid x_{12}(\vartheta, \tau) \mid d\tau \Big]. \tag{31.51}$$

For $t_* = t^{[2]}$ we derive

$$\rho_u^a(t_*, x[t^{[1]}], x_*, \varepsilon) = e_{(3)}(t_*, w^{[u]0}[t^{[1]}], w_*^{[u]0}) =$$

$$= \max_{\substack{|l^{[1]}| \le 1, \\ |l^{[2]}| + |l^{[3]}| \le 1, \\ |l^{[4]}| \le 1}} \Big[l^{[1]} x_1[t^{[1]}] + l^{[2]} x_{*2} +$$
$$+ l^{[3]}\Big(x_{21}(t^{[3]}, t_*) x_{*3} + x_{22}(t^{[3]}, t_*) x_{*4} \Big) +$$

$$+l^{[4]}\Big(\ x_{11}(\vartheta,t_*)x_{*3}+x_{12}(\vartheta,t_*)x_{*4}\ \Big)-\mid l^{[1]}\mid\varepsilon[t^{[1]}]-$$

$$-\Big(\ (l^{[2]})^2+(l^{[3]}x_{21}(t^{[3]},t_*)+l^{[4]}x_{11}(\vartheta,t_*))^2+$$

$$+(l^{[3]}x_{22}(t^{[3]},t_*)+l^{[4]}x_{12}(\vartheta,t_*))^2\ \Big)^{1/2}\varepsilon[t_*]+$$

$$+\int_{t_*}^{t^{[3]}}(g[\tau]-f[\tau])\mid l^{[3]}x_{22}(t^{[3]},\tau)+l^{[4]}x_{12}(\vartheta,\tau)\mid d\tau+$$

$$+\int_{t^{[3]}}^{\vartheta}(g[\tau]-f[\tau])\mid x_{12}(\vartheta,\tau)\mid d\tau\ \Big]. \qquad (31.52)$$

For $t^{[1]}<t_*<t^{[2]}$ we come to

$$\rho_u^a(t_*,x[t^{[1]}],x_*,\varepsilon)=e_{(3)}(t_*,w^{[u]0}[t^{[1]}],w_*^{[u]0})=$$

$$=\max_{\substack{|l^{[1]}|\le 1,\\ |l^{[2]}|+|l^{[3]}|\le 1,\\ |l^{[4]}|\le 1}}\Big[\ l^{[1]}x_1[t^{[1]}]+$$

$$+l^{[2]}\Big(\ x_{21}(t^{[2]},t_*)x_{*1}+x_{22}(t^{[2]},t_*)x_{*2}\ \Big)+$$

$$+l^{[3]}\Big(\ x_{21}(t^{[3]},t_*)x_{*3}+x_{22}(t^{[3]},t_*)x_{*4}\ \Big)+$$

$$+l^{[4]}\Big(\ x_{11}(\vartheta,t_*)x_{*3}+x_{12}(\vartheta,t_*)x_{*4}\ \Big)-\mid l^{[1]}\mid\varepsilon[t^{[1]}]-$$

$$-\Big(\ (l^{[2]}x_{21}(t^{[2]},t_*))^2+(l^{[2]}x_{22}(t^{[2]},t_*))^2+$$

$$+(l^{[3]}x_{21}(t^{[3]},t_*)+l^{[4]}x_{11}(\vartheta,t_*))^2+$$

$$+(l^{[3]}x_{22}(t^{[3]},t_*)+l^{[4]}x_{12}(\vartheta,t_*))^2\ \Big)^{1/2}\varepsilon[t_*]+$$

$$+\int_{t_*}^{t^{[2]}}(g[\tau]-f[\tau])((l^{[2]}x_{12}(t^{[2]},\tau))^2+$$

$$+(l^{[3]}x_{22}(t^{[3]},\tau)+l^{[4]}x_{12}(\vartheta,\tau))^2)^{1/2}d\tau+$$

$$+\int_{t^{[2]}}^{t^{[3]}}(g[\tau]-f[\tau])\mid l^{[3]}x_{22}(t^{[3]},\tau)+l^{[4]}x_{12}(\vartheta,\tau)\mid d\tau+$$

$$+\int_{t^{[3]}}^{\vartheta}(g[\tau]-f[\tau])\mid x_{12}(\vartheta,\tau)\mid d\tau\ \Big]. \qquad (31.53)$$

For $t_*=t^{[1]}$ we have

$$\rho_u^a(t_*,x_*,\varepsilon)=e_{(3)}(t_*,w_*^{[u]0})=$$

$$= \max_{\substack{|l^{[1]}|\leq 1,\\ |l^{[2]}|+|l^{[3]}|\leq 1,\\ |l^{[4]}|\leq 1}} \Big[l^{[1]}x_{*1}+$$

$$+l^{[2]}\Big(x_{21}(t^{[2]},t_*)x_{*1}+x_{22}(t^{[2]},t_*)x_{*2} \Big)+$$

$$+l^{[3]}\Big(x_{21}(t^{[3]},t_*)x_{*3}+x_{22}(t^{[3]},t_*)x_{*4} \Big)+$$

$$+l^{[4]}\Big(x_{11}(\vartheta,t_*)x_{*3}+x_{12}(\vartheta,t_*)x_{*4} \Big)-$$

$$-\Big((l^{[1]}+l^{[2]}x_{21}(t^{[2]},t_*))^2+(l^{[2]}x_{22}(t^{[2]},t_*))^2+$$

$$+(l^{[3]}x_{21}(t^{[3]},t_*)+l^{[4]}x_{11}(\vartheta,t_*))^2+$$

$$+(l^{[3]}x_{22}(t^{[3]},t_*)+l^{[4]}x_{12}(\vartheta,t_*))^2 \Big)^{1/2}\varepsilon[t_*]+$$

$$+\int_{t_*}^{t^{[2]}}(g[\tau]-f[\tau])((l^{[2]}x_{12}(t^{[2]},\tau))^2+$$

$$+(l^{[3]}x_{22}(t^{[3]},\tau)+l^{[4]}x_{12}(\vartheta,\tau))^2)^{1/2}d\tau+$$

$$+\int_{t^{[2]}}^{t^{[3]}}(g[\tau]-f[\tau])\mid l^{[3]}x_{22}(t^{[3]},\tau)+l^{[4]}x_{12}(\vartheta,\tau)\mid d\tau+$$

$$+\int_{t^{[3]}}^{\vartheta}(g[\tau]-f[\tau])\mid x_{12}(\vartheta,\tau)\mid d\tau \Big]. \tag{31.54}$$

For $t_*^0 \leq t_* < t^{[1]}$ we deduce

$$\rho_u^a(t_*,x_*,\varepsilon)=e_{(3)}(t_*,w_*^{[u]0})=$$

$$= \max_{\substack{|l^{[1]}|\leq 1,\\ |l^{[2]}|+|l^{[3]}|\leq 1,\\ |l^{[4]}|\leq 1}} \Big[l^{[1]}\Big(x_{11}(t^{[1]},t_*)x_{*1}+x_{12}(t^{[1]},t_*)x_{*2} \Big)+$$

$$+l^{[2]}\Big(x_{21}(t^{[2]},t_*)x_{*1}+x_{22}(t^{[2]},t_*)x_{*2} \Big)+$$

$$+l^{[3]}\Big(x_{21}(t^{[3]},t_*)x_{*3}+x_{22}(t^{[3]},t_*)x_{*4} \Big)+$$

$$+l^{[4]}\Big(x_{11}(\vartheta,t_*)x_{*3}+x_{12}(\vartheta,t_*)x_{*4} \Big)-$$

$$-\Big((l^{[1]}x_{11}(t^{[1]},t_*)+l^{[2]}x_{21}(t^{[2]},t_*))^2+$$

$$+(l^{[1]}x_{12}(t^{[1]},t_*)+l^{[2]}x_{22}(t^{[2]},t_*))^2+$$

$$+(l^{[3]}x_{21}(t^{[3]},t_*)+l^{[4]}x_{11}(\vartheta,t_*))^2+$$

$$+(l^{[3]}x_{22}(t^{[3]},t_*)+l^{[4]}x_{12}(\vartheta,t_*))^2\Big)^{1/2}\varepsilon[t_*]+$$

$$+\int_{t_*}^{t^{[1]}}(g[\tau]-f[\tau])((l^{[1]}x_{12}(t^{[1]},\tau)+l^{[2]}x_{22}(t^{[2]},\tau))^2+$$

$$+(l^{[3]}x_{22}(t^{[3]},\tau)+l^{[4]}x_{12}(\vartheta,\tau))^2)^{1/2}d\tau+$$

$$+\int_{t^{[1]}}^{t^{[2]}}(g[\tau]-f[\tau])((l^{[2]}x_{22}(t^{[2]},\tau))^2+$$

$$+(l^{[3]}x_{22}(t^{[3]},\tau)+l^{[4]}x_{12}(\vartheta,\tau))^2)^{1/2}d\tau+$$

$$+\int_{t^{[2]}}^{t^{[3]}}(g[\tau]-f[\tau])\mid l^{[3]}x_{22}(t^{[3]},\tau)+l^{[4]}x_{12}(\vartheta,\tau)\mid d\tau+$$

$$+\int_{t^{[3]}}^{\vartheta}(g[\tau]-f[\tau])\mid x_{12}(\vartheta,\tau)\mid d\tau\Big]. \tag{31.55}$$

We denote by the symbols $l_u^{[i]0}$ the maximizing values in (31.49)–(31.55) for $t_* = t_i$. Then according to (25.12), (25.15)–(25.20), (31.47) the optimal control actions $u^a[t] = u^a[t_i]$, $t_i \le t < t_{i+1}$ are determined by the equalities given below. These equalities are obtained here from conditions similar to the conditions (31.34), (31.35) with understandable changes connected with differences in the dimensions of the differential equations (31.6) and (31.41) and differences in the quality indices (31.3) and (31.42).

To economize space we give below the values $u^a[t_i]$ and $v^a[t_i]$ only for the intervals $t^{[j]} < t_i < t^{[j+1]}$, $j = 0, 1, 2, 3$. We have

$$\begin{aligned} u_1^a[t_i] &= 0 \\ u_2^a[t_i] &= -\ \mathrm{sign}\ (\ l_u^{[4]0}x_{12}(\vartheta,t_i)\ ) \end{aligned} \tag{31.56}$$

for the interval $t^{[3]} < t_i < \vartheta$,

$$\begin{aligned} u_1^a[t_i] &= 0 \\ u_2^a[t_i] &= -\ \mathrm{sign}\ [\ l_u^{[3]0}x_{22}(t^{[3]},t_i)+l_u^{[4]0}x_{12}(\vartheta,t_i)\ ] \end{aligned} \tag{31.57}$$

when $t_i \in (t^{[2]},\ t^{[3]})$,

$$d_{12} = [\ (l_u^{[2]0}x_{22}(t^{[2]},t_i))^2+$$
$$+\ (l_u^{[3]0}x_{22}(t^{[3]},t_i)+l_u^{[4]0}x_{12}(\vartheta,t_i))^2\ ]^{1/2} \tag{31.58}$$

$$
\begin{aligned}
u_1^a[t_i] &= -\frac{l_u^{[2]0} x_{22}(t^{[2]}, t_i)}{d_{12}} \\
u_2^a[t_i] &= -\frac{(l_u^{[3]0} x_{22}(t^{[3]}, t_i) + l_u^{[4]0} x_{12}(\vartheta, t_i))}{d_{12}}
\end{aligned}
\qquad (31.59)
$$

if $t_i \in (t^{[1]},\ t^{[2]})$ and

$$
d_{01} = [\ (l_u^{[1]0} x_{12}(t^{[1]}, t_i) + l_u^{[2]0} x_{22}(t^{[2]}, t_i))^2 +
$$

$$
+ (l_u^{[3]0} x_{22}(t^{[3]}, t_i) + l_u^{[4]0} x_{12}(\vartheta, t_i))^2\ ]^{1/2} \qquad (31.60)
$$

$$
\begin{aligned}
u_1^a[t_i] &= -\frac{(l_u^{[1]0} x_{12}(t^{[1]}, t_i) + l_u^{[2]0} x_{22}(t^{[2]}, t_i))}{d_{01}} \\
u_2^a[t_i] &= -\frac{(l_u^{[3]0} x_{22}(t^{[3]}, t_i) + l_u^{[4]0} x_{12}(\vartheta, t_i))}{d_{01}}
\end{aligned}
\qquad (31.61)
$$

for $t_*^0 < t_i < t^{[1]}$.

In order to calculate the approximating optimal disturbances $v^a[t] = v^a[t_i]$, $t_i \le t < t_{i+1}$ we need first to solve maximum problems similar to the problems (31.49)–(31.55). We have only to change in (31.49)–(31.55) the minus signs to plus in the terms that contain $\varepsilon[t^{[1]}]$, $\varepsilon[t^{[2]}]$, $\varepsilon[t^{[3]}]$, $\varepsilon[t_*]$). Let $l_v^{[i]0}$, $i = 1, 2, 3, 4$ be the maximizing values that solve these problems. Then according to (25.12), (25.15)–(25.20) we have

$$
\begin{aligned}
v_1^a[t_i] &= 0 \\
v_2^a[t_i] &= \text{sign}\ (\ l_v^{[4]0} x_{12}(\vartheta, t_i)\ )
\end{aligned}
\qquad (31.62)
$$

for the interval $t^{[3]} < t_i < \vartheta$,

$$
\begin{aligned}
v_1^a[t_i] &= 0 \\
v_2^a[t_i] &= \text{sign}\ [\ l_v^{[3]0} x_{22}(t^{[3]}, t_i) + l_v^{[4]0} x_{12}(\vartheta, t_i)\ ]
\end{aligned}
\qquad (31.63)
$$

when $t_i \in (t^{[2]},\ t^{[3]})$,

$$d_{12} = [\ (l_v^{[2]0} x_{22}(t^{[2]}, t_i))^2 + \\ + (l_v^{[3]0} x_{22}(t^{[3]}, t_i) + l_v^{[4]0} x_{12}(\vartheta, t_i))^2\ ]^{1/2} \qquad (31.64)$$

$$v_1^a[t_i] = \frac{l_v^{[2]0} x_{22}(t^{[2]}, t_i)}{d_{12}}$$

$$v_2^a[t_i] = \frac{(l_v^{[3]0} x_{22}(t^{[3]}, t_i) + l_v^{[4]0} x_{12}(\vartheta, t_i))}{d_{12}} \qquad (31.65)$$

if $t_i \in (t^{[1]},\ t^{[2]})$ and

$$d_{01} = [\ (l_v^{[1]0} x_{12}(t^{[1]}, t_i) + l_v^{[2]0} x_{22}(t^{[2]}, t_i))^2 +$$

$$+ (l_v^{[3]0} x_{22}(t^{[3]}, t_i) + l_v^{[4]0} x_{12}(\vartheta, t_i))^2\ ]^{1/2} \qquad (31.66)$$

$$v_1^a[t_i] = \frac{(l_v^{[1]0} x_{12}(t^{[1]}, t_i) + l_v^{[2]0} x_{22}(t^{[2]}, t_i))}{d_{01}}$$

$$v_2^a[t_i] = \frac{(l_v^{[3]0} x_{22}(t^{[3]}, t_i) + l_v^{[4]0} x_{12}(\vartheta, t_i))}{d_{01}} \qquad (31.67)$$

for $(t_*^0,\ t^{[1]})$.

The solution of the first example (31.1)–(31.5) described in this section was simulated for the following data:

$$t_0 = 0, \qquad \vartheta = 4,$$

$$f[t] = 4 - t, \qquad g[t] = 2,$$
$$\alpha[t] = 0.2, \qquad \beta[t] = -16,$$
$$t^{[1]} = t_*^0 = t_0 = 0, \qquad t^{[2]} = \tilde{t} = 2, \qquad t^{[3]} = \vartheta = 4.$$

For the initial data $r_0 = 0.2,\ \dot{r}_0 = 2$ the following results were obtained. The a priori calculated approximating optimal value was

$$\rho^a(t_0, r_0, \dot{r}_0) = 0.93.$$

In the case of the approximating optimal control actions u^a and the approximating optimal disturbances v^a the simulation of the process gave

$$\gamma_{u^a,v^a} = 0.91 \approx \rho^a(t_0, r_0, \dot{r}_0).$$

The corresponding motion is shown in Figure 31.1.

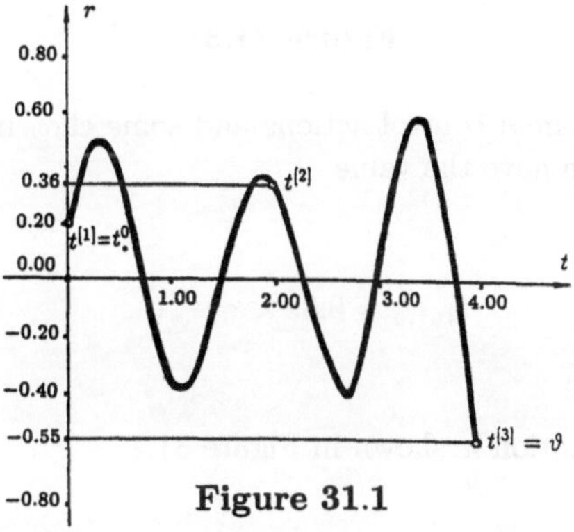

Figure 31.1

The graph of approximating optimal control actions is shown in Figure 31.2 and the graph of approximating optimal disturbances in Figure 31.3.

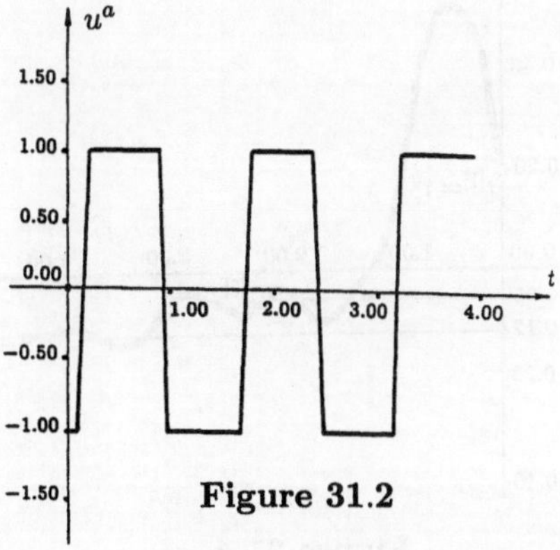

Figure 31.2

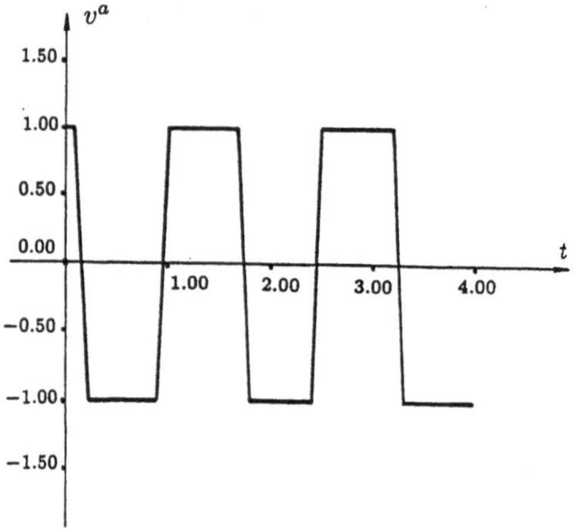

Figure 31.3

In the case of optimal control actions and some chosen admissible disturbances the simulation gave the value

$$\gamma_{u^a,v} = 0.32 < \gamma_{u^a,v^a}.$$

The corresponding motion is shown in Figure 31.4.

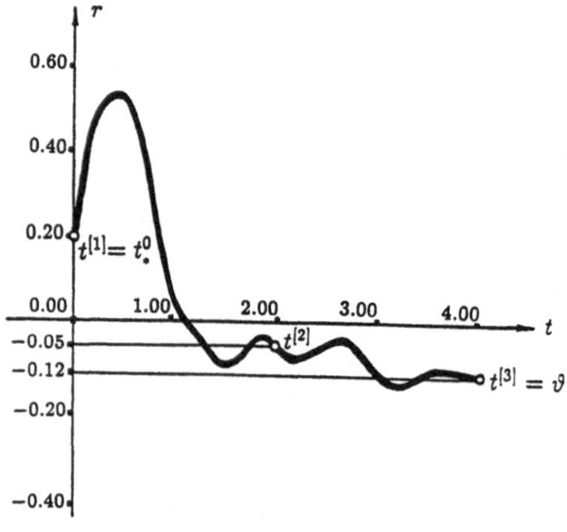

Figure 31.4

In the case of the optimal approximating disturbances and some chosen admissible control actions the simulation gave the value

$$\gamma_{u,v^a} = 3.1 > \gamma_{u^a,v^a}.$$

The corresponding motion is given in Figure 31.5.

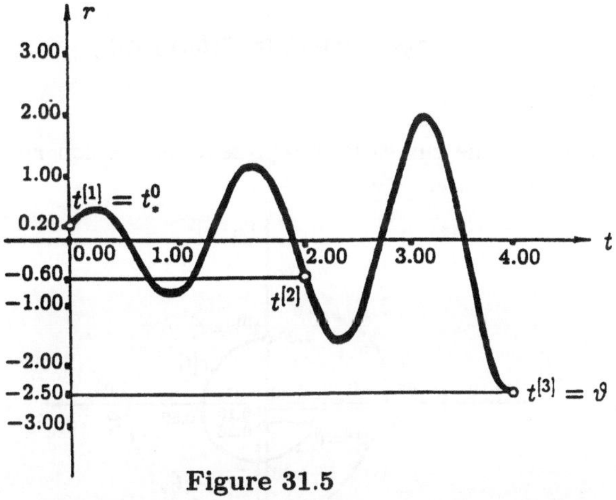

Figure 31.5

Now let us consider simulation of the control process on the basis of the solution for the second example considered in this section. Again the evolution of a vector r satisfies the equation (31.1) but now r is 2-dimensional and the quality index $\gamma_{(3)}$ has the form (31.39).

The process was simulated for the following data:

$$t_0 = 0, \quad \vartheta = 2.1,$$

$$f[t] = 2.1 - t, \quad g[t] = 0.7,$$

$$\alpha[t] = -0.2, \quad \beta[t] = -4,$$

$$t^{[1]} = t_0 = 0, \quad t^{[2]} = 0.7, \quad t^{[3]} = \tilde{t} = 1.4, \quad t^{[4]} = \vartheta = 2.1.$$

The initial conditions were

$$t_*^0 = t_0 = 0, \quad r_{01} = 0, \quad \dot{r}_{01} = 1, \quad r_{02} = 1, \quad \dot{r}_{02} = 0.$$

The a priori calculated approximating optimal value $\rho^a(t_0, r_0, \dot{r}_0)$ was

$$\rho^a(t_*, r_*, \dot{r}_*) = 0.36.$$

In the case of the approximating optimal control actions u^a and approximating optimal disturbances v^a simulation gave

$$\gamma_{u^a,v^a} = 0.38 \approx \rho^a(t_0, r_0, \dot{r}_0).$$

Figure 31.6 shows the curves that represent the motion $r[t]$ and the vector-function $\dot{r}[t]$ in this case.

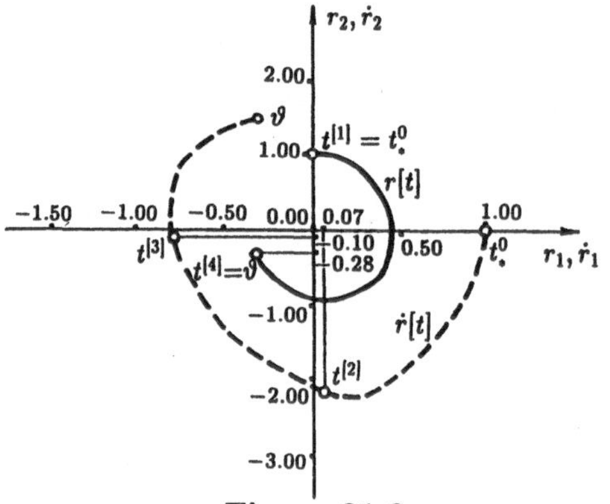

Figure 31.6

In Figures 31.7, 31.8, 31.9, and 31.10 we show the graphs of the functions $r_1[t]$, $\dot{r}_1[t]$, $r_2[t]$, and $\dot{r}_2[t]$ in the case when the control actions and disturbances are approximating optimal.

In the case of approximating optimal control actions and some chosen admissible disturbances the simulation gave the value

$$\gamma_{u^a,v} = 0.15 < \gamma_{u^a,v^a}.$$

The corresponding curves $r[t]$, $\dot{r}[t]$ are shown in Figure 31.11.

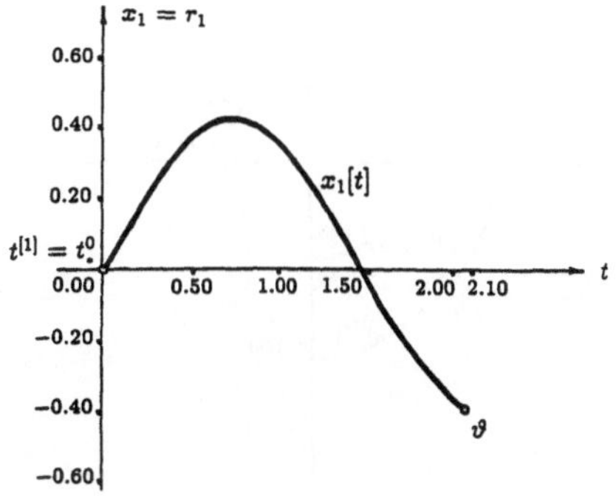

Figure 31.7

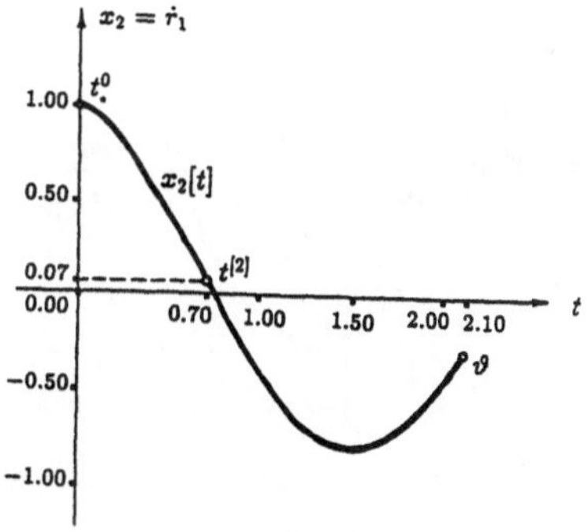

Figure 31.8

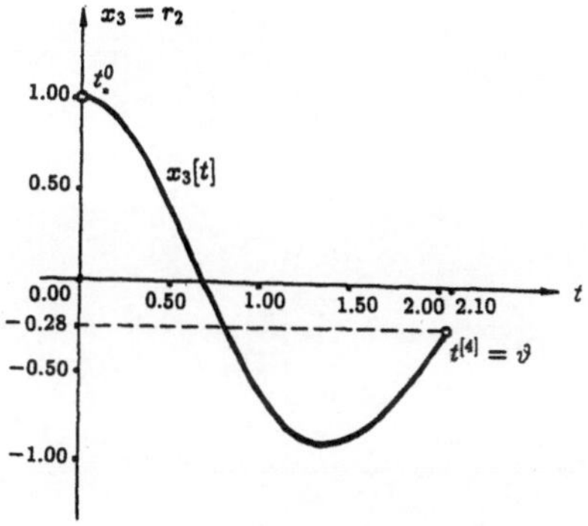

Figure 31.9

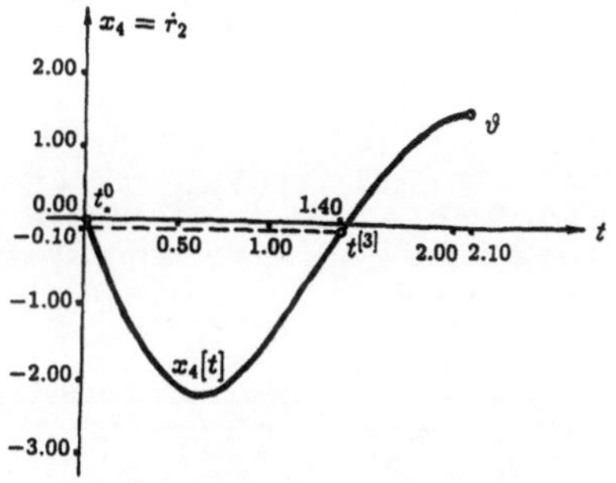

Figure 31.10

In the case of the approximating optimal disturbances and some chosen admissible control actions the simulation gave the value

$$\gamma_{u,v^a} = 2.4 > \gamma_{u^a,v^a}.$$

The corresponding curves $r[t]$, $\dot{r}[t]$ are shown in Figure 31.12.

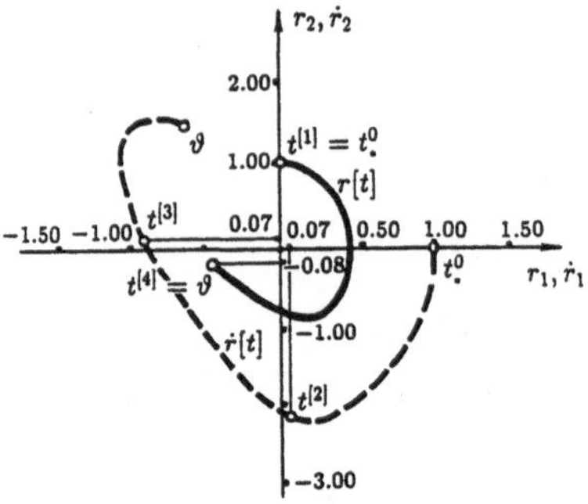

Figure 31.11

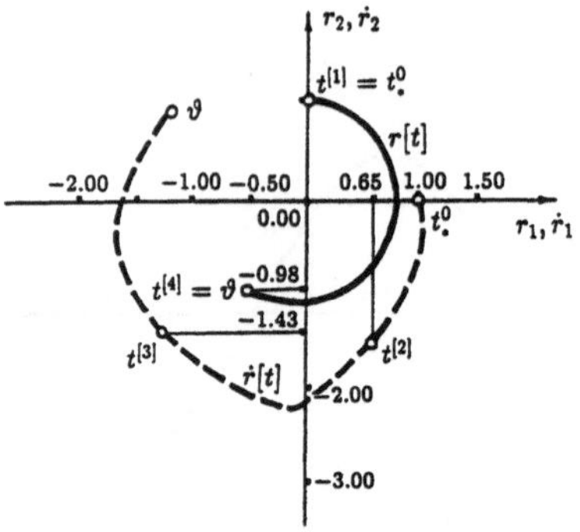

Figure 31.12

Chapter IV
Mixed strategies for positional and quasi-positional functionals

In this chapter we consider the same feedback control problem on the minimax of the quality index γ. However, now we will consider the case when the saddle point condition in a small game (see (5.1) and (10.2)) for the controlled system (3.1) or (10.1) is not valid. In this case the problem can be solved effectively within the framework of the *mixed* strategies (see the particular case in Section 2). The description of this solution is the subject of the following sections.

32. Mixed strategies S^u and S^v

Consider an object described by the differential equation (3.1), i.e., $\dot{x} = f(t, x, u, v)$, $t_0 \leq t < \vartheta$ under the conditions (3.2), (3.4) and for control actions u and disturbances v restricted by the inclusions (3.5), (3.6), i.e., $u \in P$ and $v \in Q$, where $P \subset \mathbf{R}^r$ and $Q \subset \mathbf{R}^s$ are the compact sets.

In addition we will consider the case when for the function $f(t, x, u, v)$ in (3.1) the saddle point condition in a small game (5.1) is not valid, i.e., there exists the vector $l^* \in \mathbf{R}^n$ such that the inequality

$$\max_{v \in Q} \min_{u \in P} \langle\, l^*,\ f(t, x, u, v)\,\rangle \neq \min_{u \in P} \max_{v \in Q} \langle\, l^*,\ f(t, x, u, v)\,\rangle \tag{32.1}$$

holds.

According to what was said above in Section 2 we will consider the same problem as in Chapters I and III but now in the class of *mixed strategies* [61], [62], [79]. It is known that in constructions with mixed strategies a certain computer simulated y-model (for the *first* player) or z-model (for the *second* player) could be essentially employed. This model is called the "leader" or "pilot" according to the terminology of control optimization (see [79]). And this model does satisfy the saddle point condition in a small game (5.1). Thus, all the corresponding constructions given in Chapters I–III can be adapted to this model after a suitable transformation. In particular, one can use the corresponding procedure of the construction of the approximating value of the game $\rho^a_{(i)}(\cdot)$ based on the upper convex hulls for certain auxiliary functions, (see Chapters II, III). The control of the real x-system (10.1) is implemented by following the leader-model in a suitable sense. Below we present the precise mathematical formulations and a description of these expressions.

The control scheme for the first player, together with the *x-object* (3.1), contains the *y-model.* The current phase state of this model is determined by the vector $y[t]$ whose dimension coincides with the dimension of vector x. The motion of y-model on any time interval $\tau_* \leq t \leq \tau^*$ is defined by the differential equation

$$\dot{y}[t] = \sum_{r=1}^{L}\sum_{s=1}^{M} f(t, y[t], u^{[r]}, v^{[s]})p_r^* q_s^*,$$

$$\tau_* \leq t < \tau^*, \quad y[\tau_*] = y_* \tag{32.2}$$

where $\{\tau_*, y_*\} \in G$ (3.12) is a given initial position.

In (32.2) $u^{[r]}$, $r = 1, \ldots, L$ and $v^{[s]}$, $s = 1, \ldots, M$ are the elements of some finite sets $U^* \subset P$ and $V^* \subset Q$, respectively, where P and Q are the compacta (3.5), (3.6). The vectors $u^{[r]} \in U^*$ and $v^{[s]} \in V^*$ are scattered sufficiently densely in P and Q. The *numbers* p_r^* and q_s^* in (32.2) satisfy the following conditions:

$$p_r^* \geq 0, \quad r = 1, \ldots, L, \quad \sum_{r=1}^{L} p_r^* = 1, \tag{32.3}$$

$$q_s^* \geq 0, \quad s = 1, \ldots, M, \quad \sum_{s=1}^{M} q_s^* = 1 \tag{32.4}$$

where the numbers L and M are sufficiently large.

We assume that the motion of the y-model (32.2) is simulated by a computer included in the regulator U. As has been said above, in the proposed scheme of control the motion of the y-model (32.2) leads the motion of the x-object (3.1).

We consider the case of positional functional γ (4.2), (4.9). In this case the mixed strategy S^u is determined by the following totality of components:

$$S^u = \{U_y(\cdot),\ p_y(\cdot);\ U_y^*(\cdot),\ p_y^*(\cdot);\ V_y^*(\cdot),\ q_y^*(\cdot)\} \tag{32.5}$$

where the sets

$$U_y(\cdot) = \{U_y(\varepsilon_u),\ \varepsilon_u > 0\}, \quad U_y^*(\cdot) = U_y(\cdot), \tag{32.6}$$

$$V_y^*(\cdot) = \{V_y^*(\varepsilon_u),\ \varepsilon_u > 0\} \tag{32.7}$$

and the functions

$$p_y(\cdot) = \Big\{p_y(t, x, y, \varepsilon_u), \quad t \in [t_0,\ \vartheta], \quad x \in \mathbf{R}^n, \quad y \in \mathbf{R}^n, \quad \varepsilon_u > 0\Big\}, \tag{32.8}$$

$$p_y^*(\cdot) = \{p_y^*(t, x, y, \varepsilon)\}, \quad q_y^*(\cdot) = \{q_y^*(t, x, y, \varepsilon)\} \tag{32.9}$$

are defined as follows. We assume

$$U_y(\varepsilon_u) = U_y^*(\varepsilon_u) = \{u^{[r]} \in P, \quad r = 1, \ldots, L_{\varepsilon_u}\}, \tag{32.10}$$

$$p_y(t, x, y, \varepsilon_u) = \Big\{p_{yr}(t, x, y, \varepsilon_u) \geq 0, \quad r = 1, \ldots, L_{\varepsilon_u}, \\ \sum_{r=1}^{L_{\varepsilon_u}} p_{yr}(t, x, y, \varepsilon_u) = 1\Big\}. \tag{32.11}$$

Let us explain the sense of the components $U_y(\cdot)$ (32.6), (32.10) and $p_y(\cdot)$ (32.8), (32.11) in (32.5). The employment of a mixed strategy S^u involves a certain probability mechanism of forming the control action u for an x-object. As was mentioned above, $U_y(\varepsilon)$ (32.10) is the set of control-vectors $\{u^{[r]}\}$ scattered sufficiently densely in the given compactum P (3.5). The function $p_y(\cdot)$ (32.8), (32.11) determines for every fixed $\{t, x[t], y[t], \varepsilon_u\}$ the collection of numbers $p_r \geq 0$, $\sum_{r=1}^{L_{\varepsilon_u}} p_r = 1$. These numbers p_r are understood as the *probabilities* of the corresponding events $u = u^{[r]}$. The exact mathematical treatment of this assertion will be given below.

Similarly, we assume

$$p_y^*(t, x, y, \varepsilon_u) = \Big\{p_{yr}^*(t, x, y, \varepsilon_u) \geq 0, \quad r = 1, \ldots, L_{\varepsilon_u}, \\ \sum_{r=1}^{L_{\varepsilon_u}} p_{yr}^*(t, x, y, \varepsilon_u) = 1\Big\}, \tag{32.12}$$

$$V_y^*(\varepsilon_u) = \{v^{[s]} \in Q, \quad s = 1, \ldots, M_{\varepsilon_u}\}, \tag{32.13}$$

$$q_y^*(t, x, y, \varepsilon_u) = \Big\{q_{ys}^*(t, x, y, \varepsilon_u) \geq 0, \quad s = 1, \ldots, M_{\varepsilon_u}, \\ \sum_{r=1}^{M_{\varepsilon_u}} q_{ys}^*(t, x, y, \varepsilon_u) = 1\Big\}. \tag{32.14}$$

The sets $U_y^*(\varepsilon_u)$ (32.10), $V_y^*(\varepsilon_u)$ (32.13) and the functions $p_y^*(\cdot)$ (32.9), (32.12), $q_y^*(\cdot)$ (32.9), (32.14) are similar to the set $U_y(\varepsilon_u)$ and the function $p_y(\cdot)$, and they are related to the y-model (32.2). The upper index $*$ marks here the corresponding quantities for the y-model (32.2), while the lower index y in (32.5)–(32.14) shows that we are dealing with the mixed strategy S^u of the *first* player. We assume that in (32.11), (32.12), (32.14) the functions $p_{yr}(t, x, y, \varepsilon_u)$, $p_{yr}^*(t, x, y, \varepsilon_u)$, $r = 1, \ldots, L_{\varepsilon_u}$ and $q_{ys}^*(t, x, y, \varepsilon_u)$, $s = 1, \ldots, M_{\varepsilon_u}$ are measurable in x, y for every fixed t and ε_u .

We note that in (32.5)–(32.14) the quantity $\varepsilon_u > 0$ is a parameter of accuracy similar to the parameter in the constructions of the pure strategies $u_{(i)}(\cdot)$ and $v_{(i)}(\cdot)$ which were considered earlier.

In the case of *quasi-positional* functionals $\gamma_{(i)}$, $i = 3 - 8$ the information image $\{t, x, y\}$ for the strategy S^u should be substituted by the image $\{x[t_*^0[\cdot]t], y[t_*^0[\cdot]t]\}$, $\{t, \widehat{x}, \widehat{y}\}$ or $\{t, x[t_*^0[\cdot]t],\ \widehat{x}_{n+1}, y[t_*^0[\cdot]t], \widehat{y}_{n+1}\}$ corresponding to the concrete quasi-positional functional $\gamma_{(i)}$ (see Section 26).

The mixed strategy S^v of the *second* player is defined similarly. Namely, the corresponding z-model-leader is described by the differential equation

$$\dot{z}[t] = \sum_{r=1}^{L}\sum_{s=1}^{M} f(t, z[t], u^{[r]}, v^{[s]}) p_r^* q_s^*,$$

$$\tau_* \le t < \tau^*, \qquad z[\tau_*] = z_* \tag{32.15}$$

where $z \in \mathbf{R}^n$ and all the quantities are defined similarly to (32.2).

We call

$$S^v = \Big\{V_z(\cdot),\ q_z(\cdot);\ U_z^*(\cdot),\ p_z^*(\cdot);\ V_z^*(\cdot),\ q_z^*(\cdot)\Big\} \tag{32.16}$$

a mixed strategy of the second player. Here the sets $V_z(\cdot)$, $U_z^*(\cdot)$, $V_z^*(\cdot) = V_z(\cdot)$ and the functions $q_z(\cdot)$, $p_z^*(\cdot)$, $q_z^*(\cdot)$ are defined by the equalities and the conditions that are obtained from (32.6)–(32.14) if we substitute z for y.

The realization of the mixed strategies S^u (32.5) and S^v (32.16) will be explaned in more detail in the following section.

33. The motions $x_{S^u,v}[\cdot]$ and $x_{S^v,u}[\cdot]$

The motion $x_{S^u,v}[\cdot]$ generated by the strategy S^u (32.5) together with some disturbances $v[\cdot]$ is determined in the following way.

The accurate mathematical construction of the motions of the x-object (3.1) and the y-model (32.2) is based on a certain probability space $\{\Omega_*, \mathcal{F}_*, \mathbf{P}_*\}$. This probability space, in its turn, is determined by the components in S^u (32.5) and the properties of random disturbances

$$v[\cdot] = v[t_*[\cdot]\vartheta, \cdot) = \Big\{v[t, \omega_*^{[v]}] \in Q, \quad t_* \le t < \vartheta, \quad \omega_*^{[v]} \in \Omega_*^{[v]}\Big\}. \tag{33.1}$$

Here $\{\Omega_*^{[v]}, \mathcal{F}_*^{[v]}, \mathbf{P}_*^{[v]}\}$ is some probability space given or constructed in this or that way. And the random disturbances $v[t, \omega_*^{[v]}]$ are the values of a certain random function determined on this probability space for $t_* \le t < \vartheta$. We assume here that the space $\{\Omega_*, \mathcal{F}_*, \mathbf{P}_*\}$ and subspace $\{\Omega_*^{[v]}, \mathcal{F}_*^{[v]}, \mathbf{P}_*^{[v]}\}$ are given. These objects are employed only in theoretical considerations and their concrete form is inessential.

For any initial position $\{t_*, x_*, y_*\}$, a given value of $\varepsilon_u > 0$, a partition $\Delta\{t_i\} = \{t_1 = t_*,\ t_i < t_{i+1},\ i = 1,\dots,k,\ t_{k+1} = \vartheta\}$ and some random disturbances $v[\cdot]$ (33.1), the strategy S^u (32.5) will generate a *random motion*

$$x_{S^u,v}[\cdot] = x_{S^u,v}[t_*[\cdot]\vartheta, \cdot] = \Big\{x_{S^u,v}[t,\omega_*], \quad t_* \le t \le \vartheta, \quad \omega_* \in \Omega_*\Big\} \tag{33.2}$$

of the x-object (3.1) and a random motion

$$y[\cdot] = y[t_*[\cdot]\vartheta, \cdot] = \{y[t,\omega_*], \quad t_* \le t \le \vartheta, \quad \omega_* \in \Omega_*\} \tag{33.3}$$

of the y-model as solutions of step-by-step differential equations

$$\dot{x}_{S^u,v}[t,\omega_*] = f(t, x_{S^u,v}[t,\omega_*], u[t_i,\omega_*], v[t,\omega_*]),$$

$$t_i \le t < t_{i+1}, \quad i = 1,\dots,k, \quad x_{S^u,v}[t_1,\omega_*] = x_* \tag{33.4}$$

$$\dot{y}_{S^u,v}[t,\omega_*] = \sum_{r=1}^{L\varepsilon_u}\sum_{s=1}^{M\varepsilon_u} f(t, y_{S^u,v}[t,\omega_*], u^{[r]}, v^{[s]}) *$$

$$* p^*_{yr}(t_i, x_{S^u,v}[t_i,\omega_*], y_{S^u,v}[t_i,\omega_*], \varepsilon_u)\, q^*_{ys}(t_i, x_{S^u,v}[t_i,\omega_*], y_{S^u,v}[t_i,\omega_*], \varepsilon_u)$$

$$t_i \le t < t_{i+1}, \quad i = 1,\dots,k, \quad y_{S^u,v}[t_1,\omega_*] = y_*. \tag{33.5}$$

Here in (33.5) $u^{[r]} \in U^*_y(\varepsilon_u)$ (32.10), $v^{[s]} \in V^*_y(\varepsilon_u)$ (32.13) and the collections $p^*_{yr}(\cdot)$, $r = 1,\dots,L_{\varepsilon_u}$, $q^*_{ys}(\cdot)$, $s = 1,\dots,M_{\varepsilon_u}$ are determined by the functions $p^*_y(\cdot)$ (32.12) and $q^*_y(\cdot)$ (32.14) in (32.5).

In (33.4) $u[t_i,\omega_*] \in P$ with a fixed ω_* is a realization of the random variable $u[t_i,\cdot]$ under the condition

$$P(u[t_i,\omega_*] = u^{[r]} \mid x_{S^u,v}[t_i,\omega_*], y_{S^u,v}[t_i,\omega_*]) =$$

$$= p_{yr}(t_i, x_{S^u,v}[t_i,\omega_*], y_{S^u,v}[t_i,\omega_*], \varepsilon_u) \tag{33.6}$$

where $P(\cdots \mid \cdots)$ is the conditional probability. In (33.6) $u^{[r]} \in U_y(\varepsilon_u)$ (32.10). The sense of the condition (33.6) is the following. In order to form some realization of the motion $x_{S^u,v}[t_i[\cdot]t_{i+1}]$ for the time interval $[t_i,\ t_{i+1}]$ one has to choose a vector $u[t_i] = u^{[r]}$ from the collection $U_y(\varepsilon_u)$ (33.10) of the vectors $u^{[r]}$, $r = 1,\dots,L_{\varepsilon_u}$ and take $u[t] = u[t_i]$, $t_i \le t < t_{i+1}$. This choice is determined by a random test which is stipulated by the probabilities (33.6) that correspond to the realizations $x_{S^u,v}[t_i,\omega_*]$, $y_{S^u,v}[t_i,\omega_*]$.

It is assumed that the disturbances are stochastically independent of the control actions at each step, i.e.,

$$P(v[t,\omega_*] \in C \mid x_{S^u,v}[t_i,\omega_*], y_{S^u,v}[t_i,\omega_*], u[t_i,\omega_*]) =$$

$$= P(v[t,\omega_*] \in C \mid x_{S^u,v}[t_i,\omega_*], y_{S^u,v}[t_i,\omega_*]) \tag{33.7}$$

where $C \in \mathcal{B}^s$ is any event in the Borel σ-algebra in $\mathbf{R}^s$. This condition can be interpreted in the following way. On the one hand, the disturbances can be formed by the environment, i.e., without any special calculations, in order to counteract our concrete control actions $u[t] = u[t_i,\omega_*]$, $t_i \le t < t_{i+1}$. On the other hand, the disturbances can be formed by a player with the help of special calculations in order to counteract our control actions $u[t] = u[t_i,\omega_*]$, $t_i \le t < t_{i+1}$.In the second case we assume that even if at the moment t_i this player can find out our action $u[t] = u[t_i,\omega_*]$, $t_i \le t < t_{i+1}$, he is not able to use this information effectively on the small time interval $[t_i,\ t_{i+1}]$. Of course, if the space $\{\Omega_*, \mathcal{F}_*, \mathbf{P}_*\}$ were rather complicated, then the random motions $x[\cdot]$, $y[\cdot]$ and stochastic differential equations (33.4), (33.5) would require a strict description. However, we do not need here such a detailed strict description because in our cases we need to keep in mind only rather simple probability spaces.

We have described the process of forming the control action $u[t]$ of the first player and will now consider the problem on behalf of the second player.

The motions $x_{S^v,u}[\cdot]$ and $z_{S^v,u}[\cdot]$ for the second player are determined similarly. Namely, the random motion

$$x_{S^v,u}[t_*[\cdot]\vartheta,\cdot] = \{x_{S^v,u}[t,\omega_*],\quad t_* \le t \le \vartheta,\quad \omega_* \in \Omega_*\} \tag{33.8}$$

of the x-object (3.1) and the random motion

$$z_{S^v,u}[t_*[\cdot]\vartheta,\cdot] = \{z_{S^v,u}[t,\omega_*],\quad t_* \le t \le \vartheta,\quad \omega_* \in \Omega_*\} \tag{33.9}$$

of the z-model (32.15), generated from the initial position $\{t_*, x_*, z_*\}$ by the strategy S^v (32.16) together with some random control actions

$$u[t_*[\cdot]\vartheta, \cdot) = \{u[t, \omega_*] \in P, \quad t_* \leq t < \vartheta, \quad \omega_* \in \Omega_*\} \tag{33.10}$$

are defined as step-by-step solutions of the differential equations

$$\dot{x}_{S^v,u}[t, \omega_*] = f(t, x_{S^v,u}[t], u[t, \omega_*], v[t_i, \omega_*]),$$

$$t_i \leq t < t_{i+1}, \quad i = 1, \ldots, k_v, \quad x[t_1, \omega_*] = x_*, \tag{33.11}$$

$$\dot{z}_{S^v,u}[t, \omega_*] = \sum_{r=1}^{L_{\mathcal{E}_v}} \sum_{s=1}^{M_{\mathcal{E}_v}} f(t, z_{S^v,u}[t, \omega_*], u^{[r]}, v^{[s]}) *$$

$$* p^*_{zr}(t_i, x_{S^v,u}[t_i, \omega_*], z_{S^v,u}[t_i, \omega_*], \varepsilon_v)\, q^*_{zs}(t_i, x_{S^v,u}[t_i, \omega_*], z_{S^v,u}[t_i, \omega_*], \varepsilon_v),$$

$$t_i \leq t < t_{i+1}, \quad i = 1, \ldots, k_v, \quad z_{S^v,u}[t_1, \omega_*] = z_* \tag{33.12}$$

under the assumption that in (33.11) we have

$$P\Big(v[t_i, \omega] = v^{[s]} \mid x_{S^v,u}[t_i, \omega_*], z_{S^v,u}[t_i, \omega_*] \Big) =$$

$$= q_{zs}(t_i, x_{S^v,u}[t_i, \omega_*], z_{S^v,u}[t_i, \omega_*]), \tag{33.13}$$

$$P\Big(u[t, \omega_*] \in B \mid x_{S^v,u}[t_i, \omega_*], z_{S^v,u}[t_i, \omega_*], v[t_i, \omega_*] \Big) =$$

$$= P\Big(u[t, \omega_*] \in B \mid x_{S^v,u}[t_i, \omega_*], z_{S^v,u}[t_i, \omega_*] \Big) \tag{33.14}$$

where $B \in \mathcal{B}^r$ is any event in the Borel σ-algebra in $\mathbf{R}^r$.

In (33.12) we have $u^{[r]} \in U^*_z(\varepsilon_v)$, $r = 1, \ldots, L_{\mathcal{E}_v}$ and $v^{[s]} \in V^*_z(\varepsilon_v)$, $s = 1, \ldots, M_{\mathcal{E}_v}$, where $U^*_z(\varepsilon_v)$ and $V^*_z(\varepsilon_v)$ coincide with the collection in (32.16); and $p^*_{zr}(\cdot)$ and $q^*_{zs}(\cdot)$ are the collections corresponding to the functions $p^*_z(\cdot)$ and $q^*_z(\cdot)$ in (32.16).

It should be emphasized that the basic probability space $\{\Omega_*, \mathcal{F}_*, \mathbf{P}_*\}$ in the case of the strategy S^u is constructed in accordance with the components of S^u and in accordance with the probability space $\{\Omega^{[v]}_*, \mathcal{F}^{[v]}_*, \mathbf{P}^{[v]}_*\}$ that provides the basis for the random disturbances $v[\cdot]$ (33.1). A variation of the random disturbances $v[\cdot]$ may imply the corresponding variation of probability space $\{\Omega^{[v]}_*, \mathcal{F}^{[v]}_*, \mathbf{P}^{[v]}_*\}$ and, consequently, the variation of the basic probability space $\{\Omega_*, \mathcal{F}_*, \mathbf{P}_*\}$. On the other hand, in the case of the strategy S^v the basic probability space $\{\Omega_*, \mathcal{F}_*, \mathbf{P}_*\}$ is constructed taking into account the components of S^v and taking into account the probability space $\{\Omega_*^{[u]}, \mathcal{F}_*^{[u]}, \mathbf{P}_*^{[u]}\}$ that gives the basis for the random control actions $u[\cdot]$ (33.10). It is worthwhile to note also that the elementary events $\omega_*^{[v]}$, $\omega_*^{[u]}$ in the spaces $\Omega_*^{[v]}$, $\Omega_*^{[u]}$ become some corresponding events in the spaces Ω_* for the cases of the strategies S^u, S^v, respectively.

The schemes of control processes that correspond to the mixed strategies S^u and S^v are shown in Figures 33.1 to 33.4.

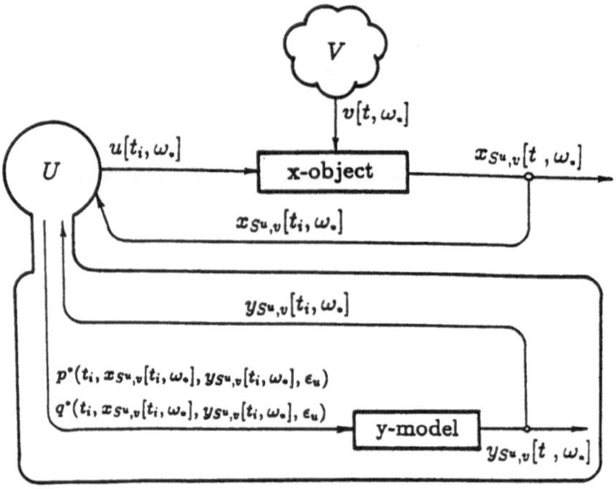

Figure 33.1

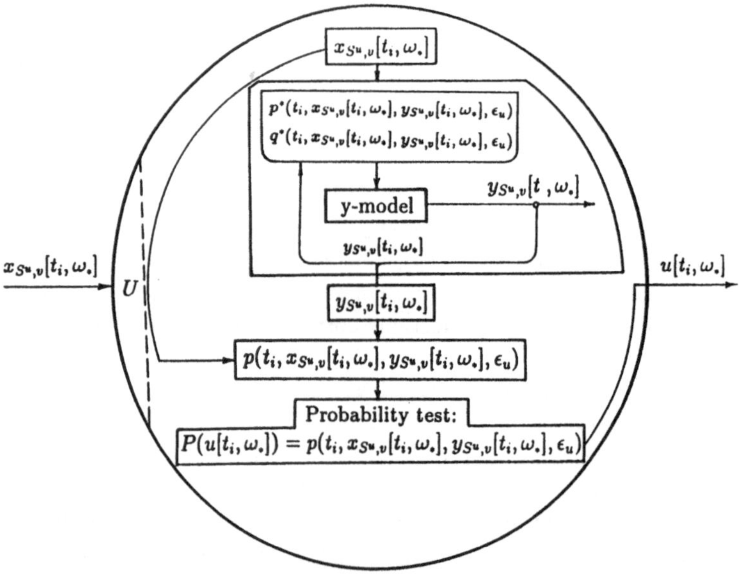

Figure 33.2

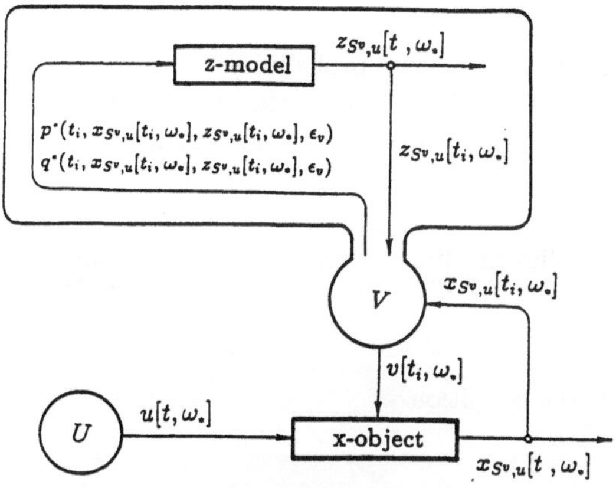

Figure 33.3

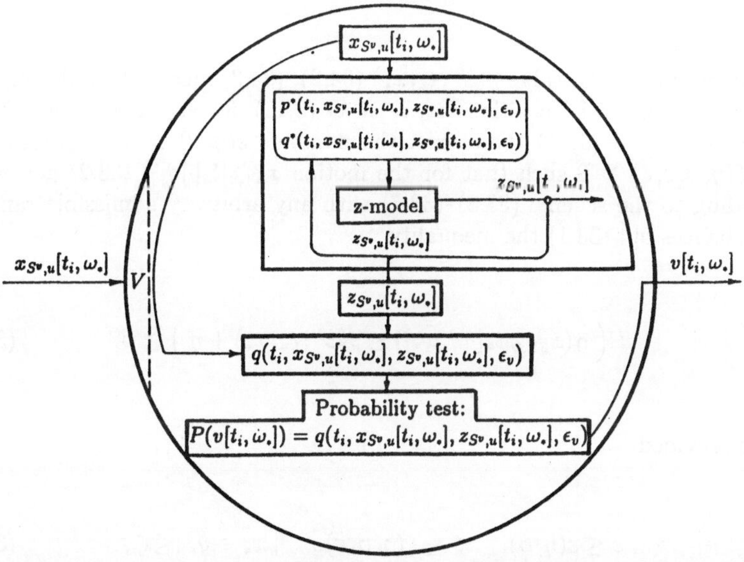

Figure 33.4

Let us remark that the schemes of control processes that correspond to the mixed strategies S^u and S^v are shown also in Section 2 in concrete cases in Figures 2.3–2.9.

34. Statement of the problem in the class of mixed strategies. Existence of the solution

Let us consider the case of *positional* quality index γ (see Section 4). Due to the stochastic structure of the motion $x_{S^u,v}[\cdot]$ (33.2)–(33.5), (or $x_{S^v,u}[\cdot]$ (33.8)–(33.12)) the quality index γ (4.2), (4.9) is a random variable. Thus, the value of quality index

$$\gamma = \gamma(x_{S^u,v}[t_*[\cdot]\vartheta, \omega_*]) \tag{34.1}$$

can be estimated only with some probability.

The ensured result $\rho(S^u; t_*, x_*)$ for the strategy S^u (32.5) and the initial position $\{t_*, x_*\} \in G$ (3.12) is defined by the equality

$$\rho(S^u; t_*, x_*) = \lim_{\beta \to 1} \overline{\lim_{\varepsilon \to 0}} \lim_{\zeta \to 0} \sup_{|x_* - y_*| \leq \zeta} \limsup_{\delta \to 0} \sup_{\Delta_\delta} \sup_{v[t_*[\cdot]\vartheta, \cdot)} (\min \alpha) \tag{34.2}$$

where α satisfies the condition

$$P\Big(\gamma(x_{S^u, v}[t_*[\cdot]\vartheta, \omega_*]) \leq \alpha \Big) \geq \beta. \tag{34.3}$$

The definition of the value $\rho(S^u; t_*, x_*)$ (34.2), (34.3) means the following.

For a given initial position $\{t_*, x_*\} \in G$ (3.12), for any number $\eta > 0$ and any number $\beta \in [0,\ 1)$ there exist the numbers $\varepsilon(\eta, \beta) > 0$, $\zeta(\eta, \beta, \varepsilon) > 0$ and $\delta(\eta, \beta, \varepsilon, \zeta) > 0$ such that for the motion $x_{S^u, v}[t_*[\cdot]\vartheta, \cdot]$ (33.2) generated according to the scheme (33.3)–(33.7) with any arbitrary admissible random disturbances $v[\cdot]$ (33.1) the inequality

$$P\Big(\gamma(x_{S^u, v}[t_*[\cdot]\vartheta, \omega_*]) \leq \rho(S^u; t_*, x_*) + \eta \Big) \geq \beta \tag{34.4}$$

holds provided

$$\varepsilon \leq \varepsilon(\eta, \beta), \qquad \zeta \leq \zeta(\eta, \beta, \varepsilon), \qquad | x_* - y_* | \leq \zeta, \tag{34.5}$$

$$\max_i | t_{i+1} - t_i | \leq \delta(\eta, \beta, \varepsilon, \zeta). \tag{34.6}$$

We call the strategy S_0^u and the ensured result $\rho_{S^u}^0(t_*, x_*)$ *optimal* if the equality

$$\rho(S_0^u; t_*, x_*) = \min_{S^u} \rho(S^u; t_*, x_*) = \rho_{S^u}^0(t_*, x_*) \tag{34.7}$$

is valid for any position $\{t_*, x_*\} \in G$ (3.12).

We call the optimal ensured result $\rho(S_0^u; t_*, x_*)$ and the optimal strategy S_0^u *uniform* in the region G if the values $\varepsilon(\eta, \beta)$, $\zeta(\eta, \beta, \varepsilon)$ and $\delta(\eta, \beta, \varepsilon, \zeta)$ can be chosen independently of the initial position $\{t_*, x_*\} \in G$.

The value $\rho(S^v; t_*, x_*)$ and *optimal* S_0^v and $\rho_{S^v}^0(t_*, x_*)$ are defined similarly.

Namely, we have

$$\rho(S^v; t_*, x_*) = \lim_{\beta\to 1} \underline{\lim_{\varepsilon\to 0}} \lim_{\zeta\to 0} \inf_{|x_*-y_*|\le\zeta} \liminf_{\delta\to 0} \inf_{\Delta_\delta} \inf_{u[t_*[\cdot]\vartheta,\cdot)} (\max\alpha) \tag{34.8}$$

where α satisfies the condition

$$P\Big(\gamma(x_{S^v,u}[t_*[\cdot]\vartheta, \omega_*]) \ge \alpha \Big) \ge \beta \,. \tag{34.9}$$

And we have also

$$\rho(S_0^v; t_*, x_*) = \max_{S^v} \rho(S^v; t_*, x_*) = \rho^0_{S^v}(t_*, x_*)$$

for every

$$\{t_*, x_*\} \in G. \tag{34.10}$$

The definitions (34.8)–(34.10) mean that for the motion $x_{S_0^v,u}[\cdot]$ (33.8), (33.9), (33.11)– (33.14) generated by the optimal strategy together with any arbitrary admissible random control actions $u[\cdot]$ (33.10) the inequality

$$P\Big(\gamma(x_{S_0^v,u}[t_*[\cdot]\vartheta, \omega_*]) \ge \rho^0_{S^v}(t_*, x_*) - \eta \Big) \ge \beta \tag{34.11}$$

holds under the conditions (34.5), (34.6).

The *uniform optimal* ensured result $\rho^0(S_0^v; t_*, x_*)$ and the *uniform* optimal strategy S_0^v are defined similarly to the above definition of $\rho^0(S_0^u; t_*, x_*)$ and S_0^u.

Using the definitions of the *value of the game* $\rho^0(t_*, x_*)$ and the *saddle point* $\{S_0^u, S_0^v\}$ that correspond to the ones given in Section 7 we can formulate the following result.

Theorem 34.1. *The differential game for the system* (3.1) *with the positional quality index* γ (4.9) *has the value* $\rho^0(t_*, x_*) = \rho^0(S_0^u, t_*, x_*) = \rho^0(S_0^v, t_*, x_*)$ *and the saddle point* $\{S_0^u, S_0^v\}$*, which is a pair of uniform (in G) optimal mixed positional strategies* S_0^u (32.5), (34.7) *and* S_0^v (32.16), (34.10).

The proof of this theorem is similar to the proof of Theorem 9.1 with some understandable changes. The optimal strategies S_0^u and S_0^v are constructed again on the basis of the value of the game $\rho^0(t,x)$ by the method of extremal shift to the accompanying points. We do not prove Theorem 34.1 here. This theorem is published in [56], [61], [62], [65]. Let us give only the corresponding conditions of extremal shift for $S^u_{(e)}$ (for the first player) and for $S^v_{(e)}$ (for the second player), similar to the conditions (8.11), (8.12) and (8.16), (8.17) given above for the pure positional extremal strategies $u_{(e)}(\cdot)$ and $v_{(e)}(\cdot)$.

According to the definitions (32.5), (32.16) of the mixed strategies S^u and S^v we now have to define the corresponding components $U_{(e)}(\cdot)$, $V_{(e)}(\cdot)$, $p_{(e)}(\cdot)$, $q_{(e)}(\cdot)$, $p^*_{(e)}(\cdot)$, $q^*_{(e)}(\cdot)$ for the extremal strategies $S^u_{(e)}$ and $S^v_{(e)}$. We note that now the extremal shift to the accompanying points is applied to the y-model-leader and z-model-leader. This shift is defined by the following condition:

$$\max_q \sum_{r=1}^{L_{\varepsilon_u}} \sum_{s=1}^{M_{\varepsilon_u}} \langle\, (y - w^{[u]0}),\ f(t,y,u^{[r]},v^{[s]})\,\rangle\, p^*_{(e)r}(t,x,y,\varepsilon_u)\, q_s =$$

$$= \min_p \max_q \sum_{r=1}^{L_{\varepsilon_u}} \sum_{s=1}^{M_{\varepsilon_u}} \langle\, (y - w^{[u]0}),\ f(t,y,u^{[r]},v^{[s]})\,\rangle p_r q_s \tag{34.12}$$

for $S^u_{(e)}$ and by the condition

$$\min_p \sum_{r=1}^{L_{\varepsilon_v}} \sum_{s=1}^{M_{\varepsilon_v}} \langle\, (w^{[v]0} - z),\ f(t,z,u^{[r]},v^{[s]})\,\rangle\, p_r q^*_{(e)s}(t,x,z,\varepsilon_v) =$$

$$= \max_q \min_p \sum_{r=1}^{L_{\varepsilon_v}} \sum_{s=1}^{M_{\varepsilon_v}} \langle\, (w^{[v]0} - z),\ f(t,z,u^{[r]},v^{[s]})\,\rangle p_r q_s \tag{34.13}$$

for $S^v_{(e)}$. In (34.12), (34.13) $w^{[u]0} = w^{[u]0}(t,y,\varepsilon_u)$ and $w^{[v]0} = w^{[v]0}(t,z,\varepsilon_v)$ are accompanying points which are defined by the conditions (8.9), (8.10) and (8.14), (8.15) if we substitute the symbol x by y or by z, respectively.

The conditions (34.12) and (34.13) define the components $p^*_{y(e)}(\cdot)$ of the extremal strategy $S^u_{(e)}$ and $q^*_{z(e)}(\cdot)$ of the extremal strategy $S^v_{(e)}$. The other components are defined from conditions similar to the ones from Lemma 7.1 and Lemma 7.2 in Section 7. Namely, the conditions that define the elements $p_{y(e)}(t,x,y,\varepsilon_u)$, $q^*_{y(e)}(t,x,y,\varepsilon_u)$ of the strategy $S^u_{(e)}$ are

$$\sum_{r=1}^{L_{\varepsilon_u}} \sum_{s=1}^{M_{\varepsilon_u}} \langle\, (x - y),\ f(t,x,u^{[r]},v^{[s]})\,\rangle\, p_{(e)r}(t,x,y,\varepsilon_u) q^*_{(e)s}(t,x,y,\varepsilon_u) =$$

$$= \min_p \max_q \sum_{r=1}^{L\varepsilon_u} \sum_{s=1}^{M\varepsilon_u} \langle (x-y),\ f(t,x,u^{[r]},v^{[s]}) \rangle p_r q_s. \qquad (34.14)$$

Choosing $p_{(e)}(t,x,y,\varepsilon_u)$, $q^*_{(e)}(t,x,y,\varepsilon_u)$, $U_{(e)}(\varepsilon_u)$ and $V_{(e)}(\varepsilon_u)$ for $S^u_{(e)}$ from conditions (34.14) we ensure *closeness* of the motion $x_{S^u_{(e)},v}[\cdot]$ to the motion $y_{S^u_{(e)},v}[\cdot]$ in the required sense. This closeness is similar to that of the motions $x_{U_e,v}[t_*[\cdot]\vartheta]$ and $w^{[u]}_a[t_*[\cdot]\vartheta]$ in Section 9 (see 9.14). The only difference is that here the desired closeness is ensured only with some probability P. However, this probability can be made as near to 1 as we desire by choosing the step δ of the partition $\Delta\{t_i\}$ sufficiently small.

Similar operations are to be done for the strategy $S^v_{(e)}$ in order to guarantee the closeness of the motion $x_{S^v_{(e)},u}[\cdot]$ to the motion $z_{S^v_{(e)},u}[\cdot]$. Namely, the conditions that define the elements $q_{z(e)}(t,x,y,\varepsilon_v)$, $p^*_{z(e)}(t,x,y,\varepsilon_v)$ of the strategy $S^v_{(e)}$ are

$$\sum_{r=1}^{L\varepsilon_v} \sum_{s=1}^{M\varepsilon_v} \langle (z-x),\ f(t,x,u^{[r]},v^{[s]}) \rangle q_{(e)s}(t,x,z,\varepsilon_v) p^*_{(e)r}(t,x,z,\varepsilon_v) =$$

$$= \max_q \min_p \sum_{r=1}^{L\varepsilon_v} \sum_{s=1}^{M\varepsilon_v} \langle (z-x),\ f(t,x,u^{[r]},v^{[s]}) \rangle q_s p_r. \qquad (34.15)$$

This concludes the constructions of the extremal strategies $S^u_{(e)}$ and $S^v_{(e)}$. Below we will consider an effective procedure for constructing the approximating values for the value of the game $\rho^0(t,x)$ in the class of mixed strategies S^u and S^v for some particular cases of the positional functional γ (4.2), (4.9).

35. Constructions of the values $e^{(s)}_{*(1)}(\cdot)$ and $e^{(s)}_{*(2)}(\cdot)$

Let us consider the cases of two typical positional functionals γ which were introduced above in Chapter II. Let us consider the quality indices $\gamma_{(1)}$ (10.3) and $\gamma_{(2)}$ (10.4) and their corresponding approximating functionals $\gamma_{*(1)}$ (11.9) and $\gamma_{*(2)}$ (11.10). In this section, for these quality indices we describe effective constructions for the quantities $e^{(s)}_{*(1)}(\tau_*,w_*,\Delta)$ and $e^{(s)}_{*(2)}(\tau_*,w_*,\Delta)$ which approximate the values of the game $\rho^{(s)0}_{*(1)}(t_*,x_*)$ and $\rho^{(s)0}_{*(2)}(t_*,x_*)$ in the class of mixed strategies S^u and S^v. Here the upper index (s) indicates that now we deal with the solutions in this class of mixed strategies. We note that the constructions of $e^{(s)}_{*(1)}(\cdot)$ and $e^{(s)}_{*(2)}(\cdot)$ are similar to the corresponding constructions for $e_{*(1)}(\cdot)$ and $e_{*(2)}(\cdot)$ which were described in Sections 13 and 14. Some changes are connected with the difference between the pure strategies $u(\cdot)$ ($v(\cdot)$) and the mixed S^u (S^v) strategies. In accordance with Sections 32 to 34 we can say that now the role of the *control actions* $u \in P$ and the *disturbances* $v \in Q$ will be played by the corresponding collections of *probabilities* p_r : $p_r \geq 0, \sum_{r=1}^{L\varepsilon} p_r = 1$ and q_s : $q_s \geq 0, \sum_{s=1}^{M\varepsilon} q_s = 1$ for the vectors $u^{[r]} \in U(\varepsilon)$ and $v^{[s]} \in V(\varepsilon)$. Therefore, we follow Sections 12 to 14 and describe the constructions of values $e_{*(i)}^{(s)}(\cdot)$ $i = 1, 2$ rather briefly.

Let us first consider the case of the quality index $\gamma_{(1)}$ (10.3). According to the constructions in Section 11 the corresponding approximating functional $\gamma_{*(1)}$ for $\gamma_{(1)}$ has a form (11.9). In our case of mixed strategies S^u and S^v the w-model (see (12.1)) is described now by the differential equation

$$\dot{w} = A(\tau)w + \sum_{r=1}^{L_\eta} \sum_{s=1}^{M_\eta} f(\tau, u^{[r]}, v^{[s]}) p_r q_s \tag{35.1}$$

where

$$p_r \geq 0, \quad \sum_{r=1}^{L_\eta} p_r = 1, \quad q_s \geq 0, \quad \sum_{s=1}^{M_\eta} q_s = 1, \tag{35.2}$$

$$u^{[r]} \in U_\eta = \Big\{u^{[r]} \in P, \qquad r = 1, \ldots, L_\eta\Big\}, \tag{35.3}$$

$$v^{[s]} \in V_\eta = \Big\{v^{[s]} \in Q, \qquad s = 1, \ldots, M_\eta\Big\}. \tag{35.4}$$

The set U_η in (35.3) has the property that for any $u \in P$ there exists $u^{[r]} \in U_\eta$ such that $| u^{[r]} - u | \leq \eta$. Similarly, V_η is a set such that for any $v \in Q$ there exists $v^{[s]} \in V_\eta : | v^{[s]} - v | \leq \eta$. Let $\{\tau_*, w_*\}$ be the initial position for the stochastic model (35.1). The probability basis for this model is the same as in Section 12. The values p_r and q_s are determined now in (35.1)–(35.4) by the stochastic nonanticipating programs $p_r[\tau, \omega]$ and $q_s[\tau, \omega]$ where

$$p_r[\tau, \omega] = p_r[\tau, \xi_1, \ldots, \xi_j],$$

$$q_s[\tau, \omega] = q_s[\tau, \xi_1, \ldots, \xi_j],$$

$$\tau_j \leq \tau < \tau_{j+1}, \quad j = 1, \ldots, k, \quad r = 1, \ldots, L_\eta, \quad s = 1, \ldots, M_\eta. \tag{35.5}$$

The quantity $e_{*(1)}^{(s)}(\cdot)$ unlike $e_{*(1)}(\cdot)$ (12.14) is now a function not only of τ_*, w_*, Δ but also of the parameter η, where η is a number that defines sets U_η and V_η of vectors $u^{[r]}$ and $v^{[s]}$. If $\eta > 0$ is sufficiently small, then the vectors $u^{[r]}$ and $v^{[s]}$ are distributed fairly densely in the compact sets P and Q. Thus, the quantity $e^{(s)}_{*(1)}(\cdot) = e^{(s)}_{*(1)}(\tau_*, w_*, \Delta, \eta)$ is defined by

$$e^{(s)}_{*(1)}(\tau_*, w_*, \Delta, \eta) = \max_{\|l_{(1)}(\cdot)\|^* \le 1} \Big[\sum_{i=g}^{N} M\{\, l^{(i)T}_{(1)}(\omega)\,\} *$$

$$* D^{[i]} X(t^{[i]}, \tau_*) w_* + M\Big\{\sum_{j=1}^{k} (\tau_{j+1} - \tau_j) *$$

$$* \max_q \min_p M\Big\{ \sum_{i=d(j)}^{N} l^{(i)T}_{(1)}(\omega)\, D^{[i]} X(t^{[i]}, \tau_j) *$$

$$* \sum_{r=1}^{L_\eta} \sum_{s=1}^{M_\eta} f(\tau_j, u^{[r]}, v^{[s]}) p_r q_s \mid \xi_1, \ldots, \xi_j \Big\}\Big\}\Big] \tag{35.6}$$

under the conditions (35.2)–(35.5). In (35.6) all the corresponding quantities are determined similarly to the ones in (12.14). According to what was said above in Chapter II it can be verified that

$$\rho^{(s)0}_{*(1)}(\tau_*, w_*) = \lim_{\eta \to 0,\ \delta \to 0} e^{(s)}_{*(1)}(\tau_*, w_*, \Delta, \eta)\,, \tag{35.7}$$

$$\delta = \max_j\, (\tau_{j+1} - \tau_j) \tag{35.8}$$

where $\rho^{(s)0}_{*(1)}(\cdot)$ is the value of the differential game (in the class of mixed strategies) for the system (10.1) with the approximating quality index $\gamma_{*(1)}$ (11.9). And more than that,

$$\rho^{(s)0}_{(1)}(\tau_*, w_*) = \lim_{\eta \to 0,\ \delta \to 0} e^{(s)}_{*(1)}(\tau_*, w_*, \Delta, \eta)\,, \tag{35.9}$$

$$\delta = \max_{j,h}\, [\ (\tau_{j+1} - \tau_j),\ (\tau^{[h+1]}_* - \tau^{[h]}_*)\ ] \tag{35.10}$$

where $\rho^{(s)0}_{(1)}(\cdot)$ is the value of the differential game for the initial functional $\gamma_{(1)}$. In (35.6)–(35.10) the numbers τ_j are the elements of the partition for $e^{(s)}_{*(1)}$, and $\tau^{[h]}_*$ are the elements of the partition (11.1) for $\gamma_{*(1)}$.

Using the notation (13.1)–(13.4), (13.6) we obtain

$$e^{(s)}_{*(1)}(\tau_*, w_*, \Delta, \eta) = \max_{m^*, m_*} \left[m^{*T} \widetilde{w}_* + \kappa^{(s)}_{*(1)}(\tau_*, m_*, \Delta, \eta) \right] \qquad (35.11)$$

where

$$\kappa^{(s)}_{*(1)}(\tau_*, m_*, \Delta, \eta) = \max_{\|l_{(1)}(\cdot)\|^* \leq 1} \Big[M \Big\{ \sum_{j=1}^{k} (\tau_{j+1} - \tau_j) *$$

$$* \max_q \min_p \Big(\sum_{i=d(j)}^{N} m_j^{(i)T} \Big) \Big(\sum_{r=1}^{L_\eta} \sum_{s=1}^{M_\eta} \widetilde{f}(\tau_j, u^{[r]}, v^{[s]}) p_r q_s \Big) \Big\} \Big] \qquad (35.12)$$

and m_*, m^* are given by (13.7)–(13.9). The minimax is taken over the arguments that satisfy the conditions (35.2)–(35.4). As in Sections 13 and 14 we will employ here the deterministic vectors $m^{[i]}$ and $l^{[i]}$ determined by the conditions (13.6), (14.2)–(14.4).

The values $\kappa^{(s)}_{*(1)}(\cdot)$ (35.12) and $e^{(s)}_{*(1)}(\cdot)$ (35.11) are calculated similarly to the constructions in Section 13 (see (13.16)–(13.42)). We suppose

$$\varphi_{k+1}(\tau_*, m) = 0, \qquad \Delta\psi_k(\tau_*, m) = J(\tau_k, \tau_{k+1}, m),$$

$$\psi_k(\tau_*, m) = \Delta\psi_k(\tau_*, m), \qquad \varphi_k(\tau_*, m) = \overline{\psi}_k(\tau_*, m), \qquad m \in G_k(\tau_*) \quad (35.13)$$

where

$$J(\tau_j, \tau_{j+1}, m) = (\tau_{j+1} - \tau_j) \max_q \min_p \Big[m^T \sum_{r=1, s=1}^{L_\eta, M_\eta} \widetilde{f}(\tau_j, u^{[r]}, v^{[s]}) p_r q_s \Big] \quad (35.14)$$

and

$$G_k(\tau_*) = \Big\{ m : \; m = D^{[N]T} l, \quad \mu^{[N]*}(l) \leq 1 \Big\}. \qquad (35.15)$$

The justification of this step is the same as in Section 13.

We assume that $j \geq 1$. Let us consider two cases. In the first case

$$\tau_{j+1} < t^{[i]}, \quad i = d(j) \qquad (35.16)$$

$t^{[i]}$ is the moment from (11.9). In this case we define $G_j(\tau_*) = G_{j+1}(\tau_*)$ and

$$\Delta\psi_j(\tau_*, m) = J(\tau_j, \tau_{j+1}, m), \qquad \psi_j(\tau_*, m) = \varphi_{j+1}(\tau_*, m) + \Delta\psi_j(\tau_*, m),$$

$$\varphi_j(\tau_*, m) = \overline{\psi}_j(\tau_*, m), \qquad m \in G_j(\tau_*). \tag{35.17}$$

In the second case

$$\tau_{j+1} = t^{[i]}, \qquad i = d(j). \tag{35.18}$$

Then

$$G_j(\tau_*) = \Big\{ m^* : \; m^* = m_* + \widehat{m}; \qquad m_* \in G_{j+1}(\tau_*), \\ \widehat{m} = X^T(t^{[i]}, \vartheta)D^{[i]T}l, \qquad \mu^{[i]*}(l) \leq 1 \Big\} \tag{35.19}$$

and we assume

$$\psi_j(\tau_*, m^*) = \max_{m_*} \Big[\Delta\psi_j(\tau_*, m^*) + \varphi_{j+1}(\tau_*, m_*) \Big]$$

$$\varphi_j(\tau_*, m^*) = \overline{\psi}_j(\tau_*, m^*), \qquad m^* \in G_j(\tau_*) \tag{35.20}$$

where $\Delta\psi_j(\tau_*, m^*) = J(\tau_j, \tau_{j+1}, m^*)$ and $G_j(\tau_*)$ is the set (35.19). We continue the construction until the time $\tau_1 = \tau_*$. Here two cases may occur. In the *first* case $\tau_* < t^{[g]}$. We then obtain a set $G_1(\tau_*) = G^{(g)}$ and a function $\varphi_1(\tau_*, m)$ for $m \in G^{(g)}$ which determines $\kappa^{(s)}_{*(1)}(\cdot)$ in (35.11) so that

$$\kappa^{(s)}_{*(1)}(\tau_*, m_*, \Delta, \eta) = \varphi_1(\tau_*, m_*), \qquad m_* \in G^{(g)}. \tag{35.21}$$

In the *second* case $\tau_* = t^{[g]}$. We then obtain a set $G_1(\tau_*) = G^{(g+1)}$ (13.37) and a function $\varphi_1(\tau_*, m)$, $m \in G^{(g+1)}$ such that

$$\kappa^{(s)}_{*(1)}(\tau_*, m_*, \Delta, \eta) = \varphi_1(\tau_*, m_*), \qquad m_* \in G^{(g+1)}. \tag{35.22}$$

In the *first* case we have $m^* = m_* \in G^{(g)}$ in (35.11). In the *second* case $m^* = m_* + \widehat{m}$, where $m_* \in G^{(g+1)}$, $\widehat{m} = X^T(\tau_*, \vartheta)D^{[g]T}l$, $\mu^{[g]*}(l) \leq 1$.

This concludes the construction of value $e^{(s)}_{*(1)}(\tau_*, w_*, \Delta, \eta)$ (35.6), (35.11).

Now we consider the quality index $\gamma_{(2)}$ (10.4). The approximating functional $\gamma_{*(2)}$ (11.10) constructed in Section 11 corresponds to $\gamma_{(2)}$. Let us describe the procedure for the calculation of the value $e^{(s)}_{*(2)}(\tau_*, w_*, \Delta, \eta)$ that determines the value of the game $\rho^{(s)0}_{*(2)}(t, x)$ for the approximating functional $\gamma_{*(2)}$ so that

$$\rho^{(s)0}_{*(2)}(\tau_*, w_*) = \lim_{\eta\to 0,\ \delta\to 0} e^{(s)}_{*(2)}(\tau_*, w_*, \Delta, \eta) \tag{35.23}$$

under the conditions (35.8).

Moreover,

$$\rho^{(s)0}_{(2)}(\tau_*, w_*) = \lim_{\eta\to 0,\ \delta\to 0} e^{(s)}_{*(2)}(\tau_*, w_*, \Delta, \eta) \tag{35.24}$$

where δ is given by (35.10).

According to (12.23), (13.1)–(13.4), (13.6)–(13.9), (14.2)–(14.4) and taking into account the above constructions in this chapter we obtain

$$e^{(s)}_{*(2)}(\tau_*, w_*, \Delta, \eta) =$$

$$= \max_{m^*, m_*, \nu} \Big[m^{*T}\widetilde{w}_* + \kappa^{(s)}_{*(2)}(\tau_*, m_*, \Delta, \eta, \nu) \Big] \tag{35.25}$$

where

$$\kappa^{(s)}_{*(2)}(\tau_*, m_*, \Delta, \eta, \nu) = \max_{\|l_{(2)}(\cdot)\|^* \le 1} \Big[M\Big\{ \sum_{j=1}^{k} (\tau_{j+1} - \tau_j) *$$

$$* \max_q \min_p \Big(\sum_{i=d(j)}^{N} m_j^{(i)T} \Big) \Big(\sum_{r=1}^{L_\mathcal{E}} \sum_{s=1}^{M_\mathcal{E}} \tilde{f}(\tau_j, u^{[r]}, v^{[s]}) p_r q_s \Big) \Big\} \Big] \tag{35.26}$$

and ν is a parameter that is similar to the parameter in the constructions of $e_{*(2)}(\cdot)$ in Section 14. In (35.25), (35.26) all the variables are defined similarly to the corresponding ones in (35.11), (35.12). Following the constructions in Section 14 (see (14.6)–(14.27)) we will obtain now the procedure of constructing the values $e^{(s)}_{*(2)}(\cdot)$ (35.25) and $\kappa^{(s)}_{*(2)}(\cdot)$ (35.26).

We take

$$\varphi_{k+1}(\tau_*, m, \nu) = 0, \qquad \nu \in [0,\ 1], \tag{35.27}$$

$$\Delta\psi_k(\tau_*, m, \nu) = J(\tau_k, \tau_{k+1}, m, \nu) \tag{35.28}$$

where $J(\tau_k, \tau_{k+1}, m, \nu)$ is analogous to $J(\tau_k, \tau_{k+1}, m)$ in (35.13), (35.14).

We assume

$$\psi_k(\tau_*, m, \nu) = \Delta\psi_k(\tau_*, m, \nu),$$

$$\varphi_k(\tau_*, m, \nu) = \overline{\psi}_k(\tau_*, m, \nu), \qquad m \in G_{k,\nu}(\tau_*) \tag{35.29}$$

where $G_{k,\nu}(\tau_*)$ is the set (14.11).

The following constructions of the quantities $\varphi_j(\tau_*, m, \nu)$, $m \in G_{j,\nu}(\tau_*)$ and the sets $G_{j,\nu}(\tau_*)$ completely repeat (14.12)–(14.20) with $J(\tau_j, \tau_{j+1}, m, \nu)$ that is analogous to $J(\tau_j, \tau_{j+1}, m)$ in (35.14).

Finally we obtain the equalities

$$\kappa^{(s)}_{*(2)}(\tau_*, m_*, \Delta, \eta, \nu) = \varphi_1(\tau_*, m_*, \nu), \qquad m_* \in G_{1,\nu}(\tau_*), \qquad \nu = 1 \tag{35.30}$$

if $\tau_* < t^{[g]}$, and

$$\kappa^{(s)}_{*(2)}(\tau_*, m_*, \Delta, \eta, \nu) = \varphi_1(\tau_*, m_*, \nu),$$
$$m_* \in G_{1,\nu}(\tau_*), \qquad \nu \le 1 - \mu^{[g]*}(l^{[g]}) \tag{35.31}$$

if $\tau_* = t^{[g]}$ (see the corresponding equalities (14.23), (14.24)). In the case $\tau_* < t^{[g]}$ we have the condition (14.24).

From (35.25), (35.30), (35.31) we deduce the equality

$$e^{(s)}_{*(2)}(\tau_*, w_*, \Delta, \eta) = \max_{m^*, m_*, \nu} \Big[\, m^{*T} \widetilde{w}_* \mid \varphi_1(\tau_*, m_*, \nu) \Big] \tag{35.32}$$

where in the case $\tau_* < t^{[g]}$ we have the condition (14.26) and in the case $\tau_* = t^{[g]}$ we have the condition (14.27).

This concludes the construction of the value $e^{(s)}_{*(2)}(\tau_*, w_*, \Delta, \eta)$ for $\gamma_{*(2)}$ (11.9).

As in Sections 15, 16, and 19 we can prove that the quantities $e^{(s)}_{(i)}(\tau_*, w_*, \Delta, \eta)$ constructed in this section satisfy conditions similar to $1^0_{*(i)}$, $2^0_{*(i)}$, $3^0_{*(i)}u$ and $4^0_{*(i)}v$, $i = 1, 2$. Therefore, the functions $e^{(s)}_{*(i)}(\cdot)$, $i = 1, 2$ can be used again as the basis for constructing the approximating optimal strategies $S^u_{a(i)}$ and $S^v_{a(i)}$, $i = 1, 2$ which form the saddle point in the differential game with $\gamma_{*(i)}$, $i = 1, 2$. And these functions enable us to construct an approximating solution of the considered game in the class of mixed strategies. In the next section we will describe these constructions.

36. Construction of approximating optimal strategies S^u_a and S^v_a

To construct the optimal mixed strategy $S^u_{0(1)}$ (34.7) for functional $\gamma_{(1)}$ we, according to the definition of mixed strategy S^u given in Section 32, have to find the elements (32.5)

$$U^0_y(\cdot),\ U^{*0}_y(\cdot),\ V^{*0}_y(\cdot),\ p^0_y(\cdot),\ p^{*0}_y(\cdot),\ q^{*0}_y(\cdot)\,. \tag{36.1}$$

As was noted in the end of Section 34 the sets

$$U_y^0(\varepsilon) = \{u^{[r]} \in P, \quad r = 1, \dots, L_\mathcal{E}\}, \tag{36.2}$$

$$V_y^{*0}(\varepsilon) = \{v^{[s]} \in Q, \quad s = 1, \dots, M_\mathcal{E}\} \tag{36.3}$$

and functions

$$p_y^0(t,x,y,\varepsilon) = \{p_{yr}^0(t,x,y,\varepsilon) \geq 0, \quad \sum_{r=1}^{L_\mathcal{E}} p_{yr}^0(t,x,y,\varepsilon) = 1\}, \tag{36.4}$$

$$q_y^{*0}(t,x,y,\varepsilon) = \{q_{ys}^{*0}(t,x,y,\varepsilon) \geq 0, \quad \sum_{s=1}^{M_\mathcal{E}} q_{ys}^{*0}(t,x,y,\varepsilon) = 1\} \tag{36.5}$$

are determined by the conditions (34.14) which guarantee that the motion (33.2), (33.4) of the x-object (10.1) and the motion (33.3), (33.5) of the y-model-leader (32.2) are close to each other. According to these conditions the sets of numbers $p_{yr}^0 = p_{yr}^0(t,x,y,\varepsilon)$ in (36.4) and $q_{ys}^{*0} = q_{ys}^{*0}(t,x,y,\varepsilon)$ in (36.5) are chosen subject to the conditions

$$\max_q \sum_{r,s=1}^{L_\eta, M_\eta} \langle\ (x-y),\ f(t,u^{[r]},v^{[s]}) p_{yr}^0 q_s\ \rangle =$$

$$= \min_p \max_q \sum_{r,s=1}^{L_\eta, M_\eta} \langle\ (x-y),\ f(t,u^{[r]},v^{[s]}) p_r q_s\ \rangle, \tag{36.6}$$

$$\min_p \sum_{r,s=1}^{L_\eta, M_\eta} \langle\ (x-y),\ f(t,u^{[r]},v^{[s]}) p_r q_{ys}^{*0}\ \rangle =$$

$$= \max_q \min_p \sum_{r,s=1}^{L_\eta, M_\eta} \langle\ (x-y),\ f(t,u^{[r]},v^{[s]}) p_r q_s\ \rangle \tag{36.7}$$

under the assumption that the relations (35.2)–(35.4) hold with $\eta = \eta(\varepsilon)$.

We define $U_y^0(\varepsilon) = U_y^{*0}(\varepsilon) = U_{\eta(\mathcal{E})}$, $V_y^{*0}(\varepsilon) = V_{\eta(\mathcal{E})}$, where $U_{\eta(\mathcal{E})}$ and $V_{\eta(\mathcal{E})}$ are the sets of vectors $\{u^{[r]}\}$ and $\{v^{[s]}\}$ in (35.1), (36.6), (36.7).

It remains to find the function

$$p_y^{*0}(\cdot) = \Big\{p_y^{*0}(t,x,y,\varepsilon) = \{p_{yr}^{*0}(t,x,y,\varepsilon) \geq 0,$$

$$\sum_{r=1}^{L_\eta} p_{yr}^{*0}(t,x,y,\varepsilon) = 1\}\Big\}. \tag{36.8}$$

According to the construction of the extremal strategy $S_{(e)}^u$ which was described in Section 34, using the *accompanying point* $w^{[u]0}$ we can determine the optimal set of numbers $p_{yr}^{*0} = p_{yr}^{*0}(t,x,y,\varepsilon)$ in (36.8) by the condition (34.12) where $p_{(e)}^* = p^{*0}$. However, the problem of finding the accompanying point $w^{[u]0}$ in the general case is far from simple. But in many cases we can use the approximating construction of value $\rho_{(1)}^{(s)0}(t,x)$ based on the quantity $e_{*(1)}^{(s)}(\cdot)$, (see Section 35). This approximating solution can be obtained similarly to the corresponding solution for the pure approximating optimal strategy $u^a(\cdot)$, (see (18.1)–(18.8)). In some cases the component $p_y^{*a}(\cdot)$ that substitutes the component $p^{*0}(\cdot)$ (36.8) can be constructed effectively in the following way.

The approximating optimal set p_{yr}^{*a} is determined by

$$\max_q \Big[m_u^{*0T}(t,y,\varepsilon)X(\vartheta,t)\sum_{r,s=1}^{L_\eta,M_\eta} f(t,u^{[r]},v^{[s]})p_{yr}^{*a}q_s\Big] =$$

$$= \min_p \max_q \Big[m_u^{*0T}(t,y,\varepsilon)X(\vartheta,t)\sum_{r,s=1}^{L_\eta,M_\eta} f(t,u^{[r]},v^{[s]})p_r q_s\Big] \tag{36.9}$$

where $m_u^{*0}(t,y,\varepsilon)$ is a solution of the following problem:

$$\langle\, m_u^{*0}(t,y,\varepsilon),\ X(\vartheta,t)y\,\rangle + \varphi_1(t,m_{*u}^0) - R^{[u]}(\varepsilon,t)\mid X^T(\vartheta,t)m_u^{*0}(t,y,\varepsilon)\mid =$$

$$= \max_{m^*,m_*}\Big[\,\langle\, m^*,\ X(\vartheta,t)y\,\rangle + \varphi_1(t,m_*) - R^{[u]}(\varepsilon,t)\mid X^T(\vartheta,t)m^*\mid\Big]. \tag{36.10}$$

Here $\varphi_1(t,m_*)$ is the function (35.21), (35.22) at $\tau_* = t$; $R^{[u]}(t,\varepsilon)$ is the quantity (8.8).

The maximum in (36.10) is taken over the vectors m^* in the set $G_1(t)$ defined in Section 35, (see the construction of $e_{*(1)}^{(s)}(\cdot)$). The employed relations are similar to (13.35), (13.42) in Section 13.

This completes the construction of the approximating optimal mixed strategy

$$S^u_{a(1)} = \{U^0_y(\cdot),\ p^0_y(\cdot);\ V^{*0}_y(\cdot),\ q^{*0}_y(\cdot);\ U^{*0}_y(\cdot),\ p^{*a}_y(\cdot)\}. \tag{36.11}$$

The approximating optimal strategy

$$S^v_{a(1)} = \{V^0_z(\cdot),\ q^0_z(\cdot);\ U^{*0}_z(\cdot),\ p^{*0}_z(\cdot);\ V^{*0}_z(\cdot),\ q^{*a}_z(\cdot)\} \tag{36.12}$$

is constructed in a similar way. The necessary changes in (36.6), (36.7) are to replace y by z, interchange p and q and replace the minus sign in (36.10) by a plus. The optimal set $q^{*a}_{zs} = q^{*a}_{zs}(t, x, z, \varepsilon)$ is defined by

$$\min_p \left[m^{*0T}_v(t, z, \varepsilon) X(\vartheta, t) \sum_{r,s=1}^{L_\eta, M_\eta} f(t, u^{[r]}, v^{[s]}) p_r q^{*a}_{zs}\right] =$$

$$= \max_q \min_p \left[m^{*0T}_v(t, z, \varepsilon) X(\vartheta, t) \sum_{r,s=1}^{L_\eta, M_\eta} f(t, u^{[r]}, v^{[s]}) p_r q_s\right] \tag{36.13}$$

where $m^{*0}_v = m^{*0}_v(t, z, \varepsilon)$ is the corresponding maximizing vector.

The approximating optimal strategies $S^u_{a(2)}$ and $S^v_{a(2)}$, which approximate the saddle point $\{S^u_{a(2)}, S^v_{a(2)}\}$ in the differential game with the quality index $\gamma_{(2)}$ (10.4), are constructed similarly using the corresponding constructions of value $e^{(s)}_{*(2)}(\cdot)$ in Section 35. We have only to replace the lower index (1) by (2).

It should be noted that the terms "approximating solution" and "approximating strategies S^u_a, S^v_a" are understood here similarly to the corresponding expressions in the case of pure strategies, (see Section 18).

37. Example with mixed strategies

In this section we illustrate the computation of the approximating value of the game $\rho^{(s)a}_{(1)}(t_*, x_*)$ (35.6) (see (35.7)–(35.10)) and the construction of mixed strategies.

Let the differential equation of the controlled object be

$$\ddot{h} = a(t)u + b(t)(u+v)^2 + c(t)v =$$
$$= F(t, u, v), \qquad t_0 \le t < \vartheta \tag{37.1}$$

where h is scalar, $a(t)$, $b(t)$ and $c(t)$ are piecewise-continuous functions and continuous from the right. The following restrictions

$$u \in P, \quad P = \{u: \ u^{[1]} = -1, \ \ u^{[2]} = 1\}, \tag{37.2}$$

$$v \in Q, \quad Q = \{v: \ v^{[1]} = -1, \ \ v^{[2]} = 1\} \tag{37.3}$$

are valid.

The quality index $\gamma_{(1)}$ is

$$\gamma_{(1)} = \mid h[t^{[1]}] \mid + \mid h[\vartheta] \mid, \quad t^{[1]} \in [t_0, \ \vartheta) \,. \tag{37.4}$$

Differential equation (37.1) can be reduced to the normal form (10.1)

$$\dot{x} = Ax + f(t, u, v) \tag{37.5}$$

where

$$x = \begin{bmatrix} x_1 \\ x_2 \end{bmatrix}, \quad A = \begin{pmatrix} 0 & 1 \\ 0 & 0 \end{pmatrix}, \quad f(t, u, v) = \begin{bmatrix} 0 \\ F(t, u, v) \end{bmatrix}. \tag{37.6}$$

Then in accordance with (10.3), (11.9) we have

$$\gamma_{(1)} = \mid Dx[t^{[1]}] \mid + \mid Dx[t^{[2]}] \mid \tag{37.7}$$

where D is a row-vector such that

$$D = [1 \ \ 0] \,. \tag{37.8}$$

In (37.7) according to (37.4) we have $t^{[2]} = \vartheta$ and $t^{[1]}$ is a given moment in the time half-interval $[t_0, \ \vartheta)$.

We follow the constructions in Section 35 and introduce the scalars $l^{[i]}$, $i = 1, 2$, $| l^{[i]} | \leq 1$ and values

$$p_r = P\Big(u = u^{[r]} \Big), \quad r = 1, 2, \quad \sum_{r=1}^{2} p_r = 1\,, \tag{37.9}$$

$$q_s = P\Big(v = v^{[s]} \Big), \quad s = 1, 2, \quad \sum_{s=1}^{2} q_s = 1\,. \tag{37.10}$$

As above, $P\Big(\cdots \Big)$ denotes the probability of the corresponding event. In (37.9), (37.10) we have

$$u^{[r]} \in U_\eta = P, \qquad v^{[s]} \in V_\eta = Q \tag{37.11}$$

where P and Q are the sets (37.2) and (37.3), respectively.

Define the vectors

$$m^{[i]} = X^T(t^{[i]}, \vartheta) D^T l^{[i]}, \quad i = 1, 2 \tag{37.12}$$

where $l^{[i]} \in R^1$, $|l^{[i]}| \leq 1$ and

$$X(t, \tau) = \begin{pmatrix} 1 & t - \tau \\ 0 & 1 \end{pmatrix} \tag{37.13}$$

is the fundamental matrix for the equation $dx/dt = Ax$ that corresponds to (37.5), (37.6). We remind the reader that in accordance with the remark in Section 13 we now consider vectors $m^{[i]}$ and $l^{[i]}$ simply as deterministic arguments.

Let $\{t_*, x_*\}$, $t_* < t^{[1]}$ be the initial position of the object (37.5). We are going to determine the approximating value $\rho_{(1)}^{(s)a}(t_*, x_*)$ that is determined by the program extremum $e_{(1)}^{(s)}(t_*, w_*, \Delta)$.

Let us choose

$$t_0 = 0, \quad \vartheta = 4, \quad t^{[1]} = 1, \quad b(t) \equiv 1/2 \tag{37.14}$$

and the functions $a(t)$ and $c(t)$ in (37.1), (37.5), which are step functions

$$a(t) = \begin{cases} 4, & t_0 \le t < 2 \\ 0, & 2 \le t \le \vartheta \end{cases} \qquad c(t) = \begin{cases} 0, & t_0 \le t < 3 \\ 2, & 3 \le t \le \vartheta \end{cases}. \tag{37.15}$$

Let us take $\tau_* < t^{[1]}$. Following the construction in Section 35 (see (35.11)–(35.22)) we compute the function

$$\Delta\psi_k(\tau_*, m^{[2]}) = J(\tau_k, \tau_{k+1}, m^{[2]}) =$$

$$= (\tau_{k+1} - \tau_k)\Big[\max_q \min_p \; m^{[2]T}\Big(\sum_{r=1}^{2}\sum_{s=1}^{2} X(\vartheta, \tau_k) f(\tau_k, u^{[r]}, v^{[s]}) p_r q_s\Big)\Big] =$$

$$= (\tau_{k+1} - \tau_k) \mid l^{[2]} \mid (\vartheta - \tau_k) \,\mathrm{extr}\,(\tau_k, \xi) \tag{37.16}$$

where $\zeta = \mathrm{sign}(l^{[2]})$,

$$\mathrm{extr}\,(\tau_k, \zeta) = \max_q \min_p \; \zeta \sum_{r=1}^{2}\sum_{s=1}^{2} F(\tau, u^{[r]}, v^{[s]}) p_r q_s \,. \tag{37.17}$$

Solving the minimax problem in (37.17) under the conditions (37.9), (37.10) we obtain the following result

$$\mathrm{extr}\,(\tau_k, 1) = 2 \qquad \mathrm{extr}(\tau_k, -1) = 0 \tag{37.18}$$

with minimizing values $p_1^{[1]} = 1$, $p_1^{[-1]} = 1$ and maximizing values $q_1^{[1]} = 0$, $q_1^{[-1]} = 1$. Then $\Delta\psi_k(\tau_*, m^{[2]}) = (|l^{[2]}| + l^{[2]})(\tau_{k+1} - \tau_k)(\vartheta - \tau_k)$, and

$\varphi_k(\tau_*, m^{[2]}) = \overline{\psi}_k(\tau_*, m^{[2]})$ in the region $| l^{[2]} | \leq 1$ is

$$\varphi_k(\tau_*, m^{[2]}) = (l^{[2]} + 1)(\tau_{k+1} - \tau_k)(\vartheta - \tau_k) . \tag{37.19}$$

Now taking $\tau_j \in [3, 4)$ $(\vartheta = 4)$ we have

$$\text{extr}\,(\tau_j, 1) = 2 \quad \text{extr}\,(\tau_j, -1) = 0 \tag{37.20}$$

with $p_1^{[1]} = 1$, $q_1^{[1]} = 1$; $p_1^{[-1]} = 0$, $q_1^{[-1]} = 1$.

For $\tau_j \in [2, 3)$ we have

$$\text{extr}\,(\tau_j, 1) = 1, \quad \text{extr}(\tau_j, -1) = -1 \tag{37.21}$$

with $p_1^{[1]} = q_1^{[1]} = p_1^{[-1]} = q_1^{[-1]} = 1/2$.

Therefore, by induction in j from $j = k$ to $j \geq j[2]$, where $\tau_{j[2]} = 2$, $\tau_{j[3]} = 3$, we obtain

$$\varphi_j(\tau_*, m^{[2]}) = (l^{[2]} + 1) \sum_{s=j[3]}^{k} (\tau_{s+1} - \tau_s) + (\vartheta - \tau_s) + l^{[2]} \sum_{s=j}^{j[3]-1} (\tau_{s+1} - \tau_s)(\vartheta - \tau_s) . \tag{37.22}$$

If $\tau_j \in [1, 2)$, we have

$$\text{extr}\,(\tau_j, 1) = -2 \quad \text{extr}\,(\tau_j, -1) = -4 \tag{37.23}$$

with $p_1^{[1]} = 1$, $q_1^{[1]} = 1$; $p_1^{[-1]} = 0$, $q_1^{[-1]} = 1$. Therefore, by induction in j from $j = j[2]$ to $j \geq j[1]$, where $\tau_{j[1]} = t^{[1]} = 1$, we obtain

$$\varphi_j(\tau_*, m^{[2]}) = \varphi_{j[2]}(\tau_*, m^{[2]}) +$$

$$+ \mid l^{[2]} \mid \sum_{s=j}^{j[2]-1} (\tau_{s+1} - \tau_s)(\vartheta - \tau_s)\text{extr}(\tau_j, \text{sign}(l^{[2]}))\,. \qquad (37.24)$$

If $\tau_j < t^{[1]}$, then

$$\text{extr}\,(\tau_j, 1) = -2, \quad \text{extr}(\tau_j, -1) = -4 \qquad (37.25)$$

with $p_1^{[1]} = 1$, $q_1^{[1]} = 0$; $p_1^{[-1]} = 0$, $q_1^{[-1]} = 1$. Proceeding by induction in j from $j = j[1]$ to $j = 1$ ($\tau_1 = \tau_* < t^{[1]} = 1$) we see that the function $\kappa_{(1)}(\cdot)$ (35.11), (35.22) is

$$\kappa_{(1)}(\tau_*, m^{[1]} + m^{[2]}) = \varphi_1(\tau_*, m^{[1]} + m^{[2]}) =$$

$$= \sum_{j=1}^{j[1]-1} (\tau_{j+1} - \tau_j) \mid l_j \mid \text{extr}\,(\tau_j, \text{sign}(l_j)) +$$

$$+ \max_{m^{[2]}} \varphi_{j[1]}(\tau_*, m^{[2]}) \quad l_j = l^{[1]}(t^{[1]} - \tau_j) + l^{[2]}(\vartheta - \tau_j)\,. \qquad (37.26)$$

In the case $\tau_* < t^{[1]}$ we have for the approximating value $\rho_{(1)}^{(s)a}(\tau_*, w_*) = e_{*(1)}^{(s)}(\tau_*, w_*, \Delta)$ of the game the following formula:

$$e_{(1)}^{(s)}(\tau_*, w_*, \Delta) = \max_{m^{[1]}, m^{[2]}} \Big[(m^{[1]} + m^{[2]})^T X(\vartheta, \tau_*) w_* +$$

$$+ \kappa_{(1)}(\tau_*, m^{[1]} + m^{[2]})\Big] \qquad (37.27)$$

under the condition

$$(m^{[1]} + m^{[2]}) \in G \qquad (37.28)$$

where

$$G = \Big\{ m^{[1]} + m^{[2]} = \begin{bmatrix} l^{[1]} \\ (t^{[1]} - \vartheta) l^{[1]} \end{bmatrix} + \begin{bmatrix} l^{[2]} \\ 0 \end{bmatrix} :$$

$$\mid l^{[i]} \mid \le 1, \quad i = 1, 2 \Big\}\,. \qquad (37.29)$$

If $\tau_* > t^{[1]} = 1$, then we have

$$e_{(1)}^{(s)}(\tau_*, w_*, \Delta) = \max_{m^{[2]}} \Big[\, m^{[2]T} X(\vartheta, \tau_*) w_* + \kappa_{(1)}(\tau_*, m^{[2]}) \,\Big] \qquad (37.30)$$

where

$$\kappa_{(1)}(\tau_*, m^{[2]}) = \begin{cases} |l^{[2]}| \sum\limits_{s=1}^{j[2]-1} (\tau_{s+1} - \tau_s)(\vartheta - \tau_s)\mathrm{extr}(\tau_*, \mathrm{sign}(l^{[2]}))+ & \\ +\varphi_{j[2]}(\tau_*, m^{[2]}), & 1 < \tau_* < 2 \\ |l^{[2]}| \sum\limits_{s=1}^{j[3]-1} (\tau_{s+1} - \tau_s)(\vartheta - \tau_s) + \varphi_{j[3]}(\tau_*, m^{[2]}), & 2 \leq \tau_* < 3 \\ (l^{[2]} + 1) \sum\limits_{j=1}^{k} (\tau_{j+1} - \tau_j)(\vartheta - \tau_j), & 3 \leq \tau_* < 4. \end{cases} \tag{37.31}$$

In the case $\tau_* = t^{[1]}$ we have

$$e^{(s)}_{(1)}(\tau_*, w_*, \Delta) = \max_{m^{[1]}, m^{[2]}} \Big[(m^{[1]} + m^{[2]})^T X(\vartheta, \tau_*)w_* + $$

$$+ \kappa_{(1)}(\tau_*, m^{[2]}) \Big] \tag{37.32}$$

where

$$\kappa_{(1)}(\tau_*, m^{[2]}) = \varphi_1(\tau_*, m^{[2]}). \tag{37.33}$$

Since the approximating value of the game $\rho^{(s)a}_{(1)}(t, x)$ is determined by the program extremum $e^{(s)}_{(1)}(\cdot)$ (37.26)–(37.33), we can construct the approximating optimal strategies $S^u_{a(1)}$ and $S^v_{a(1)}$ which form the approximating saddle point $\{S^u_{a(1)}, S^v_{a(1)}\}$ in the differential game for the system (37.5), (37.6) with the quality index $\gamma_{(1)}$ (37.7) (see Section 36). Thus, these strategies are determined by the conditions (36.6)–(36.13).

The corresponding control process was simulated on IBM-type computer for initial data $t_* = 0$, $x_* = \{1, 2\}$. For these data we have $\rho^{(s)a}_{(1)}(t_*, x_*) = 4.5$. The continuous curve in Figure 37.1 represents the motion of the object (37.5) with $S^u_{a(1)} = \{u^{[r]}, p^0_{yr}, p^{*a}_{yr}, q^{*0}_{ys}\}$, $S^v_{a(1)} = \{v^{[s]}, q^0_{zs}, q^{*a}_{zs}, p^{*0}_{zr}\}$. In this case the obtained value of the quality index $\gamma_{(1)}$ of (37.7) is close to the value of the game $\rho^{(s)a}_{(1)}(t_*, x_*) = 4.5$. The dashed line represents the motion of the object with $S^u_{a(1)}$, $S^v_{(1)} = \{v^{[s]}, q_{zs} = 1/2, q^{*a}_{zs}, p^{*0}_{zr}\} \neq S^v_{a(1)}$. In this case we obtain $\gamma_{(1)} = 3.5 < \rho^{(s)a}_{(1)}(t_*, x_*)$. Figure 37.2 shows the motion of the object for $S^u_{(1)} = \{u^{[r]},\ p_{y1} = 0.7,\ p_{y2} = 0.3, p^{*a}_{yr}, q^{*0}_{ys}\} \neq S^u_{a(1)}$, $S^v_{a(1)}$. In this case $\gamma_{(1)} = 13.5 > \rho^{(s)a}_{(1)}(t_*, x_*)$.

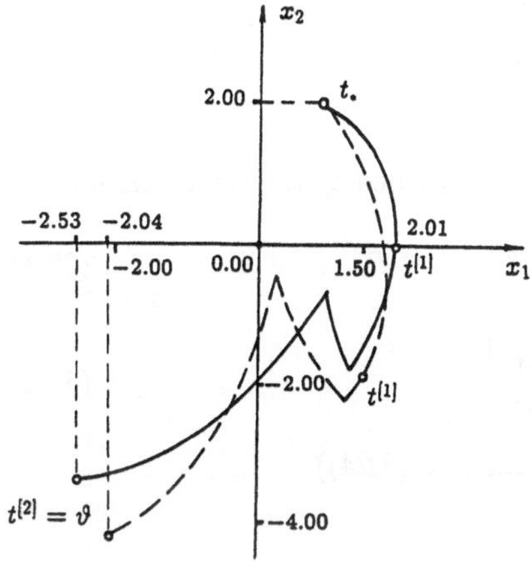

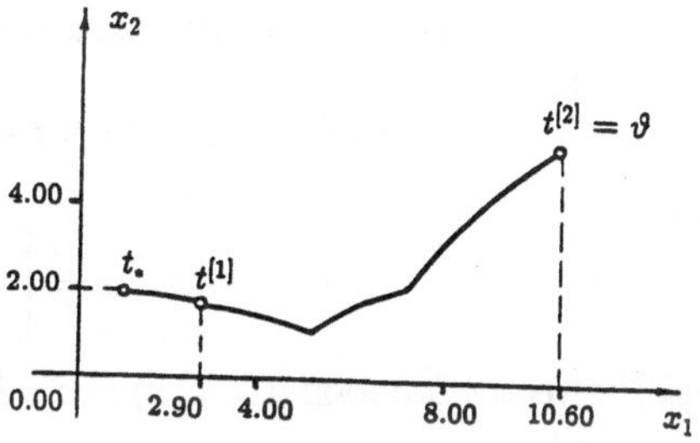

Figure 37.1

Figure 37.2

38. The second example with mixed strategies

In this section we illustrate the construction of the approximating optimal strategies by the example of the model problem from Section 2. This example is somewhat more complicated than the example considered in Section 37. And the example in this section is considered in more detail. With the help of this example we demonstrate the concrete construction of the random control law determined by the extremal shift. This shift for the x-object is aimed to the y-model-leader (for the optimal mixed strategy S^u_a) or to the z-model-leader (for S^v_a). In its turn for the y-model (for S^u_a) or for the z-model (for S^v_a) the extremal shift is aimed at the corresponding accompanying points $w^{[u]0}(t, x, y, \varepsilon)$ or $w^{[v]0}(t, x, z, \varepsilon)$. These points are calculated on the basis of the program extremum $e^{(s)}_{*(1)}(\cdot)$ (see Section 35). The construction of the upper convex hulls $\varphi_j(\tau_*, m)$ that determine the program extremum in the considered concrete case is proved to be sufficiently transparent. As a result we obtain the algorithms for forming the approximating optimal random control actions $u^a[t]$ and the approximating optimal random disturbances $v^a[t]$. These algorithms can be simulated by accessible computers. The results of this simulation were given at the end of Section 2.

In Section 2 we considered two mass points M_1 and M_2, whose motions are described by the Mestcherckii differential equations (2.1), (2.9), (2.14). In Section 2 we also indicated the transformations $x = T_t(r, \dot{r}, s, \dot{s})$ (2.18)–(2.26), which reduce the considered model problem to the game control problem for the auxiliary 2-dimensional x-object with the corresponding quality index $\gamma_{(1)}$ (2.27). This transformation does not change the corresponding restrictions for the actions u and v (see (2.28) and (2.29)). In this section we consider the construction of the approximating optimal control actions u and approximating optimal disturbances v directly for the 2-dimensional x-system of this kind with the corresponding functional $\gamma_{(1)}$ under the corresponding restrictions for the control actions and disturbances.

Thus we consider the controlled x-object described by the vector differential equation

$$\dot{x} = f(t, u, v), \qquad t_0 \le t < \vartheta, \qquad u \in P, \qquad v \in Q \tag{38.1}$$

where

$$x = \begin{bmatrix} x_1 \\ x_2 \end{bmatrix} \tag{38.2}$$

and the vector-function $f(t, u, v)$ has a form (see (2.24))

$$f(t, u, v) = B^{[1]}(t)u^{(1)}_{v^{(1)}} + B^{[2]}(t)u^{(2)} + C^{[1]}(t)v_* + C^{[2]}(t)v^*. \tag{38.3}$$

Here we have

$$B^{[i]}(t) = \begin{pmatrix} b^{[i]}(t) & 0 \\ 0 & b^{[i]}(t) \end{pmatrix}, \qquad i = 1, 2 \tag{38.4}$$

$$C^{[i]}(t) = \begin{pmatrix} c^{[i]}(t) & 0 \\ 0 & c^{[i]}(t) \end{pmatrix}, \qquad i = 1, 2. \tag{38.5}$$

In (38.3) $u^{(1)}_{v^{(1)}}$, $u^{(2)}$, v_* and v^* are the vectors (2.5), (2.10), (2.8), (2.13) respectively, i.e.,

$$u^{(1)}_{v^{(1)}} = \begin{bmatrix} u^{(1)}_1 \cos v^{(1)} - u^{(1)}_2 \sin v^{(1)} \\ u^{(1)}_1 \sin v^{(1)} + u^{(1)}_2 \cos v^{(1)} \end{bmatrix}, \tag{38.6}$$

$$u^{(1)} = \begin{bmatrix} u^{(1)}_1 \\ u^{(1)}_2 \end{bmatrix} : \quad u^{(1)}_{[1]} = \begin{bmatrix} 1 \\ 0 \end{bmatrix}, \quad u^{(1)}_{[2]} = \begin{bmatrix} 0 \\ 1 \end{bmatrix},$$

$$u^{(1)}_{[3]} = \begin{bmatrix} -1 \\ 0 \end{bmatrix}, \quad u^{(1)}_{[4]} = \begin{bmatrix} 0 \\ -1 \end{bmatrix}, \tag{38.7}$$

$$v^{(1)} : \quad -\alpha^0_1 \le v^{(1)} \le \alpha^0_2, \qquad 0 < \alpha^0_1 \le \frac{\pi}{2}, \qquad \alpha^0_2 = \alpha^0_1, \tag{38.8}$$

$$u^{(2)} = \begin{bmatrix} u_1^{(2)} \\ u_2^{(2)} \end{bmatrix} : \quad u_1^{(2)^2} + u_2^{(2)^2} \leq 1, \tag{38.9}$$

$$v_* = \begin{bmatrix} v_{1*} \\ v_{2*} \end{bmatrix} : \quad v_{1*}^2 + v_{2*}^2 \leq 1, \tag{38.10}$$

$$v^* = \begin{bmatrix} v_1^* \\ v_2^* \end{bmatrix} : \quad v_1^{*^2} + v_2^{*^2} \leq 1. \tag{38.11}$$

Thus, in (38.1) P and Q are the sets (2.28) and (2.29), respectively.

For the system (38.1)–(38.11) we consider the feedback control problem of constructing the control actions $u[t] \in P$ and disturbances $v[t] \in Q$, which are aimed to minimize and, respectively, maximize the quality index (see (2.27)).

$$\gamma = \mid x[\vartheta] \mid = \left(x_1^2[\vartheta] + x_2^2[\vartheta] \right)^{1/2}. \tag{38.12}$$

The differential equation that describes the motion of the y-model-leader (see the construction of the mixed strategy S^u in Sections 2 and 32) has a form (see (2.38))

$$\dot{y}[t] = \sum_{r=1}^{4} \sum_{s=1}^{2} B^{[1]}(t) u^{(1)}_{[r]_{v^{(1)}_{[s]}}} p^*_{yr} q^*_{ys} +$$

$$+ B^{[2]}(t) u^{(2)} + C^{[1]}(t) v_* + C^{[2]}(t) v^*, \qquad t_0 \leq t < \vartheta \tag{38.13}$$

where $\left\{u^{(1)}_{[r]}\right\}$, $r = 1, 2, 3, 4$ is the collection (38.7); $v^{(1)}_{[1]} = -\alpha_1^0$, $v^{(1)}_{[2]} = \alpha_2^0$ are the swing angles in (38.8); and the matrix-functions $B^{[i]}(t)$, $C^{[i]}(t)$, $i = 1, 2$ and the vectors $u^{(2)}$, v_* and v^* are determined by (38.4)–(38.11).

In (38.13) the numbers p^*_{yr} and q^*_{ys} satisfy the conditions

$$p^*_{yr} \geq 0, \quad r = 1, 2, 3, 4, \quad \sum_{r=1}^{4} p^*_{yr} = 1 \tag{38.14}$$

$$q^*_{ys} \geq 0, \quad s = 1, 2, \quad \sum_{s=1}^{2} q^*_{ys} = 1. \tag{38.15}$$

The z-model-leader (for the strategy S^v) is described by the differential equation (see (2.66))

$$\dot{z}[t] = \sum_{r=1}^{4} \sum_{s=1}^{2} B^{[1]}(t) u^{(1)}_{[r]_{v^{(1)}_{[s]}}} p^*_{zr} q^*_{zs} +$$

$$+ B^{[2]}(t)u^{(2)} + C^{[1]}(t)v_* + C^{[2]}(t)v^*, \qquad t_0 \le t < \vartheta \tag{38.16}$$

under the conditions

$$p^*_{zr} \ge 0, \qquad r = 1,2,3,4, \qquad \sum_{r=1}^{4} p^*_{zr} = 1 \tag{38.17}$$

$$q^*_{zs} \ge 0, \qquad s = 1,2, \qquad \sum_{s=1}^{2} q^*_{zs} = 1. \tag{38.18}$$

Let us note the following circumstance. According to the general rule for constructing the optimal mixed strategies S^u_0 and S^v_0 (see (32.5), (32.16)) or for constructing the approximating optimal mixed strategies S^u_a, S^v_a (see Section 36) we have to introduce the collections $U(\varepsilon)$ and $V(\varepsilon)$ of the vectors $u^{[r]}$ and $v^{[s]}$ (see (32.10) and (32.13) and the remark at the end of Section 32). These vectors $u^{[r]}$ and $v^{[s]}$ are admissible values for the variables u and v in the equations (38.1), (38.13), (38.16). However, in the considered case the components of strategies S^u_a and S^v_a which form the actions $u^{(2)}$, v_* and v^* work actually as pure strategies (see (2.47), (2.50) for S^u and (2.70) for S^v). Besides, the initial restrictions (38.7) and (38.8) for the variables $u^{(1)}$ and $v^{(1)}$ already determine the collections $U = U^*_y = U^*_z$ and $V = V^*_y = V^*_z$ (for S^u and S^v) which consist of four vectors $u^{(1)}_{[r]}$ in (38.7) and of two angles $v^{(1)}_{[1]} = -\alpha^0_1$, $v^{(1)}_{[2]} = \alpha^0_2$ in (38.8). We do not need to reduce the sets U and V to some other finite sets. Thus, we assume

$$U^* = U^*_y = U^*_z = \left\{ u^{(1)}_{[r]} \ (38.7), \quad r = 1,2,3,4 \right\} \tag{38.19}$$

$$V^* = V^*_y = V^*_z = \left\{ v^{(1)}_{[s]}, \quad s = 1,2, \quad v^{(1)}_{[1]} = -\alpha^0_1, \quad v^{(1)}_{[2]} = \alpha^0_2 = \alpha^0_1 \right\}. \tag{38.20}$$

That is, the approximating restricting collections U^* and V^* for the strategies S^u and S^v coincide here with the initial restricting sets.

The differential equation for $w^{[u]}$-model that determines the accompanying points $w^{[u]0}(t, x, y, \varepsilon)$ for the y-model (38.13) is identical with the equation for $w^{[z]}$-model that determines the accompanying points $w^{[v]0}(t, x, z, \varepsilon)$ for the z-model (38.16). This differential equation has the following form

$$\dot{w}[\tau] = \sum_{r=1}^{4}\sum_{s=1}^{2} B^{[1]}(\tau)u^{(1)}_{[r]_{v^{(1)}_{[s]}}} p_{*r} q_{*s} + B^{[2]}(\tau)u^{(2)} +$$

$$+ C^{[1]}(\tau)v_* + C^{[2]}(\tau)v^*, \qquad t_0 \le \tau < \vartheta. \tag{38.21}$$

The corresponding restrictions are

$$p_{*r} \ge 0, \qquad r = 1, 2, 3, 4 \qquad \sum_{r=1}^{4} p_{*r} = 1, \tag{38.22}$$

$$q_{*s} \ge 0, \qquad s = 1, 2 \qquad \sum_{s=1}^{2} q_{*s} = 1. \tag{38.23}$$

Let us describe now the construction of the program extremum $e^{(s)}_{*(1)}(\cdot)$ for the w-model (38.21). We follow the procedures given in Section 35. As is usual in such cases, we consider the interval $[\tau_*, \ \vartheta] \subset [t_0, \ \vartheta]$ of the auxiliary time τ, and introduce the partition $\Delta = \Delta\{\tau_j\} = \{\tau_* = \tau_1, \ \tau_j < \tau_{j+1}, \ j = 1, \ldots, k, \ \tau_{k+1} = \vartheta\}$. According to Section 35 in order to construct the quantity $e^{(s)}_{*(1)}(\cdot) = e^{(s)}_{*(1)}(\tau_*, w_*, \Delta)$ we have to construct the recurrent sequence of the upper convex hulls $\varphi_j(\tau_*, m)$, $m \in G_j(\tau_*)$. In the considered case for all j the following equality holds:

$$G_j(\tau_*) = G(\tau_*)$$

where

$$G(\tau_*) = \left\{ m = \begin{bmatrix} m_1 \\ m_2 \end{bmatrix} : \ | m | \le 1 \right\}. \tag{38.24}$$

Let us note that in this case we do not need to use the parameter η (in contrast to the general constructions in Section 35). Really, now the restricting set U_η described in Section 35 coincides with the union of U^* and the set (38.9). And V_η given in Section 35 is the union of V^* and the sets (38.10), (38.11).

We begin with the calculation of the quantity $\Delta\psi(\tau_*, \tau_j, m)$. We have

$$\Delta\psi(\tau_*, \tau_j, m) = (\tau_{j+1} - \tau_j)\Big[\max_{\{q_{*s}\}, v_*, v^*} \ \min_{\{p_{*r}\}, u^{(2)}} \Big[$$

$$\langle \, m, \ \Big(\sum_{r=1}^{4} \sum_{s=1}^{2} B^{[1]}(\tau_j) u^{(1)}_{[r]_{v^{(1)}_{[s]}}} p_{*r} q_{*s} + B^{[2]}(\tau_j) u^{(2)} +$$

$$+ C^{[1]}(\tau_j)v_* + C^{[2]}(\tau_j)v^* \Big) \rangle \Big] \tag{38.25}$$

under the conditions (38.22), (38.9) and (38.23), (38.10), (38.11). In (38.25) we have $m \in G(\tau_*)$ (38.24).

We introduce the following notation:

$$\zeta_1(\tau_*, \tau_j, m) = \max_{\{q_{*s}\}} \min_{\{p_{*r}\}} \langle\ m,\ \Big(\sum_{r=1}^{4} \sum_{s=1}^{2} B^{[1]}(\tau_j) u^{(1)}_{[r]_{v^{(1)}_{[s]}}} p_{*r} q_{*s} \Big) \rangle \qquad (38.26)$$

under the conditions (38.22) and (38.23);

$$\zeta_2(\tau_*, \tau_j, m) = \min_{u^{(2)}} \langle\ m,\ B^{[2]}(\tau_j) u^{[2]}\ \rangle \qquad (38.27)$$

under the condition (38.9);

$$\zeta_3(\tau_*, \tau_j, m) = \max_{v_*} \langle\ m,\ C^{[1]}(\tau_j) v_*\ \rangle \qquad (38.28)$$

under the condition (38.10);

$$\zeta_4(\tau_*, \tau_j, m) = \max_{v^*} \langle\ m,\ C^{[2]}(\tau_j) v^*\ \rangle \qquad (38.29)$$

under the condition (38.11).

Solving in (38.27)–(38.29) the minimum or maximum problems we obtain the following equalities

$$\zeta_2(\tau_*, \tau_j, m) = -\mid m \mid\mid B^{[2]}(\tau_j) \mid, \quad \mid B^{[2]} \mid = \mid b^{[2]} \mid, \qquad (38.30)$$

$$\zeta_3(\tau_*, \tau_j, m) = \mid m \mid\mid C^{[1]}(\tau_j) \mid, \quad \mid C^{[1]} \mid = \mid c^{[1]} \mid \qquad (38.31)$$

and

$$\zeta_4(\tau_*, \tau_j, m) = \mid m \mid\mid C^{[2]}(\tau_j) \mid, \quad \mid C^{[2]} \mid = \mid c^{[2]} \mid\,. \qquad (38.32)$$

According to (38.25)–(38.32) we obtain

$$\Delta\psi(\tau_*, \tau_j, m) = (\tau_{j+1} - \tau_j)\Big[\ \zeta_1(\tau_*, \tau_j, m) + \mid m \mid \Big(\mid C^{[1]}(\tau_j) \mid +$$

$$+ \mid C^{[2]}(\tau_j) \mid - \mid B^{[2]}(\tau_j) \mid \big) \Big] . \qquad (38.33)$$

Due to the symmetry it is sufficient to construct the functions $\Delta\psi(\tau_*, \tau_j, m)$, $\psi_j(\tau_*, m)$ and $\varphi_j(\tau_*, m)$ only for the values $m \in G$, where

$$G = \left\{ m = \begin{bmatrix} m_1 \\ m_2 \end{bmatrix} : \ 0 \le m_1 \le 1, \quad 0 \le m_2 \le m_1, \quad \mid m \mid \le 1 \right\}. \qquad (38.34)$$

Therefore, in the seven other parts of the circle $\mid m \mid \le 1$ the values of the function $\varphi_j(\tau_*, m)$ are determined symmetrically.

We have to emphasize that the function $\varphi_j(\tau_*, m) = \overline{\psi}_j(\tau_*, m)$ is the upper convex hull of $\psi_j(\tau_*, m)$ with respect to the whole circle. But the constructed function $\varphi_j(\tau_*, m)$, generally speaking, is not the upper convex hull of $\psi_j(\tau_*, m)$ with respect to the sector (38.34). However, the symmetry of $\Delta\psi_j$ and ψ_j makes it possible to construct this function $\varphi_j(\tau_*, m)$ dealing only with the sector (38.34).

Let us continue to construct the functions $\Delta\psi(\tau_*, \tau_j, m), \psi_j(\tau_*, m)$ and $\varphi_j(\tau_*, m)$, $m \in G$.

We consider two cases:

$$1. \ \alpha_1^0 \ge \frac{\pi}{4}, \qquad 2. \ \alpha_1^0 < \frac{\pi}{4} \qquad (38.35)$$

where α_1^0 is the angle in (38.20).

It can be proved that in the case $\alpha_1^0 \ge \pi/4$ the minimax problem in (38.26) under the conditions (38.22), (38.23) and for $m \in G$ (38.34) has a solution

$$\zeta^{(1)}(\tau_*, \tau_j, m) = -\frac{m_1^2 + m_2^2}{m_1 + m_2} \mid B^{[1]}(\tau_j) \mid \cos\alpha_1^0, \qquad \mid B^{[1]} \mid = \mid b^{[1]} \mid . \qquad (38.36)$$

And the extremal arguments p and q are

$$p_{*1}^0 = 0, \quad p_{*2}^0 = 0, \quad p_{*3}^0 = \frac{m_1}{m_1 + m_2}, \quad p_{*4}^0 = \frac{m_2}{m_1 + m_2} \qquad (38.37)$$

$$q_{*1}^0 = \frac{m_1(\sin\alpha_1^0 + \cos\alpha_1^0) + m_2(\sin\alpha_1^0 - \cos\alpha_1^0)}{2(m_1 + m_2)\sin\alpha_1^0},$$

$$q_{*2}^0 = \frac{m_1(\sin\alpha_1^0 - \cos\alpha_1^0) + m_2(\sin\alpha_1^0 + \cos\alpha_1^0)}{2(m_1 + m_2)\sin\alpha_1^0}\,. \tag{38.38}$$

In the case $\alpha_1^0 < \dfrac{\pi}{4}$ and if the inequality

$$m_1(\cos\alpha_1^0 - \sin\alpha_1^0) \le m_2(\cos\alpha_1^0 + \sin\alpha_1^0) \tag{38.39}$$

holds we obtain the same expressions (38.36)–(38.38). But if $m \in G$ (38.34) is such that

$$m_2(\cos\alpha_1^0 + \sin\alpha_1^0) \le m_1(\cos\alpha_1^0 - \sin\alpha_1^0) \tag{38.40}$$

then we obtain

$$\zeta^{(1)}(\tau_*, \tau_j, m) = -(m_1\cos\alpha_1^0 - m_2\sin\alpha_1^0) \mid B^{[1]}(\tau_j) \mid, \tag{38.41}$$

$$p_{*1}^0 = p_{*2}^0 = p_{*4}^0 = 0, \qquad p_{*3}^0 = 1\,, \tag{38.42}$$

$$q_{*1}^0 = 1, \qquad q_{*2}^0 = 0\,. \tag{38.43}$$

For the vectors m from the other sectors of the circle $\mid m \mid \le 1$ the value $\zeta^{(1)}(\tau_*, \tau_j, m)$ and the extremal values p_{*r}^0, q_{*s}^0 are obtained symmetrically.

We have determined the function $\Delta\psi(\tau_*, \tau_j, m)$ (38.33). Now we shall construct the functions $\varphi_j(\tau_*, m)$, $j = k+1, \ldots, 1$. Here the procedure of constructing these functions is similar to the corresponding procedure given in Section 28 for the case of the pure strategies. Here also the considerations are based on a sequence of certain values m_{1j}^*. As in Section 28 we start with one of the more transparent cases. Namely, we begin with the case in which this sequence $\{m_{1j}^*\}$ turns to be monotone non-increasing in $j = 1, \ldots, k$.

In this case using the expressions for $\zeta^{(1)}(\tau_*, \tau_j, m)$ (38.36), (38.41) and for $\Delta\psi(\tau_*, \tau_j, m)$ (38.33), we obtain that the desired upper convex hulls $\varphi_j(\tau_*, m)$ of the functions $\psi_j(\tau_*, m) = \Delta\psi(\tau_*, \tau_j, m) + \varphi_{j+1}(\tau_*, m)$, $m \in G$ (38.34) can be constructed according to the following formula:

$$\varphi_j(\tau_*, m) = \varphi_{j+1}(\tau_*, m) + \Delta\varphi_j(\tau_*, m) \tag{38.44}$$

where $\Delta\varphi_j(\tau_*, m)$ is the upper convex hull of the function $\Delta\psi(\tau_*, \tau_j, m)$ (38.33), (38.36), (38.41) for the region $\mid m \mid \le 1$. According to the notation of Chapter

II (see Section 13) we have

$$\Delta\varphi_j(\tau_*, m) = \overline{\Delta\psi}(\tau_*, \tau_j, m), \qquad | m | \le 1. \tag{38.45}$$

Let us construct now the functions $\Delta\varphi_j(\tau_*, m)$ (38.45). Due to the above mentioned symmetry, we again have to consider only the region G (38.34). We emphasize again that the function $\Delta\varphi_j$ (38.45) is the upper convex hull of $\Delta\psi$ (38.45) with respect to the whole circle $| m | \le 1$ (see the similar remark on p. 280).

Let us again consider two cases (38.35). In case I we assume $\alpha_1^0 \ge \frac{\pi}{4}$ and in case II we have $\alpha_1^0 < \frac{\pi}{4}$. In its turn it is convenient to distinguish two subcases in both cases I and II. Namely, we denote by I_1 and II_1 the subcases when the inequality

$$\Big(| C^{[1]}(\tau_j) | + | C^{[2]}(\tau_j) | - \\ - | B^{[2]}(\tau_j) | - | B^{[1]}(\tau_j) | \frac{\cos\alpha_1^0}{\sqrt{2}} \Big) \ge 0 \tag{38.46}$$

is valid. And let I_2 and II_2 be the subcases when the inequality (38.46) is not valid.

Let us consider at first the case I_1, i.e., $\alpha_1^0 \ge \frac{\pi}{4}$ and the inequality (38.46) holds. In this case according to (38.36) and (38.46) the maximum of the function $\Delta\psi(\tau_*, \tau_j, m)$ (38.33) is attained on the circumference $| m | = 1$ at $m_1 = 1/\sqrt{2}$ and $m_2 = 1/\sqrt{2}$. It follows that in the region

$$0 \le m_1 \le \frac{1}{\sqrt{2}}, \qquad 0 \le m_2 \le m_1 \tag{38.47}$$

the desired function $\Delta\varphi_j(\tau_*, m)$ (38.45) is defined by the equality

$$\Delta\varphi_j(\tau_*, m) = (\tau_{j+1} - \tau_j)\Big(| C^{[1]}(\tau_j) | + | C^{[2]}(\tau_j) | - \\ - | B^{[2]}(\tau_j) | - | B^{[1]}(\tau_j) | \frac{\cos\alpha_1^0}{\sqrt{2}} \Big). \tag{38.48}$$

Now let us calculate the partial derivative of $\Delta\psi(\tau_*, \tau_j, m)$ with respect to the variable m_2 at the points that belong to the circumference $| m | = 1$. We obtain

$$\nu(\tau_*,\tau_j,m) = \Big(\frac{\partial\Delta\psi(\tau_*,\tau_j,m)}{\partial m_2}\Big)_{|m|=1} = (\tau_{j+1}-\tau_j)\Big[\Big(\mid C^{[1]}(\tau_j)\mid + \mid C^{[2]}(\tau_j)\mid -$$

$$- \mid B^{[2]}(\tau_j)\mid\Big)\sqrt{1-m_1^2} - \mid B^{[1]}(\tau_j)\mid \frac{2\sqrt{1-m_1^2}\left(\sqrt{1-m_1^2}+m_1\right)-1}{\left(m_1+\sqrt{1-m_1^2}\right)^2}\cos\alpha_1^0\Big]. \tag{38.49}$$

It can be proved that under the condition (38.46) this value is non-negative if

$$\frac{1}{\sqrt{2}} \leq m_1 \leq 1. \tag{38.50}$$

Calculate the derivative

$$\frac{d\,\Delta\psi(\tau_*,\tau_j,m)}{dm_1} = \frac{d\,\Delta\psi(\tau_*,\tau_j,\{m_1,\sqrt{1-m_1^2}\,\})}{dm_1} \tag{38.51}$$

where $m^T = \{m_1, m_2\}$, $\mid m \mid = 1$, i.e., $m_2 = \sqrt{1-m_1^2}$.

It occurs that this derivative for the values m_1 (38.50) is non-positive. Moreover, the second derivative

$$\frac{d^2\Delta\psi(\tau_*,\tau_j,m)}{(dm_1)^2} = \frac{d^2\Delta\psi(\tau_*,\tau_j,\{m_1,\sqrt{1-m_1^2}\,\})}{(dm_1)^2} \tag{38.52}$$

for the values m_1 (38.50) is non-positive, too. Hence, for the values m_1 (38.50), $\mid m \mid \leq 1$, the function $\Delta\varphi_j(\tau_*,m)$ (38.45) is determined by the following equality:

$$\Delta\varphi_j(\tau_*,m) = (\tau_{j+1}-\tau_j)\Big[\mid C^{[1]}(\tau_j)\mid + \mid C^{[2]}(\tau_j)\mid -$$

$$- \mid B^{[2]}(\tau_j)\mid - \mid B^{[1]}(\tau_j)\mid \frac{1}{m_1+\sqrt{1-m_1^2}}\cos\alpha_1^0\Big]. \tag{38.53}$$

So, in the case I_1 we have determined the function $\Delta\varphi_j(\tau_*,m)$ (38.45) in the region

$$0 \leq m_1 \leq 1, \quad 0 \leq m_2 \leq m_1, \quad \mid m \mid \leq 1. \tag{38.54}$$

Thus, in accordance with the relations (38.47), (38.48), (38.50), (38.53), the set (38.54) is divided into the parts shown in Figure 38.1

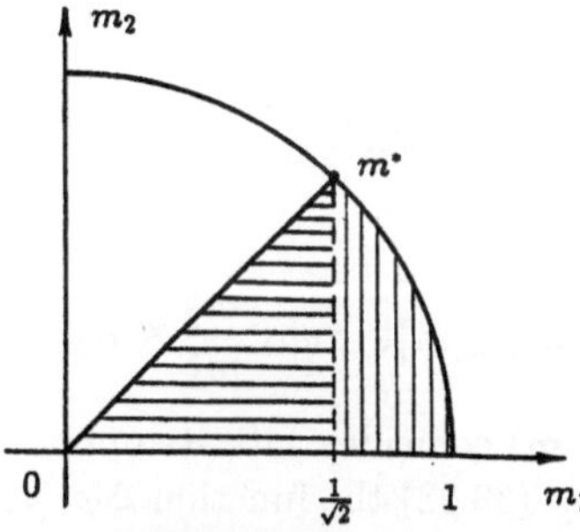

Figure 38.1

If we consider the case I_2 ($\alpha_1^0 \geq \pi/4$, the inequality (38.46) is not valid. But consider again the quantity (38.49). It can happen that the quantity (38.49) is negative in the whole interval $\dfrac{1}{\sqrt{2}} \leq m_1 < 1$. It may be if $B^{[1]}(\tau_j) = 0$. In this case we take $m_1^* = 1$ and obtain

$$\Delta\varphi_j(\tau_*, m) = \Delta\psi(\tau_*, \tau_j, m), \qquad | m | \leq 1\,. \tag{38.55}$$

Otherwise we use the following construction. Let a point $m^T = \{m_1, m_2\}$ be moving anticklockwise along the circumference $|m| = 1$ starting at the point $m_1 = 1$, $m_2 = 0$. Let m_1^* be such a value of m_1 that the quantity $\nu(\tau_*, \tau_j, m)$ (38.49) changes its sign from plus to minus when m_1 crosses m_1^* and, consequently, m_2 crosses the value

$$m_2^* = \sqrt{1 - m_1^{*2}}\,. \tag{38.56}$$

It can be proved that in the case I_2 in the situation considered now there exists a unique value m_1^* in the interval $1/\sqrt{2} < m_1 < 1$. In this interval to the left of the root m_1^* the value $(\partial\psi(\tau_*, \tau_j, m)/\partial m_2)_{|m|=1}$ is negative and to the right it is positive. For $m_1^* \leq m_1 \leq 1$, $| m |= 1$ the quantities (38.51) and (38.52) are non-positive again. Hence, in the set

$$m_1^* \leq m_1 \leq 1, \qquad | m | \leq 1 \tag{38.57}$$

the function $\Delta\varphi_j(\tau_*, m)$ is again determined by the equality (38.53).

In the region

$$0 \le m_1 \le \frac{1}{\sqrt{2}}, \qquad m_1 \frac{m_2^*}{m_1^*} \le m_2 \le m_1,$$

$$\frac{1}{\sqrt{2}} \le m_1 \le m_1^*, \qquad m_1 \frac{m_2^*}{m_1^*} \le m_2 \le \sqrt{1 - m_1^2} \tag{38.58}$$

the function $\Delta\psi(\tau_*, \tau_j, m)$ coincides with its upper convex hull.

Therefore, in the set (38.58) the function $\Delta\varphi_j(\tau_*, m)$ is determined by the equality

$$\Delta\varphi_j(\tau_*, m) = (\tau_{j+1} - \tau_j)\Big[\ | \, m \, | \ \Big(| \, C^{[1]}(\tau_j) \, | + | \, C^{[2]}(\tau_j) \, | -$$

$$- | \, B^{[2]}(\tau_j) \, | \Big) - | \, B^{[1]}(\tau_j) \, | \ \frac{m_1^2 + m_2^2}{m_1 + m_2} \cos \alpha_1^0 \Big]. \tag{38.59}$$

On the line $m_2 = m_1 \dfrac{m_2^*}{m_1^*}$ the derivative $\partial \Delta\psi(\tau_*, \tau_j, m)/\partial m_2$ is equal to zero. For $m_2 < m_1 \dfrac{m_2^*}{m_1^*}$ this derivative is non-negative. In the set

$$0 \le m_1 \le m_1^*, \qquad 0 \le m_2 \le m_1 \frac{m_2^*}{m_1^*} \tag{38.60}$$

the function $\Delta\varphi_j(\tau_*, m)$ is determined by the equality

$$\Delta\varphi_j(\tau_*, m) = (\tau_{j+1} - \tau_j) \frac{m_1}{m_1^*} \Big(\ | \, C^{[1]}(\tau_j) \, | + | \, C^{[2]}(\tau_j) \, | -$$

$$- | \, B^{[2]}(\tau_j) \, | - | \, B^{[1]}(\tau_j) \, | \ \frac{1}{m_1^* + m_2^*} \cos \alpha_1^0 \Big). \tag{38.61}$$

Thus, the function $\Delta\varphi_j(\tau_*, m)$ is determined again in the whole set (38.54). According to the given relations for $\Delta\varphi_j$, the set (38.54) is divided into the followingparts (see Figure 38.2).

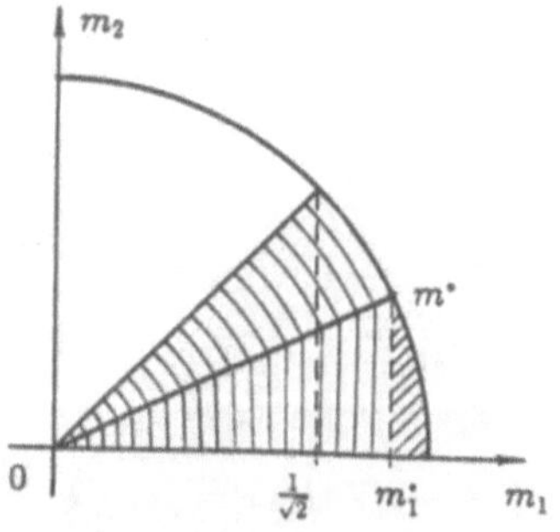

Figure 38.2

In the case I_1 the surface $\Delta\varphi = \Delta\varphi_j(\tau_*, m)$ has a form shown in Figure 38.3. In the case II_2 this surface has a form shown in Figure 38.4.

Let us consider now the case II_1 ($\alpha_1^0 < \dfrac{\pi}{4}$, and the inequality (38.46) holds). Let $\tilde{m}_1^*$ be the root of the equation

$$\sqrt{1 - m_1^2} - \frac{\cos\alpha_1^0 - \sin\alpha_1^0}{\cos\alpha_1^0 + \sin\alpha_1^0}\, m_1 = 0\,. \tag{38.62}$$

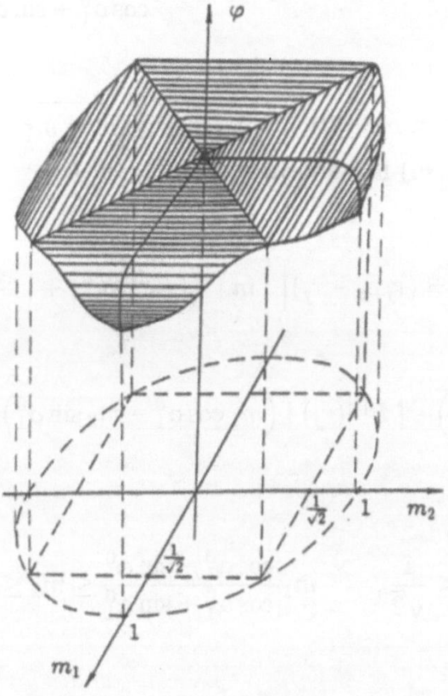

Figure 38.3

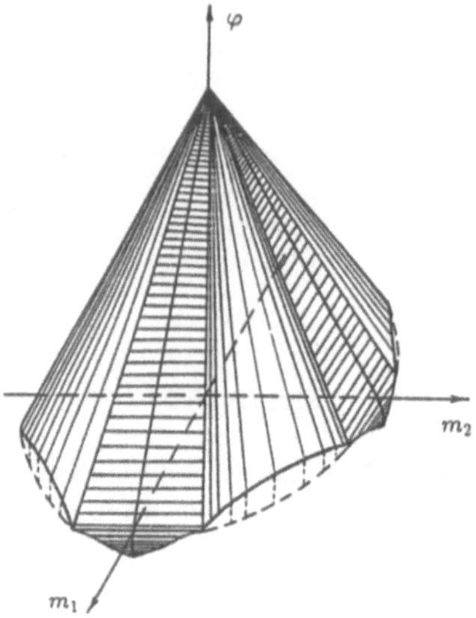

Figure 38.4

Then according to (38.40), (38.41) in the set

$$0 \le m_1 \le \tilde{m}_1^*, \qquad 0 \le m_2 \le m_1 \frac{\cos\alpha_1^0 - \sin\alpha_1^0}{\cos\alpha_1^0 + \sin\alpha_1^0},$$

$$\tilde{m}_1^* \le m_1 \le 1, \qquad 0 \le m_2 \le \sqrt{1 - m_1^2} \tag{38.63}$$

the function $\Delta\psi(\tau_*, \tau_j, m)$ has a form

$$\Delta\psi(\tau_*, \tau_j, m) = (\tau_{j+1} - \tau_j)\Big[\, |\, m \,|\, \Big(|\, C^{[1]}(\tau_j) \,| + |\, C^{[2]}(\tau_j) \,| -$$

$$- |\, B^{[2]}(\tau_j) \,|\Big) - |\, B^{[1]}(\tau_j) \,|\, \Big(m_1 \cos\alpha_1^0 - m_2 \sin\alpha_1^0\Big)\,\Big] \tag{38.64}$$

and in the set

$$0 \le m_1 \le \frac{1}{\sqrt{2}}, \qquad m_1 \frac{\cos\alpha_1^0 - \sin\alpha_1^0}{\cos\alpha_1^0 + \sin\alpha_1^0} \le m_2 \le m_1,$$

$$\frac{1}{\sqrt{2}} \le m_1 \le \tilde{m}_1^*, \qquad m_1 \frac{\cos\alpha_1^0 - \sin\alpha_1^0}{\cos\alpha_1^0 + \sin\alpha_1^0} \le m_2 \le \sqrt{1 - m_1^2} \tag{38.65}$$

this function has a form

$$\Delta\psi(\tau_*, \tau_j, m) = (\tau_{j+1} - \tau_j)\Big[\mid m \mid \Big(\mid C^{[1]}(\tau_j) \mid + \mid C^{[2]}(\tau_j) \mid -$$

$$- \mid B^{[2]}(\tau_j) \mid\Big) - \mid B^{[1]}(\tau_j) \mid \frac{\mid m \mid^2}{m_1 + m_2} \cos \alpha_1^0 \Big]. \qquad (38.66)$$

Thus, in the set (38.47) the function $\Delta\varphi_j(\tau_*, m)$ is again determined by the equality (38.48). As in the case I_1, in the considered case II_1 the quantity $(\partial\psi(\tau_*, \tau_j, m)/\partial m_2)_{|m|=1}$ is non-negative for all $m_1 \in \Big[\frac{1}{\sqrt{2}}, 1\Big]$. The derivatives (38.51) and (38.52) are non-positive for this m_1. It follows that in the set

$$\frac{1}{\sqrt{2}} \le m_1 \le \tilde{m}_1^*, \qquad \mid m \mid \le 1 \qquad (38.67)$$

the function $\Delta\varphi_j(\tau_*, m)$ is again determined by the equality (38.53), and in the set

$$\tilde{m}_1^* \le m_1 \le 1, \qquad \mid m \mid \le 1 \qquad (38.68)$$

it is determined by the equality

$$\Delta\varphi_j(\tau_*, m) = (\tau_{j+1} - \tau_j)\Big[\mid C^{[1]}(\tau_j) \mid + \mid C^{[2]}(\tau_j) \mid - \mid B^{[2]}(\tau_j) \mid -$$

$$- \mid B^{[1]}(\tau_j) \mid \Big(m_1 \cos \alpha_1^0 - \sqrt{1 - m_1^2} \sin \alpha_1^0 \Big) \Big]. \qquad (38.69)$$

Thus, the function $\Delta\varphi_j(\tau_*, m)$ is again defined in the whole set (38.54). According to the above relations for $\Delta\varphi_j$, the set (38.54) is divided into the parts depicted in Figure 38.5.

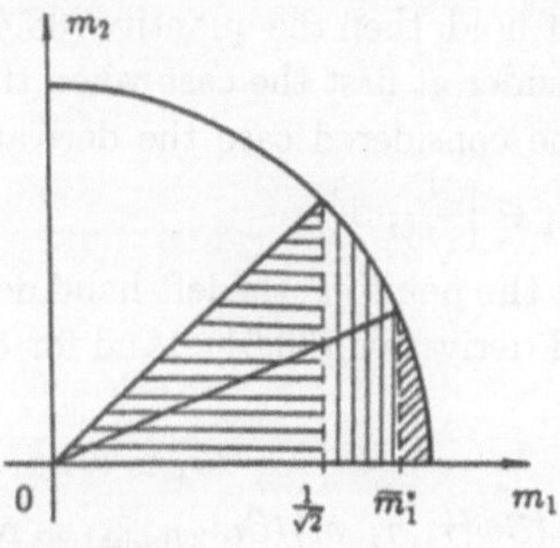

Figure 38.5

The surface $\Delta\varphi = \Delta\varphi_j(\tau_*, m)$ for the considered case is shown in Figure 38.6. The surfaces depicted in Figure 38.3 and Figure 38.6 have similar forms, though they do not coincide.

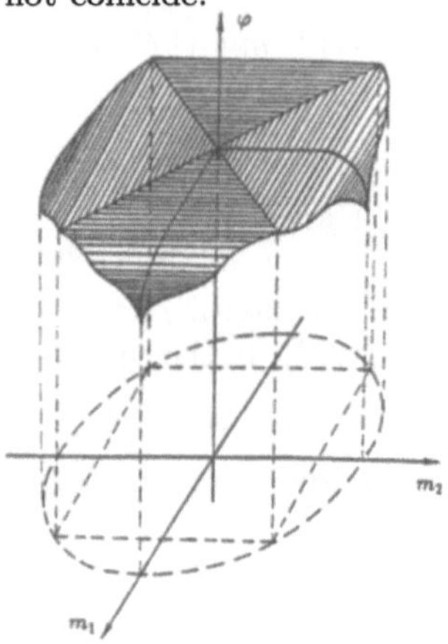

Figure 38.6

It remains only to consider the case II_2 ($\alpha_1^0 < \frac{\pi}{4}$, and the inequality (38.46) is not valid). The function $\Delta\psi(\tau_*, \tau_j, m)$ is again determined by the equalities (38.64), (38.66). Now the values of the function $\Delta\psi(\tau_*, \tau_j, m)$ in the region $0 <| m |\leq 1$ is negative. Consider the quantity (38.49) where the function $\Delta\psi(\tau_*, \tau_j, m)$ is now defined by the expressions (38.64), (38.66).

In the considered case II_2 we look for a value $m_1^* \in \left(\frac{1}{\sqrt{2}},\ 1\right)$ for which the quantity $\nu(\tau_*, \tau_j, m)$ (38.49) changes its sign from minus to plus as m_1 increases. In the interval $\tilde{m}_1^* \leq m_1 \leq 1$ the quantity (38.49) is positive. Under the imposed conditions the value m_1^* can belong only to the interval

$$\frac{1}{\sqrt{2}} < m_1 \leq \tilde{m}_1^*. \tag{38.70}$$

If this does not hold, then the quantity (38.49) for all m_1 in (38.70) is negative.

Let us consider at first the case when the value m_1^* belongs to the interval (38.70). In the considered case the derivatives (38.51) and (38.52) are non-positive for $m_1 \in \left[\frac{1}{\sqrt{2}},\ 1\right]$.

Besides, at the point $\tilde{m}_1^*$ the left-hand derivative (38.51) is not smaller than the right-hand derivative (38.51). And for the values m_1 in the interval

$$m_1^* \leq m_1 \leq \tilde{m}_1^* \tag{38.71}$$

the derivative $(\partial\psi(\tau_*, \tau_j, m)/\partial m_2)_{|m|=1}$ is non-negative. Thus, under the condition (38.71) the function $\Delta\varphi_j(\tau_*, m)$ is determined by the equality (38.53).

In the set (38.68) this function is determined by the equality (38.69); in (38.58) it is given by (38.59); in (38.60) by (38.61).

Thus, in the case II_2 and provided the root $m_1^* < \tilde{m}_1^*$ exists, the function $\Delta\varphi_j(\tau_*, m)$ is determined in the whole set (38.54). In this case in accordance with the given relations the set (38.54) is divided into the parts shown in Figure 38.7.

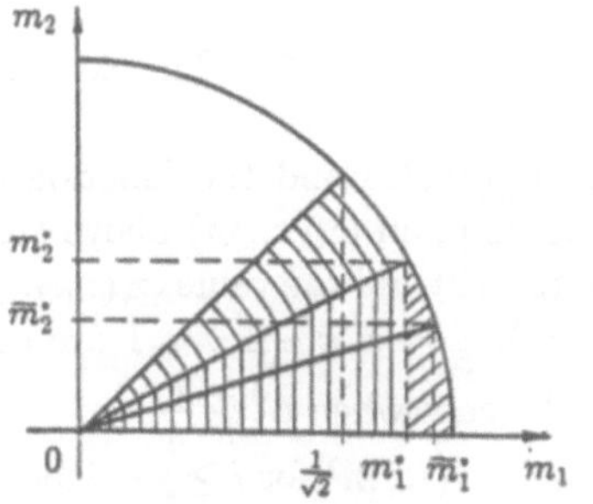

Figure 38.7

Consider now the case II_2 but under the condition that for the quantity $(\partial\Delta\psi(\tau_*, \tau_j, m)/\ \partial m_2)_{|m|=1}$ in the half interval (38.70) there exists no value m_1^*, possessing the property indicated above. Then we consider the following cases. In the first case the value (38.49) is non-positive for all m_1 in the interval $\frac{1}{\sqrt{2}} < m_1 < 1$. Then we take $m_1^* = 1$. In the second case for $\tilde{m}_1^* \leq m_1 \leq 1$ we have that the value (38.49) is non-negative. Then we take $m_1^* = \tilde{m}_1^*$. In the third case in the interval $\tilde{m}_1^* < m_1 < 1$ there exists the root m_1^*. Then this root is naturally taken for the desired value m_1^*. In these cases in accordance with the relations for $\Delta\varphi_j$ the set (38.54) is divided into the corresponding parts. We do not depict here these parts and the related surfaces $\Delta\varphi = \Delta\varphi_j$ because they have the same qualitative character as the ones considered above. In particular, one can obtain the case shown in Figure 38.8.

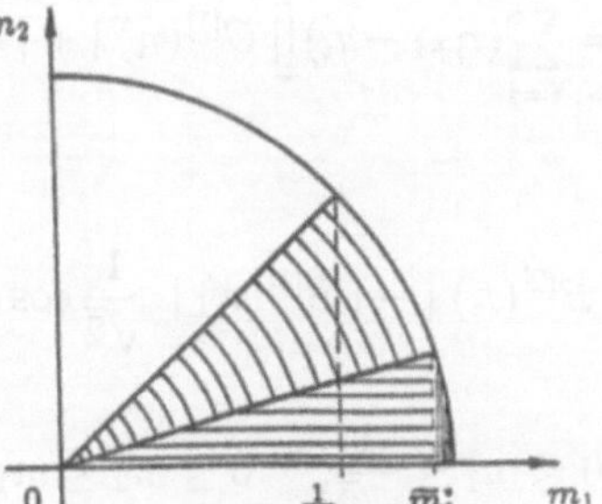

Figure 38.8

Thus the function $\Delta\varphi_j(m)$ is determined completely in the considered *monotone* case.

Let us now consider the case in which the above-formulated monotonicity condition is not valid.

In our constructions we will use the sequence of the values

$$m^*_{1i}, \quad i = k, k-1, \ldots, 1 \tag{38.72}$$

which belong to the interval $0 \leq m_1 \leq 1$. This sequence is formed recurrently in the following way.

We find the value m^*_{1k}(38.72) and the function $\varphi_k(\tau_*, m)$ the same way as the value m^*_1 and the function $\varphi_k(\tau_*, m)$ above (see p. 285). For $j = k$ we obtain the value m^*_{1k} in (38.72) and the value $\varphi_k(\tau_*, m)$. The subsequent values m^*_{1j} (38.72) and $\varphi_j(\tau_*, m)$, $j = k, k-1, \ldots, 1$ are constructed by induction. At first consider again the case when $\alpha^0_1 \geq \dfrac{\pi}{4}$.

Let the values m^*_{1i} and $\varphi_i(\tau_*, m)$ for $i \geq j+1$ be already constructed. Let us construct the values m^*_{1j} and $\varphi_j(\tau_*, m)$. Two cases are possible:

$$1^0. \quad \psi_j\Big(\tau_*, \Big\{\frac{1}{\sqrt{2}}, \frac{1}{\sqrt{2}}\Big\}\Big) \geq \varphi_{j+1}(\tau_*, \{0, 0\}) \tag{38.73}$$

$$2^0. \quad \psi_j\Big(\tau_*, \Big\{\frac{1}{\sqrt{2}}, \frac{1}{\sqrt{2}}\Big\}\Big) < \varphi_{j+1}(\tau_*, \{0, 0\}) \tag{38.74}$$

where, as usual, $\psi_j(\tau_*, m) = \Delta\psi(\tau_*, \tau_j, m) + \varphi_{j+1}(\tau_*, m)$.

In the case 1^0 (38.73) we put $m^*_{1j} = 1/\sqrt{2}$, and the function $\varphi_j(\tau_*, m)$ is constructed according to the equality

$$\varphi_j(\tau_*, m) = \sum_{i=j}^{k} (\tau_{i+1} - \tau_i)\Big[\mid C^{[1]}(\tau_i) \mid + \mid C^{[2]}(\tau_i) \mid -$$

$$- \mid B^{[2]}(\tau_i) \mid - \mid B^{[1]}(\tau_i) \mid \frac{1}{\sqrt{2}} \cos\alpha^0_1\Big] \tag{38.75}$$

if

$$0 \leq m_1 \leq \frac{1}{\sqrt{2}}, \quad 0 \leq m_2 \leq m_1$$

holds and according to the equality

$$\varphi_j(\tau_*, m) = \sum_{i=j}^{k} (\tau_{i+1} - \tau_i)\Big[\mid C^{[1]}(\tau_i) \mid + \mid C^{[2]}(\tau_i) \mid -$$

$$- \mid B^{[2]}(\tau_i) \mid - \mid B^{[1]}(\tau_i) \mid \frac{1}{m_1 + \sqrt{1 - m_1^2}} \cos \alpha_1^0 \Big] \tag{38.76}$$

if

$$\frac{1}{\sqrt{2}} \leq m_1 \leq 1, \qquad 0 \leq m_2 \leq \sqrt{1 - m_1^2}$$

is valid.

In particular, the value $\varphi_j(\tau_*, \{0,0\})$ is given by

$$\varphi_j(\tau_*, \{0,0\}) = \sum_{i=j}^{k} (\tau_{i+1} - \tau_i) \Big[\mid C^{[1]}(\tau_i) \mid + \mid C^{[2]}(\tau_i) \mid -$$

$$- \mid B^{[2]}(\tau_i) \mid - \mid B^{[1]}(\tau_i) \mid \frac{1}{\sqrt{2}} \cos \alpha_1^0 \Big] . \tag{38.77}$$

In the case 2^0 (38.74) we take some arbitrary value m_1 in the interval $1/\sqrt{2} \leq m_1 \leq 1$. For this m_1 we determine τ_l as the smallest number among the moments τ_s, $s \geq j+1$, for which $m_{1s}^* < m_1$. We consider the equation

$$\Big(\partial \Big(\sum_{i=j}^{l-1} (\tau_{i+1} - \tau_i) \Big[\Big(\mid C^{[1]}(\tau_i) \mid + \mid C^{[2]}(\tau_i) \mid - \mid B^{[2]}(\tau_i) \mid \Big) *$$

$$* \sqrt{m_1^2 + m_2^2} \mid B^{[1]}(\tau_i) \mid \frac{m_1^2 + m_2^2}{m_1 + m_2} \cos \alpha_1^0 \Big] \Big) \Big/ \partial m_2 \Big)_{|m|=1} = 0 . \tag{38.78}$$

For the chosen m_1, let the equation (38.78) have a root $m_2^* = \sqrt{1 - m_1^{*2}}$. Thus we have constructed a correspondence between m_1 and m_1^*. We need to find a value m_1 such that it coincides with the corresponding m_1^*. Let the value $m_1 = m_1^*$ be found. Let a point $m = \{m_1, m_2\}$, $|m| = 1$, be moving anticlockwise starting at the point $\{1, 0\}$. Let the quantity

$$\nu^*(\tau_*, \tau_j, m) = \partial \Big(\sum_{i=j}^{l-1} (\tau_{i+1} - \tau_i) \Big[\Big(\mid C^{[1]}(\tau_i) \mid + \mid C^{[2]}(\tau_i) \mid - \mid B^{[2]}(\tau_i) \mid \Big) *$$

$$* \sqrt{m_1^2 + m_2^2} \mid B^{[1]}(\tau_i) \mid \frac{m_1^2 + m_2^2}{m_1 + m_2} \cos \alpha_1^0 \Big] \Big) \Big/ \partial m_2 \tag{38.79}$$

change its sign from plus to minus when m_1 decreases and crosses the value m_1^*. Then we take $m_{1j}^* = m_1^*$. If for the chosen m_1 there is no value τ_s for which the conditions $s \geq j+1$ and $m_{1s}^* < m_1$ are valid, then we assume $l = k+1$. If in the interval $1/\sqrt{2} \leq m_1 \leq 1$ there exists no value m_1 that satisfies all the above-formulated conditions, then we take $m_{1j}^* = 1/\sqrt{2}$ if in the considered interval $\dfrac{\partial(\cdot)}{\partial m_2} > 0$ or $m_{1i}^* = 1$ if $\dfrac{\partial(\cdot)}{\partial m_2} < 0$ in this interval. In the case (38.74) we have

$$\varphi_j(\tau_*, \{0,0\}) = \varphi_{j+1}(\tau_*, \{0,0\}) \,. \tag{38.80}$$

Thus, the sequence of the required values m_{1j}^* and $\varphi_j(\tau_*, \{0,0\})$ for all $j = 1, \ldots, k$ is constructed by induction. The desired function $\varphi_j(\tau_*, m)$ is constructed recurrently by the following rule: Assume first that the inequality

$$m_{1j}^* > \frac{1}{\sqrt{2}} \tag{38.81}$$

is valid.

We assume

$$\varphi_j(\tau_*, m) = \varphi_{j+1}(\tau_*, m) + (\tau_{j+1} - \tau_j)\Big[\Big(\mid C^{[1]}(\tau_j) \mid + \mid C^{[2]}(\tau_j) \mid -$$

$$- \mid B^{[2]}(\tau_j) \mid\Big)\sqrt{m_1^2 + m_2^2} - \mid B^{[1]}(\tau_j) \mid \frac{m_1^2 + m_2^2}{m_1 + m_2} \cos\alpha_1^0\Big] \tag{38.82}$$

for

$$0 \leq m_1 \leq \frac{1}{\sqrt{2}}, \qquad m_1 \frac{m_{2j}^*}{m_{1j}^*} \leq m_2 \leq m_1,$$

$$\frac{1}{\sqrt{2}} \leq m_1 \leq m_{1j}^*, \qquad m_1 \frac{m_{2j}^*}{m_{1j}^*} \leq m_2 \leq \sqrt{1 - m_1^2}; \tag{38.83}$$

and

$$\varphi_j(\tau_*, m) = \varphi_{j+1}(\tau_*, m) + (\tau_{j+1} - \tau_j)\Big[\mid C^{[1]}(\tau_j) \mid + \mid C^{[2]}(\tau_j) \mid -$$

$$- \mid B^{[2]}(\tau_j) \mid - \mid B^{[1]}(\tau_j) \mid \frac{1}{m_{1j}^* + m_{2j}^*} \cos\alpha_1^0\Big] \frac{m_1}{m_{1j}^*} \tag{38.84}$$

for

$$0 \leq m_1 \leq m_{1j}^*, \qquad 0 \leq m_2 \leq m_1 \frac{m_{2j}^*}{m_{1j}^*}; \tag{38.85}$$

and

$$\varphi_j(\tau_*, m) = \sum_{i=j}^{k-1}(\tau_{i+1} - \tau_i)\Big[\mid C^{[1]}(\tau_i) \mid + \mid C^{[2]}(\tau_i) \mid -$$

$$- \mid B^{[2]}(\tau_i) \mid - \mid B^{[1]}(\tau_i) \mid \frac{1}{m_1 + \sqrt{1 - m_1^2}} \cos \alpha_1^0\Big] \tag{38.86}$$

for

$$m_{1j}^* \le m_1 \le 1, \qquad 0 \le m_2 \le \sqrt{1 - m_1^2}. \tag{38.87}$$

In the case $m_{1j}^* = 1$ the function $\varphi_j(\tau_*, m)$ is determined by the equality $\varphi_j(\tau_*, m) = \psi_j(\tau_*, m)$ in the whole set $\mid m \mid \le 1$.

In the case $m_{1j}^* = 1/\sqrt{2}$ in the set (38.81) the value $\varphi_j(\tau_*, m)$ is determined again by the expression (38.86). In the set $0 \le m_1 \le 1/\sqrt{2},\ 0 \le m_2 \le m_1$ the function $\varphi_j(\tau_*, m)$ is calculated in the following way.

If

$$\psi_j(\tau_*, \{1/\sqrt{2},\ 1/\sqrt{2}\}) \ge \varphi_j(\tau_*, \{0, 0\}) \tag{38.88}$$

then

$$\varphi_j(\tau_*, \{m_1, m_2\}) = \psi_j(\tau_*, \{1/\sqrt{2},\ 1/\sqrt{2}\}) \tag{38.89}$$

otherwise $\varphi_j(\tau_*, m)$ is calculated by the formula (38.84).

In the case $\alpha_1^0 \ge \dfrac{\pi}{4}$ the construction of the functions $\varphi_j(\tau_*, m)$ is described.

Let us consider now the case $\alpha_1^0 < \dfrac{\pi}{4}$.

We again denote by II_1 the case when $\alpha_1^0 < \dfrac{\pi}{4}$, and the inequality (38.46) is valid and denote by II_2 the case when $\alpha_1^0 < \dfrac{\pi}{4}$ and the inequality (38.46) is not valid. Here all the constructions for the similar cases considered above are repeated. The only difference is the following. Let the number $\widetilde{m}_1^*$ be determined again by the equality (38.62), and m_1^* be the number defined above (see p. 285). If m_1^* does not belong to the interval $[\widetilde{m}_1^*,\ 1]$, in the expressions corresponding to the intervals for m_1 with the right-hand end $m_1 = 1$ this upper value is substituted by the number $\widetilde{m}_1^*$. In addition, in the set

$$\widetilde{m}_1 \le m_1 \le 1, \qquad 0 \le m_2 \le \sqrt{1 - m_1^2} \tag{38.90}$$

the function $\varphi_1(\tau_*, m)$ is determined by the equality

$$\varphi_1(\tau_*, m) = \sum_{i=1}^{k}(\tau_{i+1} - \tau_i)\Big[\mid C^{[1]}(\tau_i) \mid + \mid C^{[2]}(\tau_i) \mid -$$

$$-\mid B^{[2]}(\tau_i)\mid - \mid B^{[1]}(\tau_i)\mid (m_1 \cos\alpha_1^0 - \sqrt{1-m_1^2}\sin\alpha_1^0)\Big]. \tag{38.91}$$

The construction of the function $\kappa(\tau_*, m, \Delta) = \varphi_1(\tau_*, m)$ is thus described completely.

Let us now describe the construction of the approximating optimal mixed strategies S_a^u and S_a^v. According to what has been said in this section and in Sections 2 and 36, we have

$$S_a^u = \Big\{U,\ p^0(\cdot),\ u^{(2)0}(\cdot);\ p_y^{*a}(\cdot),\ V,\ q_y^{*0}(\cdot); u_y^{*(2)a}(\cdot),\ v_{*y}^0(\cdot),\ v_y^{*0}(\cdot)\Big\}, \tag{38.92}$$

$$S_a^v = \Big\{V, q^0(\cdot),\ v_*^0(\cdot),\ v^{*0}(\cdot); U,\ p_z^{*0}(\cdot); q_z^{*a}(\cdot);\ u_z^{(2)0}(\cdot),\ v_{*z}^a(\cdot),\ v_z^{*a}(\cdot)\Big\} \tag{38.93}$$

where U and V are the collections (38.19) and (38.20) of the vectors $u_{[r]}^{(1)}$, $r = 1,2,3,4$ and of the angles $v_{[s]}^{(1)}$, $v_{[1]}^{(1)} = -\alpha_1^0$, $v_{[2]}^{(1)} = \alpha_2^0$. The other quantities which determine the strategy S_a^u (38.92) have the following form:

$$p^0(\cdot) = p^0(t,x,y,\varepsilon) = \Big\{p_r^0(t,x,y,\varepsilon) \geq 0,\ \sum_{r=1}^{4} p_r^0 = 1\Big\}, \tag{38.94}$$

$$u^{(2)0}(\cdot) = u^{(2)0}(t,x,y,\varepsilon), \tag{38.95}$$

$$p_y^{*a}(\cdot) = p_y^{*a}(t,x,y,\varepsilon) = \Big\{p_{yr}^{*a}(t,x,y,\varepsilon) \geq 0,\ \sum_{r=1}^{4} p_{yr}^{*a} = 1\Big\}, \tag{38.96}$$

$$q_y^{*0}(\cdot) = q_y^{*0}(t,x,y,\varepsilon) = \Big\{q_{ys}^{*0}(t,x,y,\varepsilon) \geq 0,\ \sum_{s=1}^{2} q_{ys}^{*0} = 1\Big\}, \tag{38.97}$$

$$u_y^{*(2)a}(\cdot) = u_y^{*(2)a}(t,x,y,\varepsilon), \qquad v_{*y}^0(\cdot) = v_{*y}^0(t,x,y,\varepsilon),$$
$$v_y^{*0}(\cdot) = v_y^{*0}(t,x,y,\varepsilon). \tag{38.98}$$

For S_a^v(38.93) we have

$$v_*^0(\cdot) = v_*^0(t,x,z,\varepsilon), \qquad v^{*0}(\cdot) = v^{*0}(t,x,z,\varepsilon) \tag{38.99}$$

and the other quantities have a form similar to the corresponding quantities in (38.95)–(38.99) if we substitute the symbol y by z.

Let us calculate the quantities (38.95)–(38.99).

Let the initial position $\{t_*, x_*\}$ of x-system (38.1)–(38.11), the parameter $\varepsilon > 0$, and the partition $\Delta\{t_i\} = \{t_* = t_1,\ t_i < t_{i+1},\ i = 1, \dots, k,\ t_{k+1} = \vartheta\}$ be chosen. Consider the current time moment $t_i \in \Delta\{t_i\}$. Let $\{t_i, x[t_i]\}$ be the current position of the x-system (38.1)–(38.11). Assume that we know the current positions $\{t_i, y[t_i]\}$ of the y-model-leader (38.13) and $\{t_i, z[t_i]\}$ of the z-model-leader (38.16). In particular, for $i = 1$ we have the given position $\{t_1, x[t_1]\} = \{t_*, x_*\}$ and we can take $\{t_1, y[t_1]\} = \{t_1, z[t_1]\} = \{t_1, x[t_1]\}$.

Consider at first the case $\alpha_1^0 \geq \frac{\pi}{4}$ where α_1^0 is the angle in (38.20).

The collections of numbers $p^0(t_i, x[t_i], y[t_i], \varepsilon) = \{p_r^0(t_i, x[t_i], y[t_i], \varepsilon) \geq 0, \sum_{r=1}^4 p_r^0 = 1\}$ is determined in the following way. Let

$$s_y[t_i] = x[t_i] - y[t_i]\,. \tag{38.100}$$

The optimal probabilities $p_r^0 = p_r^0(t_i, x[t_i], y[t_i], \varepsilon)$ are determined as the solutions of the following problem:

$$\max_{\{q_s\}} \sum_{r=1}^{4} \sum_{s=1}^{2} \langle\, s_y[t_i],\ B^{[1]}(t_i) u^{(1)}_{[r]_{v^{(1)}_{[s]}}} \,\rangle p_r^0 q_s =$$

$$= \min_{\{p_r\}} \max_{\{q_s\}} \sum_{r=1}^{4} \sum_{s=1}^{2} \langle\, s_y[t_i],\ B^{[1]}(t_i) u^{(1)}_{[r]_{v^{(1)}_{[s]}}} \,\rangle p_r q_s \tag{38.101}$$

under the conditions

$$p_r \geq 0\,, \quad \sum_{r=1}^{4} p_r = 1\,; \quad q_s \geq 0\,, \quad \sum_{s=1}^{2} q_s = 1\,. \tag{38.102}$$

It can be shown that in the case $\alpha_1^0 \geq \frac{\pi}{4}$ the desired probabilities p_r^0, $r = 1, 2, 3, 4$ are given by the following expressions (see Table 1).

$s_1 = s_{1y}[t_i] \geq 0$ $s_2 = s_{2y}[t_i] > 0$	$p_3^0 = \dfrac{\mid s_1 \mid}{\mid s_1 \mid + \mid s_2 \mid}, \quad p_4^0 = 1 - p_3^0, \quad p_1^0 = p_2^0 = 0$
$s_1 < 0$ $s_2 \geq 0$	$p_1^0 = \dfrac{\mid s_1 \mid}{\mid s_1 \mid + \mid s_2 \mid}, \quad p_4^0 = 1 - p_1^0, \quad p_2^0 = p_3^0 = 0$
$s_1 \leq 0$ $s_2 < 0$	$p_1^0 = \dfrac{\mid s_1 \mid}{\mid s_1 \mid + \mid s_2 \mid}, \quad p_2^0 = 1 - p_1^0, \quad p_3^0 = p_4^0 = 0$
$s_1 > 0$ $s_2 \leq 0$	$p_3^0 = \dfrac{\mid s_1 \mid}{\mid s_1 \mid + \mid s_2 \mid}, \quad p_2^0 = 1 - p_3^0, \quad p_1^0 = p_4^0 = 0$
$s_1 = 0, \quad s_2 = 0$	$p_r^0 = \dfrac{1}{4}, \quad r = 1, \cdots, 4$

Table 1.

According to the general control scheme described in Sections 32 to 36 the control vector $u^0[t] = u^0[t_i],\ t_i \leq t < t_{i+1}$ is determined as the result of the random test for the sampling of the vector $u[t_i] = u_{[r]}^{(1)}$ among the vectors $u_{[r]}^{(1)},\ r = 1, 2, 3, 4$ under the assumption

$$P\Big(\ u^0[t_i] = u_{[r]}^{(1)}\ \Big) = p_r^0 = p_r^0(t_i, x[t_i], y[t_i], \varepsilon) \tag{38.103}$$

where $P\Big(\ \ldots\ \Big)$ is the probability of the corresponding event. This part of the control algorithm will be described in more detail at the end of this section.

The control action $u^{(2)0}[t] = u^{(2)0}[t_i],\ t_i \leq t < t_{i+1}$ is determined by the function $u^{(2)0}(t, x, y, \varepsilon)$ (38.92), which is constructed in the following way. For t_i and $s_y[t_i]$ (38.100) the value $u^{(2)0}[t_i] = u^{(2)0}(t_i, x[t_i], y[t_i], \varepsilon)$ is a solution of the following problem:

$$\langle\ s_y[t_i],\ B^{[2]}(t_i)u^{(2)0}[t_i]\ \rangle = \min_{u^{(2)}} \langle\ s_y[t_i],\ B^{[2]}(t_i)u^{(2)}\ \rangle \tag{38.104}$$

under the condition (38.9), i.e., $\mid u^{(2)} \mid \leq 1$.

It follows from (38.104), (38.4) that the control action $u^{(2)0}[t_i]$ is determined by the formula

$$u^{(2)0}[t_i] = -\frac{s_y[t_i]}{\mid s_y[t_i] \mid} \text{ sign } b^{[2]}(t_i) \tag{38.105}$$

if $\mid s_y[t_i] \mid \neq 0$, and we choose $u^{(2)0}[t_i] = 0$ if $\mid s_y[t_i] \mid = 0$.

The control actions $p_y^{*a} = p_y^{*a}(t_i, x[t_i], y[t_i], \varepsilon)$ and $q_y^{*0} = q_y^{*0}(t_i, x[t_i], y[t_i], \varepsilon)$, which determine the motion of y-model-leader (38.13), are calculated in the following way. Let for the time moment $\tau_* = t_i$ the function $\varphi_1(\tau_*, m)$, $|m| \le 1$ be constructed as described above in this section. And for the time interval $[t_i, \ \vartheta]$ we have the corresponding partition $\Delta\{\tau_j\} = \{\tau_1 = t_i, \ \tau_j < \tau_{j+1}, \ j = 1, \ldots, k, \ \tau_{k+1} = \vartheta\}$. It is convenient to take $\Delta\{\tau_j\} = \Delta\{t_s\}$, i.e., $s = j + i - 1$, $j = 1, \ldots, k+1$, $s = i, \ldots, k+i$. We suppose that the step of the partition is sufficiently small. Further, let the maximizing vector $m_y^0 = m_y^0(t_i, x[t_i], y[t_i], \varepsilon)$ (see Figure 38.9) be determined as the solution of the problem (36.10) for $y = y[t_i]$. Then the numbers $p_{yr}^{*a} = p_{yr}^{*a}(t_i, x[t_i], y[t_i], \varepsilon)$ are the solutions of the problem (38.101), where we substitute $s_y[t_i] = m_y^0$ and $p_r^0 = p_{yr}^{*0}$. The required optimal numbers are given in Table 2.

$m_{y1}^0 = m_1^0 \ge 0$, $m_{y2}^0 = m_2^0 > 0$	$p_{y3}^{*a} = \dfrac{\mid m_1^0 \mid}{\mid m_1^0 \mid + \mid m_2^0 \mid}$, $p_{y4}^{*a} = 1 - p_{y3}^{*a}$, $p_{y1}^{*a} = p_{y2}^{*a} = 0$
$m_1^0 < 0$, $m_2^0 \ge 0$	$p_{y1}^{*a} = \dfrac{\mid m_1^0 \mid}{\mid m_1^0 \mid + \mid m_2^0 \mid}$, $p_{y4}^{*a} = 1 - p_{y1}^{*a}$, $p_{y2}^{*a} = p_{y3}^{*a} = 0$
$m_1^0 \le 0$, $m_2^0 < 0$	$p_{y1}^{*a} = \dfrac{\mid m_1^0 \mid}{\mid m_1^0 \mid + \mid m_2^0 \mid}$, $p_{y2}^{*a} = 1 - p_{y1}^{*a}$, $p_{y3}^{*a} = p_{y4}^{*a} = 0$
$m_1^0 > 0$, $m_2^0 \le 0$	$p_{y3}^{*a} = \dfrac{\mid m_1^0 \mid}{\mid m_1^0 \mid + \mid m_2^0 \mid}$, $p_{y2}^{*a} = 1 - p_{y3}^{*a}$, $p_{y1}^{*a} = p_{y4}^{*a} = 0$
$m_1^0 = 0, \ m_2^0 = 0$	$p_{yr}^a = \dfrac{1}{4}$, $r = 1, \ldots, 4$

Table 2.

The numbers $q_{ys}^{*0} = q_{ys}^{*0}(t_i, x[t_i], y[t_i], \varepsilon)$ are determined as the solutions of the problem

$$\min_{\{p_r\}} \sum_{r=1}^{4} \sum_{s=1}^{2} \langle \ s_y[t_i], \ B^{[1]}(t_i) u_{[r]_{v_{[s]}^{(1)}}}^{(1)} \ \rangle p_r q_s^{*0} =$$

$$= \max_{\{q_s\}} \min_{\{p_r\}} \sum_{r=1}^{4} \sum_{s=1}^{2} \langle \ s_y[t_i], \ B^{[1]}(t_i) u_{[r]_{v_{[s]}^{(1)}}}^{(1)} \ \rangle p_r q_s \tag{38.106}$$

under the conditions (32.3), (32.4). Here $s_y[t_i]$ is the vector (38.100). The numbers q_{yi}^{*0} are presented in Table 3.

$s_1 \geq 0$ $s_2 > 0$	$q_{y1}^{*0} = \dfrac{s_1(\sin\alpha_1^0 + \cos\alpha_1^0) + s_2(\sin\alpha_1^0 - \cos\alpha_1^0)}{2(s_1+s_2)\sin\alpha_1^0}, \quad q_{y2}^{*0} = 1 - q_{y1}^{*0}$	
$s_1 < 0$ $s_2 \geq 0$	$q_{y2}^{*0} = \dfrac{s_1(\sin\alpha_1^0 + \cos\alpha_1^0) + s_2(\sin\alpha_1^0 - \cos\alpha_1^0)}{2(s_1+s_2)\sin\alpha_1^0}, \quad q_{y1}^{*0} = 1 - q_{y2}^{*0}$	
$s_1 \leq 0$ $s_2 < 0$	$q_{y1}^{*0} = \dfrac{s_1(\sin\alpha_1^0 + \cos\alpha_1^0) + s_2(\sin\alpha_1^0 - \cos\alpha_1^0)}{2(s_1+s_2)\sin\alpha_1^0}, \quad q_{y2}^{*0} = 1 - q_{y1}^{*0}$	
$s_1 \geq 0$ $s_2 < 0$	$q_{y2}^{*0} = \dfrac{s_1(\sin\alpha_1^0 + \cos\alpha_1^0) + s_2(\sin\alpha_1^0 - \cos\alpha_1^0)}{2(s_1+s_2)\sin\alpha_1^0}, \quad q_{y1}^{*0} = 1 - q_{y2}^{*0}$	
$s_1 = 0, \quad s_2 = 0$		$q_{y1}^{*0} = q_{y2}^{*0} = \dfrac{1}{2}$

Table 3.

The vector $u_y^{*(2)a}[t_i] = u_y^{*(2)}[t_i, x[t_i], y[t_i], \varepsilon)$ is calculated as the solution of the problem

$$\langle\ m_y^0,\ B^{[2]}(t_i)u_y^{*(2)a}\ \rangle = \min_{|u_y^{*(2)}|\leq 1} \langle\ m_y^0,\ B[2](t_i)u_y^{*(2)}\ \rangle\,. \tag{38.107}$$

We note that the actions $u_y^{*(2)a}[t] = u_y^{*(2)a}[t_i],\ t_i \leq t < t_{i+1}$ provide the closeness between the motion of the y-model-leader and some imaginary motion of the accompanying w-model (see the constructions in Chapters I, II for the corresponding x-object and w-model).

The actions $v_{*y}^0[t_i] = v_{*y}^0(t_i, x[t_i], y[t_i], \varepsilon)$ and $v_y^{*0}[t_i] = v_y^{*0}(t_i, x[t_i], y[t_i], \varepsilon)$ are determined as the solutions of the following problems:

$$\langle\ s_y[t_i],\ C^{[1]}(t_i)v_{*y}^0[t_i]\ \rangle = \max_{|v_*|\leq 1} \langle\ s_y[t_i],\ C^{[1]}(t_i)v_*\ \rangle \tag{38.108}$$

and

$$\langle\ s_y[t_i],\ C^{[2]}(t_i)v_y^{*0}[t_i]\ \rangle = \max_{|v^*|\leq 1} \langle\ s_y[t_i],\ C^{[2]}(t_i)v^*\ \rangle \tag{38.109}$$

where $s_y[t_i]$ is a vector (38.100).

It follows from (38.5), (38.10), (38.109) that the actions $v^0_*[t_i]$ and $v^{*0}[t_i]$ are determined by the formulas

$$v^0_{*y}[t_i] = \frac{s_y[t_i]}{|\, s_y[t_i] \,|} \operatorname{sign} c^{[i]}(t_i) \tag{38.110}$$

$$v^{*0}_y[t_i] = \frac{s_y[t_i]}{|\, s_y[t_i] \,|} \operatorname{sign} c^{[i]}(t_i) \tag{38.111}$$

for $|\, s_y[t_i] \,| \neq 0$; and for $|\, s_y[t_i] \,| = 0$ we suppose

$$v^0_{*y}[t_i] = 0, \qquad v^{*0}_y[t_i] = 0\,. \tag{38.112}$$

Thus, the algorithm of the construction of the optimal control actions that correspond to the approximating optimal strategy S^u_a (38.92) is completely described in the case $\alpha^0_1 \geq \dfrac{\pi}{4}$.

Now let us describe the construction of the actions $v^0[t]$ that correspond to the approximating optimal mixed strategy S^v_a (38.93). We assume again that $\alpha^0_1 \geq \dfrac{\pi}{4}$. In accordance with (38.93) the second, third and fourth components of the strategy S^v_a are

$$q^0(\cdot) = q^0(t,x,z,\varepsilon) = \Big\{ q^0_s(t,x,z,\varepsilon) \geq 0,\ \sum_{s=1}^{2} q^0_s = 1 \Big\}; \tag{38.113}$$

$$v^0_*(\cdot) = v^0_*(t,x,z,\varepsilon), \qquad v^{*0}(\cdot) = v^{*0}(t,x,z,\varepsilon)\,. \tag{38.114}$$

The other components of S^v_a are determined similarly to the corresponding quantities in (38.92) if we substitute the symbol y by z.

Suppose

$$s_z[t_i] = z[t_i] - x[t_i]\,. \tag{38.115}$$

The action $v^{(1)0}[t] = v^{(1)0}[t_i]$, $t_i \leq t < t_{i+1}$ for the x-system (38.1) is obtained as a result of the random test for sampling the angle $v^{(1)}_{[r]}$ among the two admissible values $v^{(1)}_{[1]} = -\alpha^0_1$ or $v^{(1)}_{[2]} = \alpha^0_2$. The probabilities $q^0_s = q^0_s(t_i, x[t_i], z[t_i], \varepsilon)$ that determine this test are calculated as the solutions of the problem

$$\min_{p_r} \sum_{r=1}^{4} \sum_{s=1}^{2} \langle\, s_z[t_i],\ B^{[1]}(t_i) u^{(1)}_{[r]_{v^{(1)}_{[s]}}} \,\rangle p_r q^0_s =$$

$$\max_{\{q_s\}} \min_{\{p_r\}} \sum_{r=1}^{4} \sum_{s=1}^{2} \langle \; s_z[t_i], \; B^{[1]}(t_i) u^{(1)}_{[r]_{v^{(1)}_{[s]}}} \; \rangle p_r q_s \tag{38.116}$$

under the conditions (32.3), (32.4). It follows from (38.38) that in the considered case $\alpha_1^0 \geq \frac{\pi}{4}$ the optimal values q_s^0 are the numbers presented in Table 4.

$s_{z1} \geq 0$ $s_{z2} > 0$	$q_1^0 = \dfrac{s_{z1}(\sin\alpha_1^0 + \cos\alpha_1^0) + s_{z2}(\sin\alpha_1^0 - \cos\alpha_1^0)}{2(s_{z1}+s_{z2})\sin\alpha_1^0}, \quad q_2^0 = 1 - q_1^0$	
$s_{z1} < 0$ $s_{z2} \geq 0$	$q_2^0 = \dfrac{s_{z1}(\sin\alpha_1^0 + \cos\alpha_1^0) + s_{z2}(\sin\alpha_1^0 - \cos\alpha_1^0)}{2(s_{z1}+s_{z2})\sin\alpha_1^0}, \quad q_1^0 = 1 - q_2^0$	
$s_{z1} \leq 0$ $s_{z2} < 0$	$q_1^0 = \dfrac{s_{z1}(\sin\alpha_1^0 + \cos\alpha_1^0) + s_{z2}(\sin\alpha_1^0 - \cos\alpha_1^0)}{2(s_{z1}+s_{z2})\sin\alpha_1^0}, \quad q_2^0 = 1 - q_1^0$	
$s_{z1} \geq 0$ $s_{z2} < 0$	$q_2^0 = \dfrac{s_{z1}(\sin\alpha_1^0 + \cos\alpha_1^0) + s_{z2}(\sin\alpha_1^0 - \cos\alpha_1^0)}{2(s_{z1}+s_{z2})\sin\alpha_1^0}, \quad q_1^0 = 1 - q_2^0$	
$s_{z1} = 0, \quad s_{z2} = 0$		$q_1^0 = q_2^0 = \dfrac{1}{2}$

Table 4.

The optimal disturbances $v_*^0[t] = v_*^0[t_i], \; t_i \leq t < t_{i+1}$ and $v^{*0}[t] = v^{*0}[t_i], \; t_i \leq t < t_{i+1}$ are determined as the solutions of the problems

$$\langle \; s_z[t_i], \; C^{[1]}(t_i) v_*^0[t_i] \; \rangle = \max_{|v_*| \leq 1} \langle \; s_z[t_i], \; C^{[1]}(t_i) v_* \; \rangle \tag{38.117}$$

and

$$\langle \; s_z[t_i], \; C^{[2]}(t_i) v^{*0}[t_i] \; \rangle = \max_{|v^*| \leq 1} \langle \; s_z[t_i], \; C^{[2]}(t_i) v^* \; \rangle \tag{38.118}$$

where $s_z[t_i]$ is the vector (38.115).

Thus, the optimal actions according to (38.117), (38.118) are

$$v_*^0[t_i] = \frac{s_z[t_i]}{|\; s_z[t_i] \;|} \; \text{sign } c^{[1]}(t_i), \qquad |\; s_z[t_i] \;| \neq 0 \tag{38.119}$$

$$v^{*0}[t_i] = \frac{s_z[t_i]}{|\; s_z[t_i] \;|} \; \text{sign } c^{[2]}(t_i), \qquad |\; s_z[t_i] \;| \neq 0 \tag{38.120}$$

and we assume $v_*^0[t_i] = v^{*0}[t_i] \equiv 0$ if $| s_z[t_i] |= 0$.

The actions $p_{zr}^{*0} = p_{zr}^{*0}(t_i, x[t_i], z[t_i], \varepsilon)$ and $q_{zs}^{*a} = q_{zs}^{*a}(t_i, x[t_i], z[t_i], \varepsilon)$ are determined as the solution of the corresponding extremal problems

$$\max_{\{q_s\}} \sum_{r=1}^{4}\sum_{s=1}^{2} \langle\, s_z[t_i],\ B^{[1]}(t_i) u^{(1)}_{[r]_{v^{(1)}_{[s]}}}\ \rangle p_{zr}^{*0} q_s =$$

$$= \min_{\{p_r\}} \max_{\{q_s\}} \sum_{r=1}^{4}\sum_{s=1}^{2} \langle\, s_z[t_i],\ B^{[1]}(t_i) u^{(1)}_{[r]_{v^{(1)}_{[s]}}}\ \rangle p_r q_s \tag{38.121}$$

and

$$\min_{\{p_r\}} \sum_{r=1}^{4}\sum_{s=1}^{2} \langle\, m_z^0,\ B^{[1]}(t_i) u^{(1)}_{[r]_{v^{(1)}_{[s]}}}\ \rangle p_r q_{zs}^{*a} =$$

$$= \max_{\{q_s\}} \min_{\{p_r\}} \sum_{r=1}^{4}\sum_{s=1}^{2} \langle\, m_z^0,\ B^{[1]}(t_i) u^{(1)}_{[r]_{v^{(1)}_{[s]}}}\ \rangle p_r q_s \tag{38.122}$$

under the conditions (32.3), (32.4). The obtained numbers are given in Table 5 and Table 6, respectively.

$s_{z1} = [t_i] = s_{z1} \geq 0$ $s_{z2}[t_i] = s_{z2} > 0$	$p_{z3}^{*0} = \dfrac{\mid s_{z1} \mid}{\mid s_{z1} \mid + \mid s_{z2} \mid}, \quad p_{z4}^{*0} = 1 - p_{z3}^{*0}, \quad p_{z1}^{*0} = p_{z2}^{*0} = 0$
$s_{z1} < 0$ $s_{z2} \geq 0$	$p_{z1}^{*0} = \dfrac{\mid s_{z1} \mid}{\mid s_{z1} \mid + \mid s_{z2} \mid}, \quad p_{z4}^{*0} = 1 - p_{z1}^{*0}, \quad p_{z2}^{*0} = p_{z3}^{*0} = 0$
$s_{z1} \leq 0$ $s_{z2} < 0$	$p_{z1}^{*0} = \dfrac{\mid s_{z1} \mid}{\mid s_{z1} \mid + \mid s_{z2} \mid}, \quad p_{z2}^{*0} = 1 - p_{z1}^{*0}, \quad p_{z3}^{*0} = p_{z4}^{*0} = 0$
$s_{z1} > 0$ $S_{z2} \leq 0$	$p_{z3}^{*0} = \dfrac{\mid s_{z1} \mid}{\mid s_{z1} \mid + \mid s_{z2} \mid}, \quad p_{z2}^{*0} = 1 - p_{z3}^{*0}, \quad p_{z1}^{*0} = p_{z4}^{*0} = 0$
$s_{z1} = 0, \quad s_{z2} = 0$	$p_{zl}^{*0} = \dfrac{1}{4}, \quad l = 1, \ldots, 4$

Table 5.

Here the vector $m_z^0 = m_z^0(t_i, x[t_i], z[t_i], \varepsilon)$ is similar to the vector m_y^0 described above (see Figure 38.9). The only difference is that this vector is constructed for the z-model but not for the y-model.

$m_{z1}^0 \geq 0$, $m_{z2}^0 > 0$	$q_{z1}^{*a} = \frac{m_{z1}^0(\sin\alpha_1^0 + \cos\alpha_1^0) + m_{z2}^0(\sin\alpha_1^0 - \cos\alpha_1^0)}{2(m_{z1}^0 + m_{z2}^0)\sin\alpha_1^0}, \quad q_{z2}^{*a} = 1 - q_{z1}^{*a}$
$m_{z1}^0 < 0$, $m_{z2}^0 \geq 0$	$q_{z2}^{*a} = \frac{m_{z1}^0(\sin\alpha_1^0 + \cos\alpha_1^0) + m_{z2}^0(\sin\alpha_1^0 - \cos\alpha_1^0)}{2(m_{z1}^0 + m_{z2}^0)\sin\alpha_1^0}, \quad q_{z1}^{*a} = 1 - q_{z2}^{*a}$
$m_{z1}^0 \leq 0$, $m_{z2}^0 < 0$	$q_{z1}^{*a} = \frac{m_{z1}^0(\sin\alpha_1^0 + \cos\alpha_1^0) + m_{z2}^0(\sin\alpha_1^0 - \cos\alpha_1^0)}{2(m_{z1}^0 + m_{z2}^0)\sin\alpha_1^0}, \quad q_{z2}^{*a} = 1 - q_{z1}^{*a}$
$m_{z1}^0 \geq 0$, $m_{z2}^0 < 0$	$q_{z2}^{*a} = \frac{m_{z1}^0(\sin\alpha_1^0 + \cos\alpha_1^0) + m_{z2}^0(\sin\alpha_1^0 - \cos\alpha_1^0)}{2(m_{z1}^0 + m_{z2}^0)\sin\alpha_1^0}, \quad q_{z1}^{*a} = 1 - q_{z2}^{*a}$
$m_{z1}^0 = 0, \quad m_{z2}^0 = 0$	$q_{z1}^{*a} = q_{z2}^{*a} = \frac{1}{2}$

Table 6.

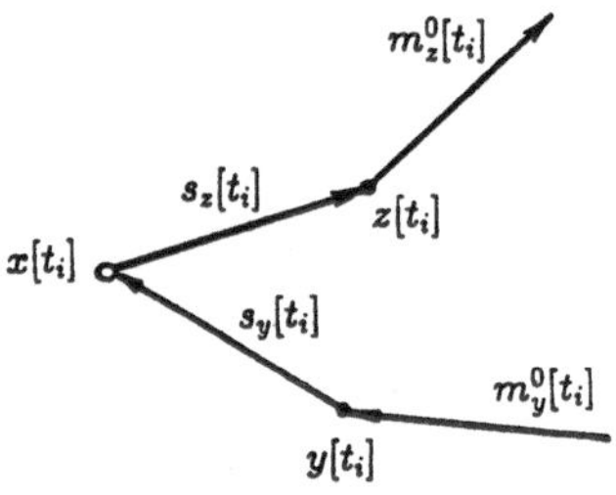

Figure 38.9

The optimal actions $u_z^{(2)0}[t_i] = u_z^{(2)0}(t_i, x[t_i], z[t_i], \varepsilon)$, $v_{*z}^a[t_i] = v_{*z}^a(t_i, x[t_i], z[t_i]$, and $v_z^{*a}[t_i] = v_z^{*a}(t_i, x[t_i], z[t_i], \varepsilon)$ for the time interval $t_i \leq t < t_{i+1}$ for the z-model are determined by the following equalities:

$$u_z^{(2)0}[t_i] = -\frac{s_z[t_i]}{\mid s_z[t_i] \mid} \text{ sign } b^{[2]}(t_i), \qquad \mid s_z[t_i] \mid \neq 0, \tag{38.123}$$

$$v_{*z}^a[t_i] = \frac{m_z^0[t_i]}{\mid m_z^0[t_i] \mid} \text{ sign } c^{[1]}(t_i), \qquad \mid m_z^0[t_i] \mid \neq 0, \tag{38.124}$$

$$v_z^{*a}[t_i] = \frac{m_z^0[t_i]}{\mid m_z^0[t_i] \mid} \text{ sign } c^{[2]}(t_i), \qquad \mid m_z^0[t_i] \mid \neq 0 \tag{38.125}$$

and $u_z^{(2)0}[t_i] = v_{*z}^0[t_i] = v_z^{*0}[t_i] \equiv 0$ if $\mid s_z[t_i] \mid = 0$ and $\mid m_z^0[t_i] \mid = 0$.

Thus, the algorithm of the construction of the actions $v_0[t]$ that correspond to the optimal strategy S_a^v (38.93) is described completely in the case $\alpha_1^0 \geq \frac{\pi}{4}$.

Let us now consider the other possible case, i.e., $\alpha_1^0 < \frac{\pi}{4}$.

If the conditions

$$| m_2^0 | > | m_1^0 | \frac{\cos \alpha_1^0 - \sin \alpha_1^0}{\cos \alpha_1^0 + \sin \alpha_1^0}, \tag{38.126}$$

are valid, then all the expressions for the optimal control actions (for S_a^u) and for the optimal disturbances (for S_a^v) are the same as the ones given above for the case $\alpha_1^0 \geq \frac{\pi}{4}$. The vector m_i^0 in (38.146) is to be treated as the appropriate vector ($s_y[t_i], m_y^0[t_i], s_z[t_i]$ or $m_z^0[t_i]$) in Tables 1–6 and in the corresponding formulas for the optimal control actions and for the optimal disturbances.

If the conditions (38.126) are not valid then the following formulas hold.

For

$$| m_2^0 | \leq | m_1^0 | \frac{\cos \alpha_1^0 - \sin \alpha_1^0}{\cos \alpha_1^0 + \sin \alpha_1^0}, \quad m_1^0 \geq 0 \tag{38.127}$$

we have

$$p_1^0 = 0, \quad p_2^0 = 0, \quad p_3^0 = 1, \quad p_4^0 = 0 . \tag{38.128}$$

For

$$| m_2^0 | \leq | m_1^0 | \frac{\cos \alpha_1^0 - \sin \alpha_1^0}{\cos \alpha_1^0 + \sin \alpha_1^0}, \quad m_1^0 < 0 \tag{38.129}$$

we obtain

$$p_1^0 = 1, \quad p_2^0 = 0, \quad p_3^0 = 0, \quad p_4^0 = 0 . \tag{38.130}$$

The formulas (38.127)–(38.130) are related to Tables 1, 2, and 5 if we understand the vector m_i^0 as the corresponding vector $s_y[t_i]$, $m_y^0[t_i]$, $s_z[t_i]$, $m_z^0[t_i]$. For example, in Table 5 for $m_1^0 = s_{z1}$, $m_2^0 = s_{z2}$ we obtain $p_{zr}^* = p_r^0$, $r = 1, 2, 3, 4$ where $\{p_r^0\}$ are the numbers in (38.128) or (38.130).

Further, for

$$| m_2^0 | \geq | m_1^0 | \frac{\cos \alpha_1^0 - \sin \alpha_1^0}{\cos \alpha_1^0 + \sin \alpha_1^0}, \quad m_2^0 \geq 0 \tag{38.131}$$

we obtain

$$p_1^0 = 0, \quad p_2^0 = 0, \quad p_3^0 = 0, \quad p_4^0 = 1 . \tag{38.132}$$

And for

$$| m_2^0 | \geq | m_1^0 | \frac{\cos \alpha_1^0 - \sin \alpha_1^0}{\cos \alpha_1^0 + \sin \alpha_1^0}, \quad m_2^0 < 0 \tag{38.133}$$

we have

$$p_1^0 = 0, \quad p_2^0 = 1, \quad p_3^0 = 0, \quad p_4^0 = 0. \tag{38.134}$$

For the optimal values in Tables 3, 4, and 6 connected with the vectors s_y, s_z, m_z^0 and with the numbers q_s^0, q_{ys}^{*0}, q_{zs}^{*0}, $s = 1, 2$ the following expressions are valid.

Under the conditions (38.127) for $m_2^0 \geq 0$ or under the conditions (38.129) for $m_2^0 < 0$ we have

$$q_1^0 = 1, \qquad q_2^0 = 0. \tag{38.135}$$

Under the conditions (38.131) for $m_1^0 \geq 0$ or under the conditions (38.133) for $m_1^0 < 0$ we obtain

$$q_1^0 = 0, \qquad q_2^0 = 1. \tag{38.136}$$

Under the conditions

$$| m_2^0 | \leq | m_1^0 | \frac{\cos\alpha_1^0 - \sin\alpha_1^0}{\cos\alpha_1^0 + \sin\alpha_1^0}, \quad m_1^0 > 0, \quad m_2^0 > 0 \tag{38.137}$$

or the conditions

$$| m_2^0 | \leq | m_1^0 | \frac{\cos\alpha_1^0 - \sin\alpha_1^0}{\cos\alpha_1^0 + \sin\alpha_1^0}, \quad m_1^0 < 0, \quad m_2^0 > 0 \tag{38.138}$$

we have the equalities (38.136).

And finally, under the conditions

$$| m_1^0 | \geq | m_2^0 | \frac{\cos\alpha_1^0 - \sin\alpha_1^0}{\cos\alpha_1^0 + \sin\alpha_1^0}, \quad m_1^0 > 0, \quad m_2^0 \leq 0 \tag{38.139}$$

or

$$| m_1^0 | \geq | m_2^0 | \frac{\cos\alpha_1^0 - \sin\alpha_1^0}{\cos\alpha_1^0 + \sin\alpha_1^0}, \quad m_1^0 \leq 0, \quad m_2^0 > 0 \tag{38.140}$$

we have the equalities (38.135).

Thus, the algorithm for the calculation of the optimal control actions (for S_a^u (38.92)) and the optimal disturbances (for S_a^v (38.93) is described for all possible cases.

In conclusion of this section we give an explicit scheme of the realization of the algorithms based on the optimal mixed strategies S_a^u (38.92) or S_a^v (38.93).

The strategy S_a^u (38.92) constructed in this section determines the control algorithm that was realized in the process of control for the material r, s-object (2.14), (2.17) considered in Section 2.

This algorithm acts in the following way.

Let

$$H(t_*) = \{t_*, r_*, \dot{r}_*, s_*, \dot{s}_*\} \tag{38.141}$$

be the measured or given initial state for the r, s-object (2.14), (2.17).

According to the constructions given in Section 2 (see (2.18)–(2.20)) the given data determine the initial state

$$\{t_*, x_*\} = \{t_*, x_{*1}, x_{*2}\} \tag{38.142}$$

for the x-object (38.1)–(38.11).

We choose some small value of the parameter $\varepsilon > 0$ and a step $\delta > 0$ of the partition $\Delta\{t_i\} = \{t_1 = t_*,\ t_i < t_{i+1},\ i = 1, \ldots, k,\ t_{k+1} = \vartheta\}$ so that

$$\delta = t_{i+1} - t_i = \frac{\vartheta - t_*}{M}, \quad i = 1, \ldots, k. \tag{38.143}$$

Here $[t_*,\ \vartheta]$ is a given time interval for the considered controlled process; $M > 0$ is a sufficiently large number.

We choose the initial position $\{t_*, y_{*1}, y_{*2}\}$ for the y-model-leader (38.13), setting for example $y_{*1} = x_{*1} - s_{*1}$ and $y_{*2} = x_{*2} - s_{*2}$, where s_{*1} and s_{*2} are sufficiently small so that $s_{*2}^2 + s_{*2}^2 \le \varepsilon^2$. It determines the vector $s_y[t_1] = s_y[t_*] = x[t_*] - y[t_*]$.

Solving the auxiliary problems considered above in this section we find the quantities $\{p_{yr}^{*a}\}$, $\{q_{ys}^{*0}\}$, $\{u_y^{*(2)a}[t_1]\}$, $\{v_{*y}^0[t_1]\}$, and $\{v_y^{0*}[t_1]\}$, which determine the motion of y-model on the time interval $[t_1,\ t_2]$ as a solution of the differential equation (38.13) for the initial state $y[t_1] = y_*$.

We also realize the random test for sampling the control vector $u_{[r]}^{(1)}$, $r = 1, 2, 3, 4$ in U (38.19). In this random sampling we suppose

$$P\Big(\ u[t_1] = u_{[r]}^{(1)}\ \Big) = p_r^0[t_1], \quad r = 1, 2, 3, 4 \tag{38.144}$$

where the collection $\{p_r^0\}$ is determined as the solution of the problem (38.101) at $t_i = t_1$.

For example, the procedure of this random sampling can be organized in the following way. A computer-realized generator of random numbers uniformly distributed in the interval [0, 1] gives the value

$$A \in [0,\ 1]\,. \tag{38.145}$$

This number A appears in one of the intervals shown in Figure 38.10

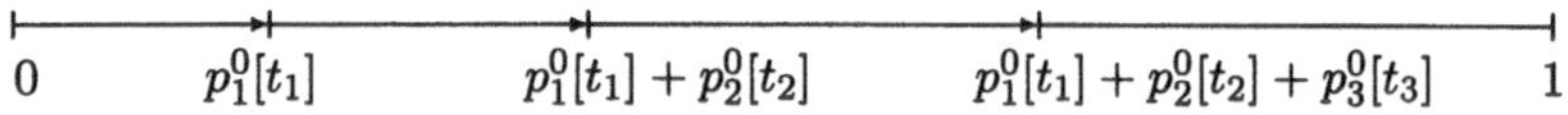

Figure 38.10

In our case only two numbers $p_{[r]}^0$ are not equal to zero. If $A \in [0,\ p_1^0]$, we choose $u^0[t_1] = u_{[1]}^{(1)} \in U$ (38.19). If $A \in (p_1^0,\ p_1^0 + p_2^0]$, then $u[t_1] = u_{[2]}^{(1)}$ and so on.

The control action $u^{(2)0}[t_1]$, $t_1 \le t \le t_2$ is determined by the formula (38.105) at $t_i = t_1$. At the same time some mechanism V_v that forms the disturbance for the r, s-object (2.14), (2.17) supplies admissible disturbances $v^{(1)}[t]$, $v_*[t]$ and $v^*[t]$, $t_1 \le t < t_2$. For example, this mechanism V_v may correspond to the optimal strategy S_v^0 (38.13). That means that the z-model-leader (38.16) is used.

Thus, the motion $r[t]$, $s[t]$, $t_1 \le t \le t_2$ of r, s-object (2.14), 2.17) is realized and a new state $\{t_2, r[t_2], s[t_2]\}$ is determined. Further, this procedure is repeated for $i = 1, \ldots, k$. We omit here the description of the steps of this procedure. At the time moment $t = \vartheta$ the position $\{\vartheta, x[\vartheta]\}$ of the x-object (38.1) and the position $\{\vartheta, r[\vartheta], s[\vartheta]\}$ of the r, s-object (2.14), (2.17) are realized. Thus, the value (38.12)

$$\gamma = \mid x[\vartheta] \mid = (x_1^2[\vartheta] + x_2^2[\vartheta])^{1/2} \tag{38.146}$$

and the value (2.15)

$$\gamma = \Big(\, (r_1[\vartheta] - s_1[\vartheta])^2 + (r_2[\vartheta] - s_2[\vartheta])^2 \,\Big)^{1/2} \tag{38.147}$$

are obtained. This value γ (38.12) must satisfy the following inequality

$$\gamma \leq \rho^0(t_*, x_*) + \zeta \tag{38.148}$$

with the probability close to unit. We can ensure that the value $\zeta > 0$ is sufficiently small if we choose the parameter $\varepsilon > 0$ and partition step $\delta > 0$ sufficiently small. Note that $\gamma \approx \rho^0(t_*, x_*)$ is valid in the case when both players use their approximating optimal mixed strategies S_a^u (38.92) and S_a^v (38.93). However, if the approximating control actions are formed by the optimal mixed strategy S_a^u but the disturbances are formed by some mechanism that differs from the optimal mixed strategy S_a^v, then, as a rule, the obtained value γ is essentially smaller than $\rho^0(t_*, x_0)$ with the probability close to 1. On the other hand, if the disturbances are formed by the approximating optimal mixed strategy S_a^v, but the control actions are formed by some mechanism that differs from S_a^u, then the obtained value γ is, as a rule, essentially greater then $\rho^0(t_*, x_*)$.

The computer simulation of these processes was given in the end of Section 2.

References

[1] Arkin, V.I. and Levin, V.L., Convexity of values of vector integrals, theorems of measurable sampling, and variational problems, *Russian Math. Surveys*, **27**(3), (1972), 21–86.

[2] Barbashin, E.A., *Introduction to the Theory of Stability*, Wolters-Noordhoff, Groningen, 1970.

[3] Barbashin, E.A. and Alimov, Yu.I., On the theory of relay differential equations, *Izvest. Vys Uch. Zav. Matematika* (Kazan State Univ. Publishing House), **1**(26), (1962), 3–13 (in Russian).

[4] Basar, T. and Olsder, J., *Dynamic Noncooperative Game Theory*, Acad. Press, New York, 1982.

[5] Bellman, R., *Introduction to the Mathematical Theory of Control Processes*, Vols. 1 & 2, Academic Press, New York, (1967), (1971).

[6] Bensoussan, A. and Friedman, A., Nonlinear variational inequalities and differential games with stopping times, *J. Funct. Anal.*, **16**(3), (1974), 305–352.

[7] Berkovitz, L.D., A variational approach to differential games, *Adv. in Game Theory, Ann. of Math. Studies,* (Princeton Univ. Press), **52** (1964), 127–174.

[8] Blaquiere, A., Gerard, F., and Leitman, G., *Quantitative and Qualitative Games,* Academic Press, New York, 1969.

[9] Bohnenblust, H.F. and Karlin, S., On a theorem of Ville, *Contributions to the theory of games*, I, *Ann. of Math. Studies*, (Princeton), I (1950), 155–160.

[10] Botkin, N.D., Evaluation of numerical construction error in differential game with fixed terminal time, *Problems of Control and Information Theory*, **11**(4), (1982), 283–295.

[11] Botkin, N.D. and Patsko, V.S., Positional control in a linear differential game, *Engineering Cybernetics*, **21**(4), (1983), 69–75.

[12] Breakwell, J.V. and Merz, A.W., Toward a complete solution of the homicidal chauffeur game, *Proc. 1st Intern. Conf. Theory and Appl. Different. Games*, Amherst, Mass., 1969, S. 1, S.a.P. 111/1–111/5.

[13] Brykalov, S.A., Equilibrium existence for some heating control systems, *Problems Control Inform. Theory*, **17**(6), (1988), 371–380.

[14] Brykalov, S.A., Nonlinear boundary value problems and the existence of stationary states of heating control systems, *Doklady Academii Nauk SSSR,* **307**(1), (1989), 11–14 (Russian); *Soviet Math. Dokl.*, **40**(1)–(4), (1990), (Math. Rev. 90k:47124) (English).

[15] Brykalov, S.A., The existence of temperature distributions close to a prescribed one in some control systems, *Problems Control Inform. Theory,* **19**(4), (1990), 279–288.

[16] Bryson, A.E., Jr. and Ho, Y.-C., *Applied Optimal Control,* Waltham, Mass., Ginn., 1969.

[17] Burton, T.A., *Stability and Periodic Solutions* of *Ordinary* and *Functional Differential Equations*, Academic Press, Orlando, Florida, 1985.

[18] Chentsov, A.G., On the structure of a game problem of convergence, *Soviet Math. Dokl.*, **16** (1975), 1404–1406.

[19] Chentsov, A.G. , On a game problem of guidance, *Soviet Mathematics,* **17**(1), (1976), 73–77.

[20] Chentsov, A.G., On a game problem of converging at a given instant of time, *Math. of the USSR Sbornik,* 28(3), (1976), 353–376.

[21] Chentsov, A.G., On the design of differential games, 1, *Problems Control Inform. Theory,* **6**(6), (1977), 5–7.

[22] Chentsov, A.G., On an alternative in the class of quasistrategies for a pursuit-evasion differential game, *J. Different. Equat.*, **16** (1980).

[23] Clarke, F., Methods of dynamic and nonsmooth optimization, *SIAM*, Philadelphia, (1989), p. 90.

[24] Coddington, E.A. and Levinson, N., *Theory of Ordinary Differential Equations*, McGraw-Hill, New York, 1955.

[25] Crandall, M.G. and Lions, P.L., Viscosity Solutions of Hamilton–Jacobi equations, *Amer. Math. Soc. Transl.*, **277**(1), (1983), 1–42.

[26] Davy, J.L., Properties of the solution set of a generalized differential equation, *Bull. Austral. Math. Soc.*, **6** (1972), 379–398.

[27] Dem'janov, V.F., *Minimax: Directional Differentiation,* Izdat. Leningrad Gos. Univ., Leningrad, 1974, (in Russian).

[28] Dem'janov, V.F. and Malozemov, V.N., *Introduction to the Minimax,* Halsted Press, New York, 1974.

[29] Dieudonne, J., *Foundations of Modern Analysis*, Academic Press, New York, 1969.

[30] Dunford, N. and Schwarz, J.T., *Linear Operators*, Part I, *General Theory*, Interscience, New York, 1958.

[31] Ekeland, I. and Temam, R., *Convex Analysis and Variational Problems; Studies in Mathematics and its Applications*, vol. 1, North-Holland, Amsterdam, 1976.

[32] Elliott, R.J. and Kalton, N.J., The existence of value in differential games of pursuit and evasion, *J. Different. Equat.*, **12**(3), (1972), 504–523.

[33] Fan, K., Minimax theorems, *Proc. Nat. Acad. Sci. USA*, **39**(1), (1953), 42–47.

[34] Filippov, A.F., On certain questions in the theory of optimal control, *SIAM J. Control*, **1** (1962), 76–84.

[35] Filippov, A.F., Differential equations with discontinuous right-hand side, *Amer. Math. Soc. Transl.*, Providence, R.I., 2:42, (1964), 199–231.

[36] Fleming, W.H., A note on differential games of prescribed durations, *Contributions to the Theory of Games, Ann. Math. Studies*, No. 3 (1957), 407–412.

[37] Fleming, W.H., The convergence problem for differential games, *J. Math. Anal. and Appl.*, **3** (1961), 102–116.

[38] Fleming, W.H., The Cauchy problem for degenerate parabolic equations, *J. of Math. and Mech.*, **13** (1964), 987–1008.

[39] Flynn, J., Pursuit in the circle: lion versus man, *Different. Games and Contr. Theory,* New York, 1974, 99–124.

[40] Friedman, A., Existence of Value and Saddle Points for Differential Games of Pursuit and Evasion, *J. Different. Equat.*, **7**(1), (1970), 92–110.

[41] Friedman, A., *Differential games*, Interscience, New York, 1971.

[42] Gamkrelidze, R.V. and Kharatishvilli, G.L., A differential game of evasion with nonlinear control, *SIAM J. Control*, Vol. 12 No. 2, (1974), 332–349.

[43] Gusjatnikov, P.B. and Nikol'skii, M.S., Optimality of pursuit time, *Soviet Math. Dokl.*, **10** (1969), 103–106.

[44] Hadwiger, H., *Vorlesungen über Inhalt, Oberfläche and und Isoperimetrie*, Springer-Verlag, Berlin, 1957.

[45] Hajek, O., *Pursuit Games*, Mathematics in Science and Engineering, vol. 20, Academic Press, New York, 1975.

[46] Hale, J.K., *Theory of Functional Differential Equation*, Springer-Verlag, New York, 1977.

[47] Ioffe, A.D. and Tikhomirov, W.M., *Theory of Extremal Problems*, North-Holland, Amsterdam, 1978.

[48] Isaacs, R., *Differential Games*, John Wiley, New York, 1965.

[49] Janin, R., On sensitivity in an optimal control problem, *J. of Math. Anal. and Appl.*, **60** (1977), 631–657.

[50] Kantorovich, L.V. and Akilov, G.P., *Functional Analysis*, 2nd ed., Pergamon Press, Oxford, 1982.

[51] Karlin, C., *Mathematical Methods and Theory in Games, Programming, and Economics,* Moscow, Mir., 1964.

[52] Kikuchi, N., On control problems for functional-differential equations, *Funkcialaj Ekvacioj.*, **14** (1971), 1–23.

[53] Kleimenov, A.F., *Nonantagonistic Positional Dfferential Games*, Nauka, Ekaterinburg, 1993, (in Russian).

[54] Kolmogorov, A.N. and Fomin, S.V., *Introductory Real Analysis*, Prentice-Hall, Englewood Cliffs, New Jersey, 1970.

[55] Krasovskii, A.N., A differential game for the positional functional, *Soviet Math. Dokl.*, **22**(1), (1980), 251–255.

[56] Krasovskii, A.N., On positional minimax control, *Jour. Appl. Math. Mech.*, **44**(4), (1980), 602–610.

[57] Krasovskii, A.N., Krasovskii N.N., and Tret'yakov, V.E., Stochastic program synthesis for a deterministic positional differential game, *Jour. Appl. Math. Mech.*, **45**(4), (1981), 579–586.

[58] Krasovskii, A.N., Nonlinear differential games with integral payoff, *Differential Equations*, **18**(2), (1982), 1306–1312.

[59] Krasovskii, A.N., Conditions for regularity of program maximin, *Izvest. AN SSSR. Tekh. Kibernetika,* No. 4 (1983), 70–77.

[60] Krasovskii, A.N. and Tret'yakov, V.E., Stochastic program synthesis in differential game with integral quality index, *Prikl. Mat. Mekh.*, **48**(5), (1984), 883–889.

[61] Krasovskii, A.N., Construction of mixed strategies based on stochastic programs, *Jour. Appl. Math. Mech.*, **51**(2), (1987), 186–192.

[62] Krasovskii, A.N., *Synthesis of Mixed Control Strategies*, Ural State Univ., Sverdlovsk, 1988, (in Russian).

[63] Krasovskii, A.N. and Reshetova, T.N., *Control under Lack of Information*, Ural State Univ., Sverdlovsk, 1990.

[64] Krasovskii, A.N., Control under minimax of an integral functional, *Dokl. Acad. Nauk SSSR*, **320**(4), (1991), 785–788. (English transl. in *Soviet Math. Dokl.*, **44**(2), (1992)).

[65] Krasovskii, A.N., Control in mixed strategies on the minimax of an integral functional, *Jour. Appl. Math. Mech.*, **56**(2), (1992), 167–175.

[66] Krasovskii, A.N., A differential game for a positional functional of the phase coordinates and control actions, *Soviet Math. Dokl.*, **322**(5), (1992), 180–183.

[67] Krasovskii, A.N., Differential games with positional functionals, *Izv. Akad. Nauk. Technicheskaya Kibernetika*, No. 1 (1993), 142–147.

[68] Krasovskii, A.N. and Krasovskii, N.N., A differential game for the minimax of a positional functional, *Advances in Nonlinear Dinamical and Control; A Report from Russia*, Birkhäuser Boston, (1993), 41–72.

[69] Krasovskii, N.N., On a problem of tracking, *Jour. Appl. Math. Mech.*, **17** (1963), 363–377.

[70] Krasovskii, N.N., *Theory of Controlling Motions*, Nauka, Moscow, (1968), (in Russian).

[71] Krasovskii, N.N., *Games Problems about Contact of Motions*, Nauka, Moscow, (1970), (in Russian).

[72] Krasovskii, N.N., Mixed strategies in differential games, *Dokl. AN SSSR*, **235**(3), (1977), 519–522.

[73] Krasovskii, N.N., Pursuit-evasion games with stochastic leader, *Dokl. AN SSSR*, **237**(5), (1977), 1020–1023.

[74] Krasovskii, N.N., Differential games, approximation and formal models, *Math. USSR Sbornic*, **35**(6), (1979), 795– 2.

[75] Krasovskii, N.N. and Tret'yakov, V.E., Stochastic program synthesis for positional differential games, *Dokl. AN SSSR*, **259**(1), (1981), 24–27.

[76] Krasovskii, N.N. and Tret'yakov, V.E., On programmed synthesis of positional control, *Dokl. Akad. Nauk SSSR*, **264**(6), (1982), 1309–1312.

[77] Krasovskii, N.N., *The Control of a Dynamic System*, Nauka, Moscow, 1985, (in Russian).

[78] Krasovskii, N.N., On synthesis for differential games, *Jour. Appl. Math. Mech.*, **50**(6), (1986), 898–902.

[79] Krasovskii, N.N. and Subbotin, A.I., *Game-Theoretical Control Problems*, Springer-Verlag, New-York, 1988.

[80] Krasovskii, N.N. and Reshetova, T.N., On the program synthesis of a guaranteed control, *Problems of Control and Information Theory*, **17**(6), (1988), 333–343.

[81] Krasovskii, N.N., Control and stabilization under lack of information, *Izv. Akad. Nauk. Tekhnicheskaya Kibernetika*, No. 1 (1993), 148–151.

[82] Kruzhkov,S.N., Non-linear equations of the first order and differential games connected with them, *Uspekhi Mat. Nauk.*, 24(2), (1969), 227–228, (in Russian).

[83] Kryazhimskii, A.V., Positional differential pursuit-evasion games, *Dokl. AN SSSR*, (1978).

[84] Kryazhimskii, A.V., Maksimov V.I., and Osipov, Yu.S., On positional modelling in dynamical systems, *Jour. Appl. Math. Mech.*, **47**(6), (1983), 883–898.

[85] Kuratowski, K. and Ryll-Nardzewski, C., A general theorem on selectors, *Bull Acad. Polon. Sci.; Ser. Sci. Math. Astr. Phys.*, **13** (1965), 397–403.

[86] Kurzhanskii, A.B., Differential games of observation, *Dokl. AN SSSR*, **207**(3), (1972), 527–530.

[87] Kurzhhanskii, A.B., On minimax control and estimation strategies under incomplete information, *Probl. Control Inform. Theory*, **4**(3), (1975), 205–218.

[88] Kurzhanskii, A.B., *Control and Observation under Conditions of Uncertainty*, Nauka, Moscow, 1977, (in Russian).

[89] Kurzhanskii, A.B., Dynamic control system estimation under uncertainty conditions I, II, *Probl. Control Inform. Theory*, **9**(6), (1980), 395–406, **10**(1), 33–42.

[90] Kurzhanskii, A.B., On stochastic filtering approximations of estimation problems for systems with uncertainty, *Stochastics*, **23** (1988), 109–130.

[91] Kurzhanskii, A.B., Identification – a theory of guaranteed estimates//From Date to Model, Berlin etc.: Springer, (1989), 135–214.

[92] Lions, P.L., *Generalized Slutions of Hamilton–Jacobi equations*, Pitman, Boston, 1982.

[93] Lions, P.L., Optimal control and viscosity solutions, *Lect. Notes Math.*, **1119** (1985), 94–112.

[94] Lions, P.L. and Souganidis, P.E., Differential games, optimal control and directional derivatives of visconty solutions of Bellman's and Isaacs equations, *SIAM J. Control and Optim.*, **23** (1985), 566–583.

[95] Lions, J.L., Exact controllability, stabilization and perturbations for distributed systems, *SIAM Review*, **30**(1), (1988), 1–68.

[96] Loeve, M., *Probability Theory*, 2nd edition, D. van Nostrand, Princeton, New Jersey, (1960).

[97] Liptser, R.Sh. and Shiryaev, A.N., *Statistics of stochastic processes*, Nauka, Moscow, (1974).

[98] Lokshin, M.D., On the optimal control of a linear system under the condition of the integral disturbance constraint, *Problems of Control and Information Theory*, **19**(2), (1990), 111–127.

[99] Lokshin, M.D., On differential games with integral constraints on control actions, *Differential Equations*, **28**(11), (1992), 1952–1961.

[100] Lusin, N., Sur la probleme de J. Hadamard d'uniformisation des ensembles, *Mathematica*, **4** (1930), 54–66.

[101] Malkin, I.G., *Theory of Stability of Motions*, Nauka, Moscow, 1965, (in Russian).

[102] Marchal, C., Analytical Study of a case of the homicidal chauffeur game problem, *Lect. Notes Comput.*, **27** (1975), 472–481.

[103] McKinsey, J.C.C., *Introduction to the Theory of Games*, McGraw–Hill, New York, 1952.

[104] Melikyan, A.A., Optimal interaction of two pursuers in a game problem, *Izvest. AN SSSR, Tech. Kibernetica*, No. 4 (1981), 10–18.

[105] Miscenko, E.F., Pursuit and evasion problems in differential games, *Engineering Cybernetics*, **9** (1971), 787–791.

[106] Natanson, I.P., *Theory of Functions of a Real Variable*, F. Ungar, ed., New York, 1961.

[107] von Neumann, J., Zur Theorie der Geselshaftsspiele, *Math. Ann.*, **100** (1928), 295–320.

[108] Neveu, J., *Mathematical Foundations of the Calculus of Probabilities*, Holden–Day, San Francisco, CA., 1965.

[109] Nikolskii, M.S., Application of the first direct method of Pontryagin in pursuit games, *Engineering Cybernetics*, **10**(6), (1972), 984–985.

[110] Olech, C., Existence theory in optimal control. Control theory and topics in functional analysis, (*Internat. Sem., Internat. Centre Theoret. Phys.*, Trieste) vol. I (1974), 291–328.

[111] Osipov, Ju.S., Alternative in a differential–difference game, *Soviet Math. Dokl.*, **12**(2), (1971), 619–624.

[112] Osipov, Ju.S., Differential games for systems with aftereffect, *Dokl. AN SSSR*, **196**(4), (1971), 779–782.

[113] Osipov, Ju.S., Differential games for systems with with distributed parameters, *Dokl. AN SSSR*, **223**(6), (1975), 1314–1317.

[114] Osipov, Ju.S., Positional control in parabolic systems, *Prikl. Mat. Mekh.*, **41**(2), (1977), 195–201.

[115] Osipov, Ju.S. and Kryazhimskii, A.V., Dynamical solving operator equations, *Dokl. AN SSSR* , **269**(3), (1983), 552–556.

[116] Parthasarathy, T. and Raghavan, T., Some Topics in Two-Person Games, *Modern Analytic and Computational Methods in Science and Mathematics*, No. 22, American Elsevier, New York, 1971.

[117] Patsko, V.S., Quality differential games of second order, *Jour. Appl. Math. Mech.*, **46**(4), (1982), 596–604.

[118] Petrosyan, L.A., Differential games of pursuit, *Leningrad, Gos. Univ.*, (1977).

[119] Petrov, N.N., On the existence of a value for pursuit games, *Soviet Math. Dokl.*, **11** (1970), 292–294.

[120] Pontryagin, L.S., *Ordinary Differential Equations*, Addison–Wesley, Reading, MA., 1962.

[121] Pontryagin, L.S., Boltyanskii, V.G., Gamkrelidze, R.V., and Mishchenko, E.F., *The Mathematical Theory of Optimal Processes,* Interscience, New York, 1962.

[122] Pontryagin, L.S., On the theory of differential games, *Russian Mathematical Surveys*, **21**(4), (1966), 193–246.

[123] Pontryagin, L.S., Linear differential games, 1, *Soviet Math. Dokl.*, **8** (1967), 769–771.

[124] Pontryagin, L.S., Linear differential games, 2, *Soviet Math. Dokl.*, **8** (1967), 910–912.

[125] Pshenichnii, B.N., The structure of differential games, *Soviet Math. Dokl.*, **10** (1969), 70–72.

[126] Pshenichnii, B.N. and Sagaidak, N.I., Differential games of prescribed duration, *Cybernetics*, **6**(2), (1970), 72–83.

[127] Pshenichnii, B.N., *Convex Analysis and Extremal Problems,* Nauka, Moscow, 1980.

[128] Pshenichnii, B.N. and Ostapenko, V.V., *Differential Games*, Naukova Dumka, Kiev, 1992.

[129] Rademacher, H., Über partielle und totale Differenzierbarkeit von Funktionen mehrerer Variablem über die Transformation der Doppelintegrale, *Math. Ann.*, **79** (1918), 340–354.

[130] Rockafellar, R.T., *Convex Analysis*, Princeton University Press, Princeton, New Jersey, 1970.

[131] Roxin, E., Axiomatic approach in differential games, *J. Optimiz. Theory & Appl.*, **3**(3), (1969), 153–163.

[132] Rudin, W., *Principles of Mathematical Analysis,* 2nd edition, McGraw-Hill, New York, 1964.

[133] Sansone, G., *Odinary Differential Equations*, Moscow, Inostr. Liter., vols. 1 & 2, 1954.

[134] Subbotin, A.I., Differential games with constraints on phase states, *Dokl. AN SSSR*, **193**(2), (1970), 294–297.

[135] Subbotin A.I., Pursuit–evasion dynamical games, *Dokl. AN SSSR*, **234**(2), (1977), 323–326.

[136] Subbotin, A.I., A generalization of the basic equation of the theory of differential games, *Soviet Math. Dokl.*, **22**(2), (1980), 358–362.

[137] Subbotin, A.I., *Minimax inequalities and Hamilton–Jacobi equations*, Nauka, Moscow, 1991, (in Russian).

[138] Subbotin, A.I. and Subbotina, N.N., Game control problems under incomplete information, *Izv.Math. Nauk. Tech. Kibern.*, No. 5 (1977), 14–23.

[139] Subbotin, A.I. and Subbotina, N.N., Nessesary and sufficient conditions for piecewise smooth value of a differential game, *Dokl. AN SSSR*, **243**(4), (1978), 862–865.

[140] Subbotin, A.I. and Chentsov, A.G., *Optimization of a Guarantee in Control Problems,* Nauka, Moscow, 1981, (in Russian).

[141] Subbotina, N.N., Universal optimal strategies in positional differential games, *Differential Equations* , **19**(11), (1983), 1377–1382.

[142] Tarlinskii, S.I., On a positional guidance problem, *Soviet Math. Dokl.*, **13**(6), (1972), 1459–1463.

[143] Tret'yakov, V.E., Program syntethesis in stochastical differential games, *Dokl. AN SSSR*, **270**(2), (1983), 273–300.

[144] Tret'yakov V.E., Synthesis of optimal guaranteeing control, *Problems of Control and Information Theory*, **17**(4), (1988), 207–221.

[145] Ushakov, V.N., Krasovskii, N.N., and Subbotin, A.I., Minimax differential games, *Dokl. AN SSSR*, **206**(2), (1972), 277–280.

[146] Ushakov, V.N., On the problem of constructing stable bridges in a differential game of approach and avoidance, *Engineering Cybernetics*, **18**(4), (1980), 29–36.

[147] Varaiya, P. and Lin, J., Existence of saddle points in differential games, *SIAM J. Control*, **7**(1), (1969), 142–157.

[148] Warga, J., *Optimal Control of Differential and Functional Equations*, Academic Press, New York, 1972.

[149] Wazewski, T., Sur une generalisation de *la* notion des solution d'une equation *au* contingent, *Bull. Acad. Polon. Sci.; Ser. Sci. Math. Astr. Phys.*, **10** (1962), 11–15.

[150] Zaremba, S.Ch., Sur les equations *au* paratingent, *Bull. Sci. Math.*, 2:60, (1936), 139–160.

[151] Zelikin, M.I., On a differential game, *Usp. Math. Nauk.*, **21**(4), (1966), 272–274.

INDEX

Systems & Control: Foundations & Applications

Series Editor
Christopher I. Byrnes
School of Engineering and Applied Science
Washington University
Campus P.O. 1040
One Brookings Drive
St. Louis, MO 63130-4899
U.S.A.

Systems & Control: Foundations & Applications publishes research monographs and advanced graduate texts dealing with areas of current research in all areas of systems and control theory and its applications to a wide variety of scientific disciplines.

We encourage the preparation of manuscripts in TEX, preferably in Plain or AMS TEX LaTeX is also acceptable—for delivery as camera-ready hard copy which leads to rapid publication, or on a diskette that can interface with laser printers or typesetters.

Proposals should be sent directly to the editor or to: Birkhäuser Boston, 675 Massachusetts Avenue, Cambridge, MA 02139, U.S.A.

Estimation Techniques for Distributed Parameter Systems
H.T. Banks and K. Kunisch

Set-Valued Analysis
Jean-Pierre Aubin and Hélène Frankowska

Weak Convergence Methods and Singularly Perturbed
Stochastic Control and Filtering Problems
Harold J. Kushner

Methods of Algebraic Geometry in Control Theory: Part I
Scalar Linear Systems and Affine Algebraic Geometry
Peter Falb

H_∞ -Optimal Control and Related Minimax Design Problems
Tamer Başar and Pierre Bernhard

Identification and Stochastic Adaptive Control
Han-Fu Chen and Lei Guo

Viability Theory
Jean-Pierre Aubin

Representation and Control of Infinite Dimensional Systems, Vol. I
A. Bensoussan, G. Da Prato, M. C. Delfour and S. K. Mitter

Representation and Control of Infinite Dimensional Systems, Vol. II
A. Bensoussan, G. Da Prato, M. C. Delfour and S. K. Mitter

Mathematical Control Theory: An Introduction
Jerzy Zabczyk

H_∞-Control for Distributed Parameter Systems: A State-Space Approach
Bert van Keulen

Disease Dynamics
Alexander Asachenkov, Guri Marchuk, Ronald Mohler, Serge Zuev

Theory of Chattering Control with Applications to Astronautics, Robotics, Economics, and Engineering
Michail I. Zelikin and Vladimir F. Borisov

Modeling, Analysis and Control of Dynamic Elastic Multi-Link Structures
J. E. Lagnese, Günter Leugering, E. J. P. G. Schmidt

First Order Representations of Linear Systems
Margreet Kuijper

Hierarchical Decision Making in Stochastic Manufacturing Systems
Suresh P. Sethi and Qing Zhang

Optimal Control Theory for Infinite Dimensional Systems
Xunjing Li and Jiongmin Yong

Generalized Solutions of First-Order PDEs: The Dynamical Optimization Process
Andreĭ I. Subbotin